Lecture Notes in Artificial Intelligence 8755

Subseries of Lecture Notes in Computer Science

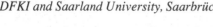

Michael Beetz   Benjamin Johnston
Mary-Anne Williams (Eds.)

Social Robotics

6th International Conference, ICSR 2014
Sydney, NSW, Australia, October 27-29, 2014
Proceedings

 Springer

Volume Editors

Michael Beetz
Technische Universität München
Department of Computer Science
Boltzmannstr. 3
85748 Garching, Germany
E-mail: michael.beetz@in.tum.de

Benjamin Johnston
University of Technology, Sydney
Faculty of Engineering and IT
Ultimo, NSW 2007, Australia
E-mail: benjamin.johnston@uts.edu.au

Mary-Anne Williams
University of Technology, Sydney
Faculty of Engineering and IT
Ultimo, NSW 2007, Australia
E-mail: mary-anne.williams@uts.edu.au

ISSN 0302-9743   e-ISSN 1611-3349
ISBN 978-3-319-11972-4   e-ISBN 978-3-319-11973-1
DOI 10.1007/978-3-319-11973-1
Springer Cham Heidelberg New York Dordrecht London

Library of Congress Control Number: 2014950402

LNCS Sublibrary: SL 7 – Artificial Intelligence

Typesetting: Camera-ready by author, data conversion by Scientific Publishing Services, Chennai, India

Printed on acid-free paper

Springer is part of Springer Science+Business Media (www.springer.com)

Preface

Welcome to the proceedings of the 6th International Conference on Social Robotics (ICSR) 2014.

The ICSR conference series brings together international researchers to discuss cutting edge developments in the field of social robotics. ICSR was first held in 2009, in Incheon, Korea as part of the FIRA RoboWorld Congress. Since then it has grown in size and scope, while continuing to maintain its reputation for its collegial, supportive, and constructive atmosphere. The University of Technology, Sydney is therefore proud to have hosted the sixth conference in the series.

The theme of the 2014 conference was Social Intelligence. Social Intelligence refers to the ability of people (and robots) to get along with others, to cooperate and to encourage cooperation. Social Intelligence is a crucial ability for people and robots to negotiate complex social relationships and environments. The theme of the conference invited researchers to explore computational models, robotic embodiments and behaviors that enable social robots to develop sophisticated levels of social intelligence.

In addition to the main theme of the conference, work was invited to two special sessions: *Social Robots for Therapeutic Purposes* and *Knowledge Representation and Reasoning in Robotics*.

Therapy, rehabilitation, medicine, and the care of the elderly will become an important application of social robotics. A robot that can empathize with patients, anticipate their pain and provide personalized care will, no doubt, be more readily accepted and achieve higher levels of satisfaction.

Research in knowledge representation and reasoning has proven useful in robotics. Social robotics is an exciting new avenue for this kind of research. Mapping and understanding complex human relationships and societies is a problem that is well suited to rich formalizations and representations. Furthermore, future generations of robots will inhabit complex social environments that will help inspire new approaches to representing and managing the complex knowledge that is involved.

The papers presented in this volume represent innovations from a global research community. Submissions came from Australasia, Asia, North and South America, and Europe. Each paper was subject to a rigorous peer review process by an esteemed International Program Committee of experts in Social Robotics.

In addition to the technical papers presented here, the conference featured invited talks by four distinguished researchers: Anthony Cohn (University of Leeds), Peter Gärdenfors (Lund University), Guy Hoffman (IDC Herzliya) and Oussama Khatib (Stanford University).

We wish to express our appreciation to the Program Committee members who generously donated their time and to our sponsors (The Journal of Artificial Intelligence, Aldebaran Robotics, The Association for the Advancement of Artificial Intelligence, The Centre for Quantum Computation & Intelligent Systems, Robohub and The Stanford Center for Legal Informatics). In addition, the conference would not have been possible without the contributions of the local Organizing Committee, workshop chairs, exhibition chairs and the Standing Committee.

October 2014

Michael Beetz
Benjamin Johnston
Mary-Anne Williams

Organization

ICSR2014 was organized by the Centre for Quantum Computation & Intelligent Systems at the University of Technology, Sydney. The conference was held at the Powerhouse Museum, Sydney.

Standing Committee

Ronald Arkin	Georgia Institute of Technology, USA
Paolo Dario	Scuola Superiore Sant'Anna, Italy
Suzhi Sam Ge	National University of Singapore, Singapore
Oussama Khatib	Stanford University, USA
Jong Hwan Kim	Korea Advanced Institute of Science and Technology, Korea
Haizhou Li	A*Star Singapore, Singapore
Maja Matarić	University of South Carolina, USA

General Chair

Mary-Anne Williams	University of Technology, Sydney, Australia

Program Chairs

Michael Beetz	University of Bremen, Germany
Benjamin Johnston	University of Technology, Sydney, Australia

Workshop Chairs

Alen Alempijevic	University of Technology, Sydney, Australia
Giuseppe Boccignone	Università degli Studi di Milano, Italy
Bruce MacDonald	University of Auckland, New Zealand

Exhibition Chairs

Gavin Paul	University of Technology, Sydney, Australia
Greg Peters	Sabre Autonomous Solutions, Australia

Special Sessions Chairs

Social Robots for Therapeutic Purposes
Ho Seok Ahn   University of Auckland, New Zealand
Bruce MacDonald   University of Auckland, New Zealand

Knowledge Representation and Reasoning in Robotics
Mohan Sridharan   Texas Tech University, USA
Subramanian Ramamoorthy   University of Edinburgh, UK
Vaishak Belle   University of Toronto, Canada

Organization Chairs

Teresa Vidal Calleja   University of Technology, Sydney, Australia
Xun Wang   University of Technology, Sydney, Australia

Organization Committee

Shaukat Abedi   University of Technology, Sydney, Australia
Sajjad Haider   Institute of Business Administration, Pakistan
Rony Novianto   University of Technology, Sydney, Australia
Pavlos Peppas   University of Technology, Sydney, Australia
Syed Ali Raza   University of Technology, Sydney, Australia
Jonathan Vitale   University of Technology, Sydney, Australia

Webmaster

Pramod Parajuli   University of Technology, Sydney, Australia

Program Committee

Shaukat Abidi   University of Technology, Sydney, Australia
Arvin Agah   University of Kansas, USA
Ho Seok Ahn   University of Auckland, New Zealand
Rini Akmeliawati   International Islamic University Malaysia,
                                Malaysia
Minoo Alemi   Sharif University of Technoloy, Iran
Alen Alempijevic   University of Technology, Sydney, Australia
Marcelo Ang   National University of Singapore, Singapore
Muhammad Anshar   University of Technology Sydney, Australia
Khelifa Baizid   UNCIAS, Italy
Ilaria Baroni   Ospedale San Raffaele—Milano, Italy
Christoph Bartneck   University of Canterbury, UK
Paul Baxter   Plymouth University, UK
Tony Belpaeme   Plymouth University, UK
Eduardo Benitez Sandoval   University of Canterbury, UK

Bingbing Liu	A*STAR Singapore
Katrin Solveig Lohan	Heriot-Watt University, UK
Manja Lohse	University of Twente, The Netherlands
David Lu	Washington University in St. Louis, USA
Andrew McDaid	University of Auckland, New Zealand
Ali Meghdari	Sharif University of Technology, Iran
Ben Mitchinson	University of Sheffield, UK
Wendy Moyle	Griffith University, Australia
Omar Mubin	University of Western Sydney, Australia
Marco Nalin	Telbios S.p.A., Italy
Rony Novianto	University of Technology, Sydney, Australia
Mohammad Obaid	Chalmers University of Technology, Sweden
Maurice Pagnucco	University of New South Wales, Australia
Marco Paleari	Italian Institute of Technology, Italy
Amit Kumar Pandey	Aldebaran Robotics, France
Pramod Parajuli	University of Technology, Sydney, Australia
Kathy Peri	University of Auckland, New Zealand
Tony Pipe	Bristol Robotics Laboratory, UK
Karola Pitsch	Bielefeld University, Germany
Nima Ramezani Taghiabadi	University of Technology, Sydney, Australia
David Rye	University of Sydney, Australia
Selma Sabanovic	Indiana University, USA
Maha Salem	University of Hertfordshire, UK
Miguel A. Salichs	Universidad Carlos III, Spain
Stefan Schiffer	RWTH Aachen University, Germany
Ruth Schulz	Queensland University of Technology, Australia
Christoph Schwering	RWTH Aachen University, Germany
Ravindra De Silva	Toyohashi University of Technology, Japan
David Silvera-Tawil	University of New South Wales, Australia
Reid Simmons	Carnegie Mellon University, USA
Wing Chee So	Chinese University of Hong Kong, Hong Kong
Mohan Sridharan	University of Auckland, New Zealand
Adriana Tapus	ENSTA-ParisTech, France
Keng Peng Tee	Institute for Infocomm Research, Singapore
Konstantinos Theofilis	University of Hertfordshire, UK
Bram Vanderborght	Vrije Universiteit, Belgium
Mari Velonaki	University of New South Wales, Australia
Manuela Veloso	Carnegie Mellon University, USA
Teresa Vidal-Calleja	University of Technology Sydney, Australia
Jonathan Vitale	University of Technology, Sydney, Australia
Frank Wallhoff	Jade University of Applied Sciences, Germany
Michael Walters	University of Hertfordshire, UK
Chen Wang	National University of Singapore, Singapore
Xun Wang	University of Technology, Sydney, Australia
Wei Wang	University of Technology, Sydney, Australia
Astrid Weiss	Vienna University of Technology, Austria

Johannes Wienke	Bielefeld University, Germany
Gregor Wolbring	University of Calgary, Canada
Alvin Wong	A*STAR Singapore
Agnieszka Wykowska	Ludwig Maximilians University Munich, Germany
Karolina Zawieska	PIAP, Poland
Jakub Zlotowski	University of Canterbury, UK

Sponsoring Organizations

Journal of Artificial Intelligence
Aldebaran Robotics
Association for the Advancement of Artificial Intelligence
Centre for Quantum Computation & Intelligent Systems
The Stanford Center for Legal Informatics (CodeX)
Robohub

Table of Contents

Development of Brain Training Games
for a Healthcare Service Robot for Older People

Ho Seok Ahn[1], Mary Pauline Grace Santos[2],
Charu Wadhwa[3], and Bruce MacDonald[4]

Department of Electrical and Computer Engineering, CARES, University of Auckland,
Auckland, New Zealand
{hs.ahn,b.macdonald,msan065,cwad018}@auckland.ac.nz

Abstract. Many people suffer from memory loss as they get older. This may lead to severe memory conditions, where everyday communication and activities for a person become much more difficult and independent living is a challenge. Brain training is one therapeutic method for people who are concerned about brain function decline. We previously found that brain training games are helpful and enjoyable for older people. In this paper, we describe the development of new computer-based brain training games based on paper-based brain training exercises created by experts on brain function. We develop four games targeting specific areas of memory, prospective memory, face recognition skills, verbal memory, and short-term memory. These games are deployed on our healthcare service robot, which is used in an individual home environment. We report a usability study with older adults to evaluate our brain training games on a healthcare service robot. Results show the games are usable and people responded positively about them, and some improvements were identified for future development.

Keywords: Brain training game, healthcare robot system, caring of older people, healthcare service, therapeutic method.

1 Introduction

Recently, the older population growth is faster than that of young people [1-2]. Some older people may suffer memory loss, a common example being memory lapses that occur almost every day as people age [3] and hinder a person's ability to remember and recall, an example being trying to remember a loved one's name. A person's concentration, memory and judgment slowly deteriorate, affecting their ability to do daily tasks, making independent living difficult for those affected [4]. Some researchers report that exercising cognitive functions mitigates cognitive decline [5]. Researchers and companies have been developing brain fitness games to exercise cognitive functions. Lumosity is a web based brain fitness application [6]. Most users of Lumosity are between the ages of 25-34; an age group that is unlikely to be affected by cognitive issues [7]. Anti-Aging is another brain training website and their games are designed to exercise and help improve memory [8]. They focus on people

M. Beetz et al. (Eds.): ICSR 2014, LNAI 8755, pp. 1–10, 2014.

who are over 35 years of age, not people with severe brain disorders. Unlike other games, Dakim Brain Fitness is a software program dedicated for users over the age of 60 and focuses on people with a variety of cognitive conditions [9-10].

In our previous research, we designed assistant robots for older people, which provide companionship as well healthcare support. We conducted several studies in the initial design and testing stages. We focused on how effective a healthcare robot is, which functions are useful, what are the important factors for healthcare robot systems, and the differences in various places and people, and found that robots are acceptable to older people [2, 11-30]. Our healthcare robots have several service applications that include a medication reminding function, a caregiver service to guide how to measure vital signs, a video chatting service with medical staff as well as family members, entertainment services to play videos and music [21]. We installed Dakim's Brain Fitness game on our healthcare robots, and older people enjoyed playing the game with the robot. Fig. 1 shows our healthcare robots with users. However some brain training software is not clearly visible on the small screens of some robots, for example on the right in Fig. 1, which is particularly important for older people who may suffer some decline in vision.

In this paper, we introduce new brain training games that are deployed on our healthcare robots, are based on clinical established brain training concepts, and are suitable for smaller screens. Dr. Allison Lamont and Gillian Eadie, founders of the Brain and Memory Foundation and the Healthy Memory Company, created paper-based brain training exercises that exercise the six key areas of memory: working memory, verbal memory, non-verbal memory, short-term memory, face recognition and prospective memory [31], and two initial software prototypes for computer games. We have developed four computer games for brain training since 2012, based on the paper-based exercises and initial software. We plan to develop two more games next year. We considered the requests and feedback from the experts (Dr Lamont and Ms. Eadie) of the paper-based brain training exercises when we designed and developed the games.

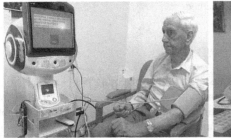

Fig. 1. Our healthcare robot systems; a nursing assistant robot system in the hospital environment (left) and a personal healthcare service robot system in the individual home environment (right)

This paper is organized as follows. In Section 2, we introduce the computer version of four brain training games. In Section 3, we present experiments and evaluations. Finally, we conclude this paper in Section 4.

2   Brain Training Games

We have chosen four brain training games from the six paper-based brain training exercises according to the advice of experts. Each brain training game targets a specific area of memory as follows.

- Night at the Movies: Prospective memory
- Cross the Bridge: Short-term memory
- Shopping Spree: Verbal memory
- Wild West Hunt: Face recognition skills

We took advice from the experts regarding the requirements for the software design, in order to maximize the therapeutic effect of the game and minimize divergence of the game from the memory theory behind the paper based exercises, which can occur during the agile, iterative software development process we used. For example, there should be 8-10 levels of increasing difficulty in each game. The difficulty for each level should slowly increase, and the next level can only be unlocked once the previous level has been completed. Hints should be provided to improve performance in each level and feedback is provided on the performance of the user in every level. Instructions on how to play games and information about brain function should be displayed at the start of each game. Sound effects should be heard depending on whether the user passes or fails at each level. The developed game should be deployable on a robot. The Graphical User Interface or the look and feel of the game should be simple and attractive for the 50-70 age group. It should function smoothly without failures or faults. Our robots interact by touch screen input, speech and audio output, so the games should enable touch screen inputs and provide audio output in addition to the screen display; there is no keyboard or mouse input.

Fig. 2. The main page of the brain training game. It shows four different games and users can start any game as well as re-select games after coming back to this page.

Fig. 2 shows the four developed brain training games. When the user executes the applications through the healthcare software on the robot, it shows the main page where a game can be selected. It is possible to go back to the main page, restart the same game again, and start different games. When each game is started, it shows information about the intended memory therapy effect and instructions. Each game has 10 different levels, and the next level is unlocked if the user clears the current level.

2.1   Night at the Movies

Night at the Movies is a game that strengthens a person's prospective memory, which is related to the ability to remember something in the future, for example, doctor appointments, birthdays, etc. The game works by having the users remember certain parts of a movie clip, indicated by a star, and being able to recall these parts when the video clip is played back the second time. At each level, star timings were randomized by splitting the duration of the clip depending on the number of stars for the level. Users should click the right timing when stars were shown, to succeed in playing the game. Fig. 3 shows sample screenshots of Night at the Movies.

Fig. 3. Screenshots of Night at the Movies; users learn how to play this game from the instructions (left) and should remember the timing star is bright (right)

2.2   Cross the Bridge

Cross the Bridge is a game that strengthens a person's short-term memory. The game works by having the users remember a bridge pattern, consisting of a number of various colored blocks. Users need to reconstruct the bridge correctly, so that the person crossing the bridge will not fall into the water. The array of blocks on the bridge is randomly generated and the numbers of blocks is increased as level increases. Fig. 4 shows screenshots of Cross the Bridge.

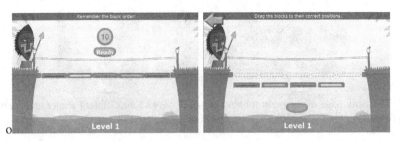

Fig. 4. Screenshots of Cross the Bridge; users should remember the color of blocks on the bridge (left) and drag and drop the correct blocks (right)

2.3   Shopping Spree

Shopping Spree is a game that strengthens a person's verbal memory, which is related to the ability to remember people's names, memorable locations, etc. The game works by having the user remember a written list of items in a shopping list. Users need to correctly identify and select items given in the list, from a shelf of items provided. The shopping list and item placement on shelves are randomly generated and the numbers of items are increased as level goes up. Fig. 5 shows screenshots of Shopping Spree.

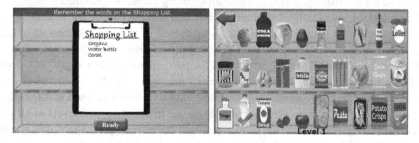

Fig. 5. Screenshots of Shopping Spree; users should remember the shopping list (left) and select the correct items (right)

2.4   Wild West Hunt

Wild West Hunt is a game that strengthens a person's face recognition skills based on features of faces. The game works by having the users remember a cartoon criminal face. Users need to identify the criminal from a line-up of cartoon faces. The criminal and a line-up of cartoon faces are randomly generated and the number of faces is increased as level goes up. Fig. 6 shows screenshots of Wild West Hunt.

Fig. 6. Screenshots of Wild West Hunt; users should remember the cartoon face (left) and select the correct face (right).

3   Experiments and Evaluations

3.1   Overview of Study

We deployed the developed four brain training games on our healthcare service robot iRobiS, shown in Fig. 1 (right), from Yujin robotics in South Korea. It is small enough to take hold with one hand, in size measuring 45x32x32cm and weigh 7kg [21]. It has an Intel Atom processor based internal computer. Physically, it has two arms, which are used mainly for getting attention, indicating emotions and gesturing. It is also equipped with a number of touch sensors at different locations on its body. This enables the programming of realistic responses when users pat, tap, touch, or nudge the robot. Especially, as it has a 7 inch touchscreen on its body, it is useful to play a game without external input devices. The games were developed by authors Santos and Wadhwa using the Adobe Flex environment which is used by our software framework for the iRobiS, and is suitable for creating animated graphical content.

We undertook two kinds of evaluation; one is a usability study for evaluating the overall design of the games, and the other is analyzing of task completion rate of each game. Prior to the study, we obtained ethics approval from the Ethics committee at the University of Auckland. We advertised for participants for this study, and a total of 10 participants were recruited between the ages of 50 - 70 who lived independently. The participants were required to carry out pre-defined tasks, such as navigating to the instruction screen, completing the first and second levels of all four games. Following the completion of the specified tasks, each participant was given a questionnaire to fill out. Table 1 shows the questionnaire for this study.

Table 1. Instructions and Wording of the brain training game Questionnaire

Scale instructions	Strongly Disagree				Strongly agree
Q1) The games have a similar look and feel	1	2	3	4	5
Q2) The navigation between the different screens is smooth	1	2	3	4	5
Q3) The color scheme is appealing for each of the games	1	2	3	4	5
Q4) The text can be clearly read for each of the games	1	2	3	4	5
Q5) The games are intuitive and simple in nature	1	2	3	4	5
Q6) The difficulty level of the games increases gradually	1	2	3	4	5
Q7) The games are fun and maintain the user's interest	1	2	3	4	5

3.2 Experimental Results

Fig. 7 shows the results of the usability study. Each participant's response was scaled from 1 to 5 for the agree/disagree statements and the sum of each response is the total rating for a particular aspect of the games. From the results, it is evident that smooth navigation between screens (Q1), text readability (Q4) and fun and maintain user's interest (Q7) were the strengths of the application whereas, the games not having a similar look and feel (Q1) was a weakness. A possible reason for this weakness may be that each game targets different areas of the memory and the theme of each game is quite different. Participants enjoyed playing games, and were satisfied about the performance of the game.

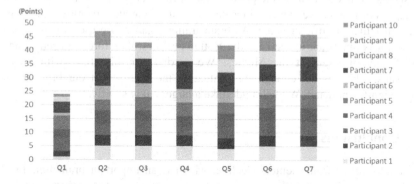

Fig. 7. The result of usability study of brain training games; sum of participants' responses scaled between 1 and 5 based on the questionnaire shown in Table 1

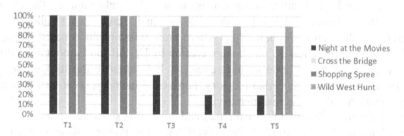

Fig. 8. The result of task completion analysis of brain training games. We used the log data of the usability study. T1: navigation to brain screen, T2: navigation to instruction screen, T3: passing of the first level, T4: passing of the second level, T5: reattempt of the first level.

We analyzed the task completion rate of each brain training game using the log data of the usability study. We analyzed five items; navigation to the brain and instruction screen, passing of the first and second level, and reattempts of the first level. Fig. 8 shows the results. None of the participants had any problem navigating to the brain and instruction screen in any of the four brain training games. There were different results on passing the first and second level in the four brain training games. The passing rate of all games was decreased when participants did the second level. It means that the level of difficulty increased well as the levels go higher, which is one of the design requirements.

The passing rate of each game was quite different. Wild West Hunt had the highest passing rate: 100% for the first level and 90% for the second level, which means that only one participant failed to clear the second level. Whereas, Night at the Movies had the lowest passing rate: 40% for the first level and 20% for the second level, which means that only two participants succeeded to clear the second level of Night at the Movies. A potential reason for this could be that Night at the Movies requires participants to multitask by watching the video, processing auditory and visual information, and guessing the appearances of stars. The reattempt rate of the first level shows similar patterns to the passing rate of the four brain training games. From the analysis of results, we can consider that participants tended to enjoy easier games again rather than a difficult game. As brain training games are used for therapeutic purposes, it is important to increase the reattempt times, therefore we need to adjust the difficulty of Night at the Movies. We carried out this study with only 10 participants, which is not enough for evaluating the usability of our brain training game in significant detail. We will undertake a larger study shortly.

4   Conclusions

Older adults suffer from memory decline, which is an important brain function for everyday communication and activities. As one of the methods to mitigate the slow progression of memory loss, various brain training games are used, and we used one commercial brain game for our research. We applied it on our healthcare service robots, and found that brain training games are a good application for the elderly. However this software did not present well on small screens of smaller robots. Therefore, we developed a computer version of brain training games based on paper-based brain training exercises, which were created by Dr. Allison Lamont and Gillian Eadie. These paper-based brain training exercises help improve the six key areas of memory: working memory, verbal memory, non-verbal memory, short-term memory, face recognition and prospective memory.

Among them, we selected four exercises and developed software versions of them on the advice of experts: Night at the Movies related to prospective memory, Cross the Bridge related to short-term memory, Shopping Spree related to verbal memory, and Wild West Hunt related to face recognition skills. We designed the games by considering the requests from the experts of the paper-based brain training exercises. We applied the developed brain training games on our healthcare service robot. We conducted a usability study to evaluate our four brain training games with 10 participants between the ages of 50 - 70 who live independently. Another aim was to find any inconsistencies in the design of the games. From the results, we confirmed that our brain training games are well designed and developed for the elderly and the healthcare robot system. Participants had fun with the games and robot, but Night at the Movies had failed to give interest and suffered an increase in the number of reattempt times due to its difficulty. Therefore, we need to adjust the difficulty of Night at the Movies in the future. We will also conduct more usability studies to find any differences for conditions such as age, gender, and cultural background. We will also develop two more games, which are related to working memory and non-verbal memory, and apply them on various platforms such as tablets and web applications. Once the software usability is improved we plan to study the effectiveness of the games for promoting memory function.

References

1. Lutz, W., Sanderson, W., Scherbov, S.: The coming acceleration of global population ageing. Nature 451(7179), 716–719 (2008)
2. Ahn, H.S., MacDonald, B.A., Kuo, I.-H., Datta, C., Stafford, R., Kerse, N., Peri, K., Broadbent, E.: Design of a Kiosk Type Healthcare Robot System for Older People in Private and Public Places. In: International Conference on Simulation, Modeling, and Programming for Autonomous Robots (2014)
3. Smith, M., Robinson, L., Segal, R.: Last Age-Related Memory Loss, helpguide.org (2013), http://www.helpguide.org/life/prevent_memory_loss.htm
4. Robinson, H., MacDonald, B.A., Kerse, N., Broadbent, E.: Suitability of Healthcare Robots for a Dementia Unit and Suggested Improvements. Journal of the American Medical Directors Association 14(1), 34–40 (2013)
5. Willis, S.L., Tennstedt, S.L., Marsiske, M., Ball, K., Elias, J., Koepke, K.M., Morris, J.N., Rebok, G.W., Unverzagt, F.W., Stoddard, A.M., Wright, E.: Long-term effects of cognitive training on everyday functional outcomes in older adults. Journal of the American Medical Association 296, 2805–2814 (2006)
6. http://www.lumosity.com
7. Alban, D.: Lumosity Brain Training Program (2012), http://bebrainfit.com/lifestyle/mental/lumosity-brain-training-program/
8. http://www.anti-aginggames.com/games.html
9. Alban, D.: Brain Training Programs Compared - Which Is Right for You? (2013), http://bebrainfit.com/lifestyle/mental/brain-training-programs-compared-which-is-right-for-you/
10. http://www.dakim.com
11. Robinson, H., MacDonald, B.A., Kerse, N., Broadbent, E.: The Psychosocial Effects of a Companion Robot: A Randomized Controlled Trial. Journal of the American Medical Directors Association 14(9), 661–667 (2013)
12. Kuo, I.H., Rabindran, J.M., Broadbent, E., Lee, Y.I., Kerse, N., Stafford, R.M.Q., MacDonald, B.A.: Age and gender factors in user acceptance of healthcare robots. In: IEEE International Symposium on Robot and Human Interactive Communication (ROMAN 2009), pp. 214–219 (2009)
13. Broadbent, E., Tamagawa, R., Kerse, N., Knock, B., Patience, A., MacDonald, B.: Retirement home staff and residents' preferences for healthcare robots. In: IEEE International Symposium on Robot and Human Interactive Communication (ROMAN 2009), pp. 645–650 (2009)
14. Stafford, R.Q., Broadbent, E., Jayawardena, C., Unger, U., Kuo, I.H., Igic, A., Wong, R., Kerse, N., Watson, C., MacDonald, B.A.: Improved robot attitudes and emotions at a retirement home after meeting a robot. In: IEEE International Symposium on Robot and Human Interactive Communication, pp. 82–87 (2010)
15. Jayawardena, C., Kuo, I., Datta, C., Stafford, R.Q., Broadbent, E., MacDonald, B.A.: Design, implementation and field tests of a socially assistive robot for the elderly: HealthBot Version 2. In: RAS/EMBS International Conference on Biomedical Robotics and Biomechatronics, pp. 1837–1842 (2012)
16. Broadbent, E., Tamagawa, R., Patience, A., Knock, B., Kerse, N., Day, K., MacDonald, B.A.: Attitudes towards health care robots in a retirement village. Australasian Journal on Ageing 31(2), 115–120 (2012)
17. Ahn, H.S., Choi, J.Y.: Can We Teach What Emotions a Robot should Express? In: IEEE International Conference on Intelligent Robots and Systems, pp. 1407–1412 (2012)

18. Ahn, H.S., Lee, D.-W., Choi, D., Lee, D.-Y., Hur, M., Lee, H.: Uses of Facial Expressions of Android Head System according to Gender and Age. In: IEEE International Conference on Systems, Man, and Cybernetics, pp. 2300–2305 (2012)

19. Ahn, H.S., Lee, D.-W., Choi, D., Lee, D.-Y., Hur, M., Lee, H.: Appropriate Emotions for Facial Expressions of 33-DOFs Android Head EveR-4 H33. In: IEEE International Symposium on Robot and Human Interactive Communication, pp. 1115–1120 (2012)

20. Ahn, H.S., Lee, D.-W., Choi, D., Lee, D.-Y., Hur, M., Lee, H.: Difference of Efficiency in Human-Robot Interaction According to Condition of Experimental Environment. In: Ge, S.S., Khatib, O., Cabibihan, J.-J., Simmons, R., Williams, M.-A. (eds.) ICSR 2012. LNCS, vol. 7621, pp. 219–227. Springer, Heidelberg (2012)

21. Datta, C., Yang, H.Y., Kuo, I.-H., Broadbent, E., MacDonald, B.A.: Software platform design for personal service robots in healthcare. In: IEEE International Conference on Robotics, Automation and Mechatronics, pp. 156–161 (2013)

22. Ahn, H.S., Choi, J.Y., Lee, D.-W., Shon, W.H.: A Behavior Combination Generating Method for Reflecting Emotional Probabilities using Simulated Annealing Algorithm. In: IEEE International Symposium on Robot and Human Interactive Communication, pp. 192–197 (2011)

23. Stafford, R.Q., MacDonald, B.A., Jayawardena, C., Wegner, D.M., Broadbent, E.: Does the Robot Have a Mind? Mind Perception and Attitudes Towards Robots Predict Use of an Eldercare Robot. International Journal of Social Robotics 6(1), 17–32 (2014)

24. Ahn, H.S.: Designing of a Personality based Emotional Decision Model for Generating Various Emotional Behavior of Social Robots. Advances in Human-Computer Interaction 2014, 1–14 (2014)

25. Ahn, H.S., Lee, D.-W., Choi, D., Lee, D.-Y., Lee, H., Baeg, M.-H.: Development of an Incarnate Announcing Robot System using Emotional Interaction with Humans. International Journal of Humanoid Robotics 10(2), 1–24 (2013)

26. Kuo, C.J.-H., Broadbent, E., MacDonald, B.A.: Socially Assistive Robot HealthBot: Design, Implementation, and Field Trials. IEEE Systems Journal (2014)

27. Ahn, H.S., Sa, I.-K., Lee, D.-W., Choi, D.: A Playmate Robot System for Playing the Rock-Paper-Scissors Game with Humans. Artificial Life and Robotics 16(2), 142–146 (2011)

28. Lee, D., Ahn, H.S., Choi, J.Y.: A General Behavior Generation Module for Emotional Robots Using Unit Behavior Combination Method. In: IEEE International Symposium on Robot and Human Interactive Communication, pp. 375–380 (2009)

29. Broadbent, E., et al.: Benefits and dis-benefits of healthcare robots in a retirement village: A comparison trial. Australasian Journal on Ageing (2014)

30. Ahn, H.S., Choi, J.Y., Lee, D.-W., Shon, W.H.: Natural Changing of Emotional Expression by Considering Correlation of Behavior History. In: IEEE International Symposium on Consumer Electronics, pp. 369–372 (2011)

31. http://www.brainfit.co.nz

Impact of a Social Humanoid Robot as a Therapy Assistant in Children Cancer Treatment

Minoo Alemi[1,3], Ali Meghdari[1,*], Ashkan Ghanbarzadeh[1],
Leila Jafari Moghadam[2], and Anooshe Ghanbarzadeh[1]

[1] Social Robotics Laboratory,
Center of Excellence in Design, Robotics, and Automation,
Sharif University of Technology, Tehran, Iran
[2] Mahak Hospital and Rehabilitation Complex, Tehran, Iran
[3] Islamic Azad University, Tehran North Branch, Tehran, Iran
meghdari@sharif.edu

Abstract. Treating cancer encompasses many invasive procedures that can be a source of distress in oncology patients. Distress itself can be a major obstruction in the path of acceptance of treatment and the patient's adaptation to it, thereby reducing its efficiency. These distress symptoms have been found to be prevalent in children suffering from cancer, in a spectrum from mild to critical. In the past years in response to this psychological suffering, researchers have proposed and tested several methods such as relaxation, hypnosis, desensitization, and distraction. This paper propounds a new approach by exploring the effect of utilizing a humanoid robot as a therapy-assistive tool in dealing with pediatric distress. Ten children, ages 6-10, diagnosed with cancer were randomly assigned into two groups of *SRAT* (5 kids) and psychotherapy (5 kids) at two specialized hospitals in Tehran. A *NAO* robot was programmed and employed as a robotic assistant to a psychologist in the SRAT group to perform various scenarios in eight intervention sessions. The promising results of this study in the level of anger, depression, and anxiety could render using social robots applicable in psychological interventions for children with cancer. Results of this study shall be beneficial to psychologists, oncologists, and robot specialists.

Keywords: Social Robot-Assisted Therapy (*SRAT*), Cancer, Anxiety, Anger, Depression, Psychotherapy.

1   Introduction

Treatment of cancer, besides its hardships and setbacks, has considerable mental and physical side effects that may be as significant as the disease itself. Psychological distress, anxiety, reduction of appetite, and weight loss are among the most prevalent symptoms in cancer patients at all stages of treatment. In the physical respect, chronic pain, post-operational pain and anticipatory nausea are commonly observed in most

---

* Corresponding author.

M. Beetz et al. (Eds.): ICSR 2014, LNAI 8755, pp. 11–22, 2014.

patients suffering from cancer [1]. In recent years, numerous studies have been conducted in this area, with the aim of alleviating the aforementioned symptoms through non-pharmacological techniques. In psychological intervention, which is the focus of this review, cognitive-behavioral therapy (CBT) and group therapy have been successful in decreasing distress and enhancing the quality of life in cancer patients. CBT has been proved to be fulfilling in appeasing the physical and psychological symptoms of distress, including depression, anxiety, pain, and post-cancer fatigue. [1, 3-8]

Various forms of group therapy have been studied in the context of cancer, all of which have shown improvement in adjustment skills, pain management, and traumatic stress among adult cancer patients. Moreover, all forms of group therapy have proven to be undeniably effective in elevating the quality of life, enhancing psychological symptoms, and assuaging pain among patients diagnosed with metastatic breast cancer [9-11].

In addition to psychological interventions, behavioral techniques have displayed effective results in the management of cancer distress. These techniques include relaxation, distraction, desensitization, and hypnosis and all have been modified with regard to age-considerations, so that they could be used in child patients. [1] Utilization of the mentioned methods has been shown to be effective in the management of anticipatory nausea and vomiting, and ameliorating anxiety and pain [12-23]. Wide adoption of behavioral methods in clinics is due to the comparative simplicity of their application, directness of their beneficial influence, and the sense of control they provide patients, in their most vulnerable moments. [12]

Alongside the advances in psychological methods, the progress in Social Robotics has prompted tremendous potentials in patient-robot interactions. Clinical and educational applications of robots are of distinct importance in cases involving children, since robots have attractive features that can increase the efficacy of communication. One of the recent research concentrations has been in the utilizations of humanoid robots in both diagnosis and treatment of Autism Spectrum Disorders (ASD). These studies have proved that the use of interactive robotics not only raises autistic children's interest in treatment sessions, but also provides more responsive feedback and engagement from them, when compared to cases in which only a human administers the treatment process [24-27].

In the study at hand, in an original and novel procedure, we tried to combine psychological methods with social applications of a humanoid robot to observe its influence on distress management of pediatric cancer patients. Several factors have been considered in this study such as encompassing the short-term changes in the children's level of anxiety, depression and anger. In this research, the robot acted as an assistant to the psychologist (or psychiatrist) in order to work on the abovementioned factors with children through several preplanned scenarios. These scenarios were performed in an interactive manner between the trainer, the psychologist, the robot, and the kids. The humanoid robot used in this study was able to exhibit sympathetic emotions with speech tone and body motions and could play a part that was close to the child being treated. Hence, it could provide the incentive for the child to express his or her feelings, and thereby engender enhancements in targeted factors.

2   Research Questions

a). What are the anxiety, anger, and depression levels of the children with cancer before and after social robot-assisted therapy (SRAT) and psychotherapy?

b). Is there any improvement in SRAT and psychotherapy groups regarding their anxiety, anger, and depression levels after their therapies?

3   Method

Participants

Participants were gathered for an 8-sessioned period (24 April to 12 May 2014), from a non-governmental hospital, MAHAK, specializing in pediatric cancer, and the oncology sector of another medical center, MARKAZ-e-TEBI-KOODAKAN (MTK). During the trial, all children were receiving active treatment and were able to attend the scheduled sessions. Participants were chosen in an age range of 7-12 years old and the data of those who didn't succeed in taking part in more than 6 sessions were not included in the results. This means that ultimately, from the initial 10 participants only 6 were included in the final results (a mean age of 9.5, and a std. of 1.26). The control group (5 members) was also selected randomly from available patients in both medical centers (average=9.4, std. =1.36).

Instrument

The core device of this study was *NAO* (renamed as *Nima*, a Persian name for better interactions with the kids), a programmable humanoid robot developed by Aldebaran Robotics Company. Its physical specifications are presented in Table.1. *NAO* is capable of displaying human-like body gestures, speaking, playing sound effects

Table 1. Specifications of the *NAO* robot

Nao Next Gen (2011)	
Height	58 centimeters (23 in)
Weight	4.3 kilograms (9.5 lb)
Autonomy	60 minutes (active use), 90 minutes (normal use)
DOF	21
CPU	Intel Atom @ 1.6 GHz
Built-in OS	Linux
Compatible OS	Windows, Mac OS, Linux
Programming languages	C++, Python, Java, MATLAB, Urbi, C, .Net
Vision	Two HD 1280x960 cameras
Connectivity	Ethernet, Wi-Fi

and music, and dancing. The scenario conversations between two trainers (one psychologist and the study investigator) and robot, and also the robot's dialogs to the children were all composed and loaded on the device before each session. The software used to design and plan NAO's speeches and animations was Choregraphe 1.14.5 developed by the original company. Choregraphe is a multi-platform desktop application that allows the user to connect to the real robot, and conveniently create animations for its joints and body parts.

A human operator took charge of sending commands from a laptop (Windows 7, SP1, 2.5 GHz processor), via a modem to NAO at suitable times during sessions. Another laptop (Windows XP, SP2, 1.0 GHz processor) was utilized to display presentation slides for each session.

Questionnaires

Three factors; anxiety, depression, and anger were assessed by three standard psychological questionnaires, scaled for children. These tests were administrated twice, prior to the first and after the eighth session. All the questionnaires were formatted in Lickret self-report and 3- or 4-point rating response choice. Anxiety was measured with the Multidimensional Anxiety Children Scale (MASC, 39 items). This test was developed by March and others (1997) for an age span of 8-18, and its validity for Iranian children was confirmed in a report by Mashhadi and others (2012) for a sample of 507 students from 3rd, 4th, and 5th grade. Depression was measured by Kovaks' (1985) Children's Depression Inventory (CDI, 27 items), developed for individuals from 7 to 17 years old. Finally, Children's Inventory of Anger (CIA, Nelson and Finch, 2000) having 39 items, was employed to measure anger level in participants. The CDI and CIA confirmatory retests in Iran were also reported in Mashhadi's study [30-35].

Intervention (Data Collection)

With respect to the study's objectives, 8 scenarios were composed, each focusing on a concept of major importance for the children who were receiving treatment. The general plot was primarily based on the conjecture that knowing about the procedure and necessity of their treatment could meaningfully influence the child's level of acceptance, cooperation, and adjustment.

The NAO robot (renamed *Nima* in this project) was programmed to play a different role in each session, and convey all the general and necessary information that could serve the purpose of reducing distress. Alongside with making children aware of their condition, Nima displayed a sense of sympathy with the situation that patients were in. In this study, Nima was introduced as a baby boy robot that had an illness similar to the patients and was mandated by a doctor to attend hospital twice every week to obtain his dose of chemotherapy. It should be noted that Nima's character as a baby robot was maintained in all sessions, i.e. he kept his baby tone and cheerfulness while playing his roles. In other words, he used different roles in order to convey his information to kids in a more enjoyable, and of course, organized way (see Figure 1). The clinical objective of each session is presented briefly in Table 2.

Table 2. Clinical goals of each intervention session

Session	Clinical Objective
Introduction	Children getting prepared to communicate with the NAO (Nima) robot, sharing fears and worries in hospital milieu
Nima as a doctor	Getting more acquainted with the hospital and its different sections, treatment and diagnostic procedures, kids' confronting their fears and stress by getting aware of the reason behind each procedure, kid's expressing/sharing their feelings and emotions about various kinds of treatment
Nima as Chemo-Hero	Establishing a positive image about chemotherapy and its adverse side effects, appreciation of kids' forbearance and bravery against the disease
Nima as a nurse	Instruction in important points about hygiene and respecting children's independence in their everyday tasks, teaching kids how to relax themselves with the "robot-spaghetti" technique while listening to a soothing music
Nima as a cook	Introducing beneficial and necessary foods while constructing an image about their advantages for health and strength, instructing methods to reduce nausea, discussing various solutions to increase appetite
Nima as an ill kid	Kid's developing a sense of sympathy with a sick and confused robot while comparing him with themselves, seeing themselves in the "sad", "cranky", and occasionally "angry" ego of Nima, and also in the state of power and wisdom
Hopes and dreams	Giving children hope for their future life, and helping them visualize themselves in inspiring, wonderful, and advantageous jobs when they grow up
Saying Goodbye	Reviewing the instructed concepts during previous sessions, preparing the kids to say farewell to Nima

The Performance of Sessions

All sessions were performed in the playroom of the medical centers. Three individuals including a psychotherapist, an operator, and a trained person in communication with children suffering from cancer were present at sessions. After warming up to the ambience, Nima pretended that he was connecting to the network by playing a special sound, and acquiring information associated with his role. Then, some costumes were put on the robot and the session continued with different discussions on important subjects. Mainly, the trained person and robot led the scenario by talking to each other (according to their ascribed lines) and to the children. Each discussion was initiated by Nima with a brief introduction in the form of instructive data or a simulated experience in 3-5 minute timeframes. As noted before, Nima kept his character during all sessions, and acted as a kid playing a role-taking game. The children were aware of this game, and sometimes joined Nima in playing by imitating his words and gestures. All the robot's actions were guided by a trained operator, whose control of the robot

(a)

(b)

Fig. 1. (a) The psychologist with the patients, (b) the NAO robot (Nima) acting ill

was not observed by the children, via an ordinary computer and a modem. Afterward, the robot remained silent, pretending he was carefully listening to the kids' opinions and sharing experiences, for example about spending time in the CT scanning machine. The psychologist and/or a trained person intervened to encourage the participants to get involved in discussions and exchange their feelings. Nima's program was designed to constitute some encouraging comments such as "bravo!" or "how interesting!" as inducements for the kids to be active and expressive. Sessions usually closed with a cheerful song accompanied with Nima's dancing, as to enhance the children's mood and instigate them to participate for the next sessions.

Some assignments were also given to the participants during sessions and Nima was designed to perform some special actions to stimulate the kids. As observed, all of these special performances elevated the kids' cooperation in doing the assigned activities and made the sessions more enjoyable and friendly. Moreover, they gave the robot a human character with whom the kids felt much more close and easy to communicate. The main purpose was to make the best use of the robot in an intervention practice (see Figure 2).

Fig. 2. The NAO robot entertaining and performing for the patients with the help of education specialists and psychologist

4   Results and Discussions

In order to answer the first research question on groups' anxiety, anger, and depression levels, first the mean and standard deviation of each were computed, based on the standard questionnaires previously introduced. Table 3 shows the descriptive statistics of the total mean of the anxiety (4-point scale), anger (4-point scale), and depression (3-point scale) scores of both groups.

Table 3. Descriptive Statistics of Experimental and Control Groups (pre/post-tests)

GROUP		Min	Max	Mean	SD
Experimental	Total Anxiety (pretest)	1.90	2.49	2.23	.227
	Total Anxiety (posttest)	1.69	2.18	1.89	.203
	Total Depression (pretest)	1.26	1.48	1.35	.093
	Total Depression (posttest)	1.07	1.26	1.00	.078
	Total Anger (pretest)	2.10	3.41	2.73	.546
	Total Anger (posttest)	1.97	2.82	2.31	.313
Control	Total Anxiety (pretest)	2.05	3.10	2.36	.440
	Total Anxiety (posttest)	2.00	3.05	2.38	.425
	Total Depression (pretest)	1.11	1.59	1.31	.195
	Total Depression (posttest)	1.04	1.52	1.30	.180
	Total Anger (pretest)	1.95	3.18	2.60	.504
	Total Anger (posttest)	2.67	3.00	2.82	.161

As shown in Table 3, the total mean for the anxiety, depression, and anger scores for the experimental-group's pretest are enhanced respectively from 2.23, 1.35, and 2.73 to 1.89, 1.00, and 2.31 in posttests (see also Graph 1). However, the control group's scores do not display any meaningful improvements.

To answer the second research question, "Is there any improvement in SRAT or psychotherapy groups regarding anxiety, depression, and anger levels after therapy?" a paired sample t-test was run as shown in Table 4.

Table 4 shows the results of the paired t-test of the overall mean of anxiety, depression, and anger of both groups. The t-test results of anxiety, depression, and anger in the experimental group are (t(4)=6.82, p=.002<.05), (t(4)=4.91, p=.039<.05), and (t(4)=3.19, p=.049<.05), respectively, revealing that there was a significant difference between the levels of the targeted factors in patients after treatment in the robotic group. However, conducting the same t-tests for the control group shows no improvements in any of the symptoms. The results clearly provide evidence that the use of a social robot as a psychologist assistant helps to lower the children's psychological problems.

Table 4. Total Paired Samples Tests for Experimental and Control Groups (pre/post-tests)

			Mean	SD	SEM	95% CI	t	df	Sig. (2-tailed)
Experimenta	Pai	Total Anxiety(pretest) Total Anxiety(posttest)	.338	.110	.049	[.200, .476]	6.826	4	.002
	Pai	Total Depression(pretest) Total Depression(posttest)	.350	.056	.032	[.019, .301]	4.914	4	.039
	Pai	Total Anger(pretest) Total Anger(posttest)	.384	.240	.120	[.001, .767]	3.198	4	.049
Control	Pai	Total Anxiety(pretest) Total Anxiety(posttest)	-.015	.150	.067	[-.201, .171]	-.229	4	.830
	Pai	Total Depression(pretest) Total Depression(posttest)	.014	.085	.038	[-.091, .120]	.389	4	.717
	Pai	Total Anger(pretest) Total Anger(posttest)	-.179	.424	.212	[-.855, .496]	-.845	4	.460

Graph1. Descriptive Statistics of the overall posttest mean of anxiety, depression, and anger of both groups

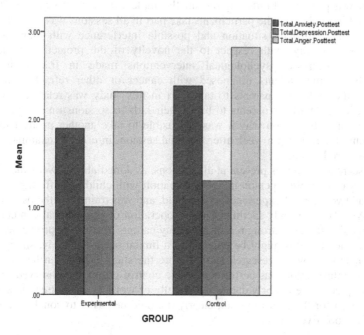

5   Conclusions and Implications

Cancer treatment consists of many invasive and painful procedures that are the main sources of distress and discontentment for patients. It is important to note that these procedures could be psychologically more influential on pediatric patients [17]. In addition, the side effects that treatment brings, such as hair loss, lack of appetite, fatigue, and inability to do many activities exacerbate this situation. So, it is apparent that seeking a method which could alleviate distress and its debilitating impacts is a necessity.

The method this study suggests can be placed alongside other behavioral techniques introduced in literature review, as an independent method for instructing the kids and as an assistant in the process of psychological intervention. From a closer view, a humanoid robot can be used as an addition in any other techniques as a device that can increase the efficiency of communication, involve kids' imaginations in learning, and induce them to be more responsive, as well as cooperative.

To conclude, utilizing a humanoid robot with different communication abilities can be beneficial, both in elevation of efficacy in interventions, and encouraging kids to be more interactive. Also, a humanoid robot was shown to be significantly useful in teaching children about their afflictions, and also instructing them the methods to confront their distress themselves, and take control of their situation.

Study Limitations

This study was primarily limited by the small sample size. This was due to a number of reasons. Firstly, having the participants take part in all sessions was not convenient, considering their difficult situation and possible interference with their treatment protocol. Additionally, with respect to the novelty of the project, and the scant number of systematic psychological interventions made in Iranian hospitals, particularly for the patients diagnosed with cancer or other refractory illnesses, persuading the children's parents to take part in this study was relatively difficult. Generally, it was hard for parents to bring their kids to sessions on a regular basis. These were also the reason why it wasn't feasible to take another posttest from the whole group, for instance a week after the final session, in order to examine whether the effects were lasting.

The researcher who was present at all sessions, as a mediator between the kids and the robot, was a certified person in communication with children suffering from cancer. He had two years of experience in this field, and was trained by the psychologists of MAHAK. This not only facilitated the cooperation of the medical centers in this project, but also it was a strong reason for many parents to trust the project team with their kids. Thereby, this might be counted as a limitation in the study, since most of the participants knew the research team, before the study began. Furthermore, inasmuch as no intervention was performed on the control group, the observed results in SRAT group could be majorly due to the novelty of work. If a new treatment method were also used for the control group, clearly it would be easier to conclude about the source of the positive outcomes.

Acknowledgements. We would like to thank the National Elites Foundation of Iran (http://www.bmn.ir) for their moral and financial support throughout this project. Furthermore, the cooperation of MAHAK and MARKAZ-e-TEBI-KOODAKAN Hospitals in Tehran during the course of this research is highly appreciated. We also thank all individuals who helped us throughout this research: Marjan Vosoughi, Dr. Leyli Koochakzadeh, Dr. Farzad, Kompani, Shadi Ansari, Saeedeh Zorofchi, Elaheh Rahimian, Fatemeh Mirdoraghi.

References

1. Holland, J.C., Alici, Y.: Management of Distress in Cancer Patients. Journal of Supportive Oncology 8, 4–12 (2010)
2. The National Comprehensive Cancer Network, Distress Management Clinical Practice Guidelines in Oncology, version 1 (2009), http://www.nccn.org/professionals/physician_gls/f_guidelines.asp# supportive (accessed May 20, 2014)
3. Kuppenheimer, W.G., Brown, R.T.: Painful procedures in pediatric cancer: A comparison of interventions. Clinical Psychology Review 22(5), 753–786 (2002)
4. Dalton, J.A., Keefe, F.J., Carlson, J., Youngblood, R.: Tailoring cognitive-behavioral treatment for cancer pain. Pain Management Nursing 5(1), 3–18 (2004)
5. Gielissen, M.F., Verhagen, S., Witjes, F., Bleijenberg, G.: Effects of Cognitive Behavior Therapy in Severely Fatigued Disease-Free Cancer Patients Compared With Patients Waiting for Cognitive Behavior Therapy: A Randomized Controlled Trial. Journal of Clinical Oncology 24(30), 4882–4887 (2006)

6. Strong, V., Waters, R., Hibberd, C., Murray, G., Wall, L., Walker, L., McHugh, G., Walker, A., Sharpe, M.: Management of depression for people with cancer (SMaRT oncology 1): a randomised trial. Lancet 372, 40–48 (2008)
7. Hopko, D.R., Bell, J.L., Armento, M., Robertson, S., Mullane, C., Wolf, N., Lejuez, C.W.: Cognitive-Behavior Therapy for Depressed Cancer Patients in a Medical Care Setting. Journal of Behavior Therapy 39(2), 126–136 (2008)
8. Jacobsen, P.B., Jim, H.S.: Psychosocial Interventions for Anxiety and Depression in Adult Cancer Patients: Achievements and Challenges. CA Cancer J. Clin. 58, 214–230 (2008)
9. Classen, C., Butler, L.D., Koopman, C., Miller, E., DiMiceli, S., Giese-Davis, J., Fobair, P., Carlson, R.W., Kraemer, H.C., Spiegel, D.: Supportive-expressive group therapy and distress in patients with metastatic breast cancer: a randomized clinical intervention trial. Archives of General psychiatry 58(5), 494–501 (2001)
10. Kissane, D.W., Bloch, S., Smith, G.C., Miach, P., Clarke, D.M., Ikin, J., Love, A., Ranieri, N., McKenzie, D.: Cognitive-existential group psychotherapy for women with primary breast cancer: a randomised controlled trial. Psychooncology 12(6), 532–546 (2003)
11. Kissane, D.W., Love, A., Hatton, A., Bloch, S., Smith, G., Clarke, D.M., Miach, P., Ikin, J., Ranieri, N., Snyder, R.D.: Effect of cognitive-existential group therapy on survival in early-stage breast cancer. Journal of Clinical Oncology 22(21), 4255–4260 (2004)
12. Redd, W.H., Montgomery, G.H., DuHamel, K.N.: Behavioral Intervention for Cancer Treatment Side Effects. Journal of the National Cancer Institute 93(11) (2001)
13. Beale, I.L.: Scholarly Literature Review: Efficacy of Psychological Interventions for Pediatric Chronic Illnesses. Journal of Pediatric Psychology 31(5), 437–451 (2006)
14. Figueroa-Moseley, C., Jean-Pierre, P., Roscoe, J.A., Ryan, J.L., Kohli, S., Palesh, O.G., Ryan, E.P., Carroll, J., Morrow, G.R.: Behavioral interventions in treating anticipatory nausea and vomiting. Journal of National Comprehensive CancerNetwork 5(1), 44–50 (2007)
15. Lotfi-Jam, K., Carey, M., Jefford, M., Schofield, P., Charleson, C., Aranda, S.: Nonpharmacologic Strategies for Managing Common Chemotherapy Adverse Effects: A Systematic Review. Journal of Clinical Oncology 26(34), 5618–5629 (2008)
16. Richardson, J., Smith, J.E., Pilkington, K.: Hypnosis for Procedure-Related Pain and Distress in Pediatric Cancer Patients: A Systematic Review of Effectiveness and Methodology Related to Hypnosis Interventions. Journal of Pain and Symptom Management 31(1), 70–84 (2006)
17. Butler, L.D., Symons, B.K., Henderson, S.L., Shortliffe, L.D., Spiegel, D.: Hypnosis reduces distress and duration of an invasive medical procedure for children. Pediatric 115, 77–85 (2005)
18. Syrjal, K.L., Donaldson, G.W., Davis, M.W., Kippes, M.E., Carr, J.E.: Relaxation and imagery and cognitive-behavioral training reduce pain during cancer treatment: a controlled clinical trial. Pain 63(2), 189–198 (1995)
19. Spiegel, D., Moore, R.: Imagery and hypnosis in the treatment of cancer patients. Oncology (Williston Park) 11(8), 1179–1189 (1997)
20. Smith, J.E., Mccall, G., Richardson, A., Pilkington, K., Kirsch, I.: Hypnosis for procedure-related pain and distress in pediatric cancer patients: a systematic review of effectiveness and methodology related to hypnosis interventions. Journal of Pain and Symptom Management 31(1), 70–84 (2006)
21. Kleibe, C., Harper, D.C.: Effects of Distraction on Children's Pain and Distress during Medical Procedures: A Meta-Analysis. Nurse Research 48(1), 44–49 (1999)

22. Gershon, J., Zimand, E., Lemos, R., Rothbaum, B.O., Hodges, L.: Use of Virtual Reality as a Distracter for Painful Procedures in a Patient with Pediatric Cancer: A Case Study. Cyber Psychology & Behavior 6(6), 657–661 (2003)

23. Windich-Biermeier, A., Sjoberg, I., Dale, J.C., Eshelman, D., Guzzetta, C.E.: Effects of Distraction on Pain, Fear, and Distress During Venous Port Access and Venipuncture in Children and Adolescents With Cancer. Journal of Pediatric Oncology Nursing 24(1), 8–19 (2007)

24. Robins, B., Dautenhahn, K., Dubowski, J.: Does appearance matter in the interaction of children with autism with a humanoid robot? Interaction Studies 7(3), 509–542 (2006)

25. Scassellati, B., Admoni, H., Matari, M.: Robots for Use in Autism Research. Annual Review of Biomedical Engineering 14, 275–294 (2012)

26. Meghdari, A., Alemi, M., Pouretemad, H.R., Taheri, A.R.: Clinical Application of a Humanoid Robot in Playing Imitation Games for Autistic Children in Iran. In: CD-Rom Proc. of the 2nd Basic Clinical and Neuroscience Congress 2013, Tehran, Iran (2013) (in Persian)

27. Meghdari, A., Alemi, M., Taheri, A.R.: The Effects of Using Humanoid Robots for Treatment of Individuals with Autism in Iran. In: 6th Neuropsychology Symposium, Tehran, Iran (2013)

28. Meghdari, A., Alemi, M., Ghazisaedy, M., Taheri, A.R., Karimian, A., Zandvakili, M.: Applying Robots as Teaching Assistant in EFL Classes at Iranian Middle-Schools. In: CD Proc. of the Int. Conf. on Education and Modern Educational Technologies (EMET 2013), Venice, Italy (2013)

29. Alemi, M., Daftarifard, P., Pashmforoosh, R.: The Impact of Language Anxiety and Language Proficiency on WTC in EFL Context. Cross-Cultural Communication Journal 7(3), 150–166 (2011)

30. March, J.S., Parker, J.D.A., Sullivan, K., Stallings, P.: The Multidimensional Anxiety Scale for Children (MASC): factor structure, reliability, and validity. Journal of the American Academy of Child and Adolescent Psychiatry 36, 554–565 (1997)

31. Chorpita, B.F., Moffitt, C.E., Gray, J.: Psychometric properties of the Revised Child Anxiety and Depression Scale in a clinical sample. Behaviour Research and Therapy 43, 309–322 (2005)

32. Mashhadi, A., Mirdoraghi, F., Bahrami, B., SoltaniShal, R.: Effect of Demographic Variables on Children Anxiety. Iranian J. Psychiatry 7, 3 (2012)

33. Kovacs, M.: The Children's Depression Inventory (CDI). Psychopharmacology Bulletin 21(4), 995–998 (1985)

34. Nelson, W.M., Finch, A.J.: Children's Inventory of Anger. Western Psychological Services, Los Angeles (2000)

35. Zibaei, A., Gholami, H., Zare, M., Mahdian, H., Yavari, M., Haresabadi, M.: The effect of offline education on anger management in guidance school girls in Mashhad. Journal of North Khorasan University of Medical Sciences 5(2), 385 (2013)

36. Motzfeldt, H.: Der Chemo-Kasper: und seine Jagd auf die bösen Krebszellen, Deutsche Kinderkrebsstiftung der Deutsche Leukämie-Forschungshilfe, Bonn (Translated by P. Lajevardi) (2010), http://www.mahak-charity.org/main/fa/about-mahak2/593

Skeleton Tracking Based Complex Human Activity Recognition Using Kinect Camera

Muhammad Latif Anjum[1], Omar Ahmad[1], Stefano Rosa[1],
Jingchun Yin[1], and Basilio Bona[2]

[1] Department of Mechanical and Aerospace Engineering (DIMEAS),
Politecnico di Torino, 10129, Torino, Italy
{muhammad.anjum,omar.ahmad,stefano.rosa,jingchun.yin}@polito.it
[2] Department of Control and Computer Engineering (DAUIN),
Politecnico di Torino, 10129, Torino, Italy
basilio.bona@polito.it

Abstract. This paper presents a new and efficient algorithm for complex human activity recognition using depth videos recorded from a single Microsoft Kinect camera. The algorithm has been implemented on videos recorded from Kinect camera in OpenNI video file format (.oni). OpenNI file format provides a combined video with both RGB and depth information. An OpenNI specific dataset of such videos has been created containing 200 videos of 8 different activities being performed by different individuals. This dataset should serve as a reference for future research involving OpenNI skeleton tracker. The algorithm is based on skeleton tracking using state of the art OpenNI skeleton tracker. Various joints and body parts in human skeleton have been tracked and the selection of these joints is made based on the nature of the activity being performed. The change in position of the selected joints and body parts during the activity has been used to construct feature vectors for each activity. Support vector machine (SVM) multi-class classifier has been used to classify and recognize the activities being performed. Experimental results show the algorithm is able to successfully classify the set of activities irrespective of the individual performing the activities and the position of the individual in front of the camera.

Keywords: OpenNI, Skeleton tracking, Multi-class SVM, Activity recognition, RGBD Dataset.

1 Introduction

With every passing day, the research in robotics is converging to make the robots more and more social. Robots today are jumping out of their once strong field, the industrialized robotics, and are about to invade the human society. The first and the foremost task at their hands is to figure out what is going on around them or to understand what activities the individuals around them are performing. We have presented one such algorithm in this paper which can be implemented

M. Beetz et al. (Eds.): ICSR 2014, LNAI 8755, pp. 23–33, 2014.

with a very low cost single Kinect sensor and enables the robots to recognize the activities being performed in front of them and respond accordingly.

Human activity recognition has been a highly sought after subject in the recent times. After a quick literature review, this research area can be divided into two major groups: (1) the activity recognition research using wearable sensors, and (2) the activity recognition research using RGB and depth cameras. The research based on wearable sensors mainly targets sports activities and athletes. A good insight into this field of activity recognition can obtained from [9], [2] and [4]. The wearable sensors based recognition is not directly applicable to robotics because, in a normal day to day environment, humans are not expected to use wearable sensor for robots' guidance.

Activity recognition using cameras is more relevant to robotics mainly because robots can be easily equipped with a camera. Most of the previous research in this area has been focused on activity recognition using videos and images created by 2D cameras. Huimin et al. [10] used a multi-class SVM classifier for human activity recognition using RGB videos. Their algorithm provides good results both on their own home-brewed dataset and public dataset provided in [11]. Jinhui et al. [5] used motion and structure change estimation in RGB videos to classify different human activities. Similar hand gesture recognition has been implemented on RGB videos by Omar et al. [1]. They have presented results using both SVM and ANN classifiers.

Ever since the emergence of Microsoft Kinect camera that can capture depth images and videos, there is more focus on using depth videos for human activity recognition. Youding et al. [12] used depth image sequences to track human body pose. They have used Bayesian framework and have been successfully able to track 3D human pose. Tayyab et al. [7] used videos from a Kinect camera mounted on a quadrocopter for gesture recognition. Based on the gesture recognition results, the quadrocopter was successfully able to follow the individual performing the gesture. More recent work was presented by Hema et al. [6]. They have used RGB-D videos for activity recognition using both skeleton tracking and object affordances where the training was done using structural support vector machines (SSVM).

The contribution of this paper to this growing field of research is two-fold: (1) provision of the public dataset of RGB-D videos (.oni file format) that can be used for OpenNI skeleton tracker, (2) implementation of an efficient activity recognition algorithm based on skeleton tracking. In summary, we provide the following:

- RGB-D activity dataset purposely built for OpenNI skeleton tracker.
- A modified OpenNI skeleton tracker (as ROS package) for offline recorded videos.
- A ROS package for extracting joints' position based features from videos.
- A ROS package for training and testing using SVM multi-class classifier.

The dataset and ROS packages have been made available on the website of our research group, LabRob (http://www.polito.it/labrob). We have selected at least three joints or body parts in human skeleton to track using OpenNI skeleton tracker. The change in position in these three joints has been used to construct the feature vectors for each activity. The selection of these joints has been made based on the nature of activity being performed. Once feature vectors for all activities have been constructed, we have used multi-class SVM classifier for training and testing of the algorithm.

2   RGB-D Activity Dataset

Looking at the increasing attention the Kinect sensor is getting from the robotics research community, it is all but necessary that a specific dataset of activities be constructed and made available to the research community. Our dataset contains 200 videos (.oni file format) of 8 different activities being performed by two different individuals. Each video in the dataset starts with a surrender / Psi pose (figure 1) required for calibration in OpenNI skeleton tracker. Before getting to features and statistics of our dataset, let's have a look at already available RGB-D datasets.

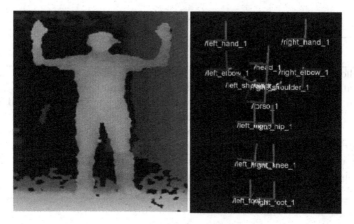

Fig. 1. Surrender / Psi pose in front of Kinect camera and its corresponding skeleton tracking out put shown in rviz

2.1   Available RGB-D Datasets

There are very few datasets publicly available containing RGB-D videos of human activities. Since most of the previous research has been done using RGB videos, most of the datasets contain only RGB videos. A comprehensive list of all available datasets (both RGB and RGB-D) can be found in [8]. Among them there are only two RGB-D datasets available. Cornell Activity Dataset (CAD-120) [6] contains 120 videos of 10 different activities. However they provide data in image format (both RGB and depth images) which require a complex process to convert it into .oni video format for OpenNI tracker. Furthermore, none of their activity starts with a necessary surrender pose. RGBD-HuDaAct [8] is the second available dataset that provides the data in required file format, but since they are not using OpenNI tracker for activity recognition, they do not start the activity with a surrender pose. Our dataset is purposely built to be used for OpenNI skeleton tracker and should serve as a reference for a vibrant ROS OpenNI community.

2.2 Features and Statistics of Our Dataset

We have used Microsoft XBOX 360 Kinect Sensor to record our videos. Each video is recorded using NiViewer[1] and has a resolution of 640x480 with a .oni file format. Each video starts with a surrender pose required for calibration in OpenNI skeleton tracker. The dataset includes 8 different activities with 25 videos recorded for each activity. We have 5 daily life activities while 3 activities consist of umpire's signals from a Cricket match. Table 1 enlists all activities contained in the dataset.

Table 1. Activities performed in dataset

Activity	Activity description	No. of videos
Wave hello	A person waves hello with his right hand	25
Check watch	A person check time from his left hand wrist watch	25
Pick from ground	A person bends and pick something lying on ground and places it on a cupboard	25
Sit stand	A person sits and stands four times in an exercise fashion	25
Sit and drink	A person sits on a chair and drinks water from a bottle	25
Four signal	The cricket umpire gives a four / boundary signal	25
Leg bye signal	The cricket umpires gives a leg bye signal	25
Dead ball signal	The cricket umpires gives a dead ball signal	25

3 Skeleton Tracking

Our algorithm has been implemented using Robot Operating System (ROS)[2]. We use a modified version of ROS wrapper package for OpenNI based skeleton tracking[3]. The original package works with online Kinect camera attached to the PC and publishes user's skeleton positions as a set of transforms (/tf). The available ROS package has been modified to work with offline recorded videos (.oni file format). The tracker can instantly detect user but requires Psi pose for calibration after which it starts tracking the user skeleton. Once calibrated,

---

[1] The program, NiViewer, is available with OpenNI SDK.
[2] http://www.ros.org/
[3] http://wiki.ros.org/openni_tracker

it starts publishing 3D positions and rotations quaternions of 15 joints or body parts with respect to a fixed frame of reference. The published skeleton joints include both feet, knees, shoulders, hands, elbows, hips, head, neck, and torso. The fact that it can track virtually every joint and part of the body signifies the potential of OpenNI tracker for human activity recognition. With a proper combination of different joints, we can recognize any movement or activity.

4   Constructing Feature Vectors

Constructing feature vectors is the most crucial step of this work, where we decide which joints or body parts are to be tracked for each activity and how to arrange them in a mathematical model for the construction of training and testing data for SVM.

4.1   Selection of Joints to be Tracked

The OpenNI tracker publishes 3D positions and rotation quaternions of 15 joints. If we track all available joints for the construction of feature vectors, the algorithm will become computationally heavy and might not work in real time. Besides, not all joints or body parts are undergoing change in position in any given activity. So, it is all but natural to select the joints and body parts to be tracked for the construction of feature vectors for each activity. The OpenNI tracker publishes positions of all joints relative to a fixed frame of reference. If we use the position relative to a fixed frame of reference, the same activity being pepformed at different positions in front of the camera may give different results. We have, therefore, used joint position relative to another joint position to account for different positions of the user in front of the camera. Table 2 summarizes the joints and their references tracked for each activity to construct the feature vectors.

The joint or body part undergoing the most distinct motion during an activity has been tracked to construct the feature vectors. For example, waving involves a continuous and distinct motion of the right hand and right elbow. Similarly it would be the most useful to track the position of the left hand and left elbow during the check watch activity. Bending down to pick something can be distinguished by tracking position of head and right hand while sit stand activity involves distinct motion of head, hip and torso. Although the selection of joints to be tracked is manual and based on the intuitive understanding of the motion scenarios, this selection is only for the training purposes. Once trained, the algorithm will not require manual selection of joints to be tracked. Figure 2 shows the sample depth images during first four activities in the dataset.

Sitting in a chair and drinking from a bottle is a fairly complex activity. We have tracked the position of right hip, right hand and left hand to construct its feature vector. Making a four signal can be distinguished by tracking right hand and right elbow along with the position of left hand. We have purposely included the leg bye signal in the dataset because it includes movement of the lower part

Table 2. Summary of the joints tracked along with their references for each activity in dataset

Activity	Joints tracked	Reference points
	/right_hand	/head
Wave hello	/right_elbow	/torso
	/right_elbow	/neck
	/left_hand	/head
Check watch	/left_elbow	/torso
	/left_hand	/torso
	/right_hand	/right_foot
Pick something from ground	/head	/right_foot
	/right_shoulder	/right_foot
	/head	/right_foot
Sit stand	/right_hip	/right_foot
	/torso	/right_foot
	/right_hip	/right_foot
Sit on a chair and drink from bottle	/right_hand	/head
	/left_hand	/head
	/right_hand	/head
Four signal	/right_elbow	/head
	/right_elbow	/neck
	/right_hand	/head
Leg bye signal	/right_knee	/head
	/right_knee	/left_foot
	/right_hand	/head
Dead ball signal	/left_hand	/head
	/head	/right_foot

of the body i.e. foot and knee. We have tracked the position of right hand, right knee and right foot for this activity. The dead ball signal involves movement of the two hands along with slight bending down of head, so we have tracked these three part to construct its feature vector. Figure 3 shows the depth images of the next 4 activities in our dataset.

4.2   Mathematical Formulation of Feature Vectors

Let us consider $j1$, $j2$ and $j3$ be the three joints we are tracking in a given activity. The 3D position of each of these joints is published in successive frames. Equation 1 gives the formulation of feature vector for activity A1.

$$FV = \{A1, \{(j1_{x,0}, j1_{y,0}, j1_{z,0}), (j2_{x,0}, j2_{y,0}, j2_{z,0}), (j3_{x,0}, j3_{y,0}, j3_{z,0})\},$$
$$\{(j1_{x,1}, j1_{y,1}, j1_{z,1}), (j2_{x,1}, j2_{y,1}, j2_{z,1}), (j3_{x,1}, j3_{y,1}, j3_{z,1})\},$$
$$..., \{(j1_{x,n}, j1_{y,n}, j1_{z,n}), (j2_{x,n}, j2_{y,n}, j2_{z,n}), (j3_{x,n}, j3_{y,n}, j3_{z,n})\}\} \quad (1)$$

where $j1_{x,0}$ indicates the x position of the joint $j1$ in frame number 0 relative to the reference joint while $j1_{x,1}$ indicates the x position of the same joint in

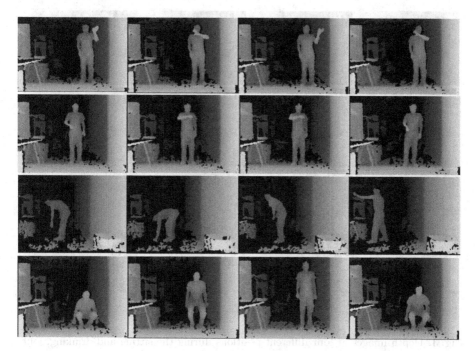

Fig. 2. Depth images at four different positions during the waving, checking watch, picking something from ground and sit stand activities respectively

frame number 1 relative to the same reference joint. The first element in each feature vector is the label of the activity, indicated as $A1$ in equation 1. Based on the length of the longest activity, we have tracked joints' positions for 2260 consecutive frames in each activity. So the dimension of each feature vector is 1x2260.

5   Training and Testing with Multi-class SVM

We now have 200 feature vectors constructed using the mathematical model presented in equation 1. All these feature vectors are put into a matrix to construct feature data as given in equation 2.

$$featureData = \begin{cases} (A_i, f_i) & i = 1, ..., n \\ & A_i \in \{1, 2, ..., 8\} \\ & f_i \in R^{\{1X2260\}} \end{cases} \qquad (2)$$

where A_i represents the activity label and f_i represents the feature sets of the activity. The number of activity videos is represented by n which is 200 in our case. Our goal now can be stated as: Given the feature vector of any activity f_i from the n videos, we have to successfully predict its label A_i.

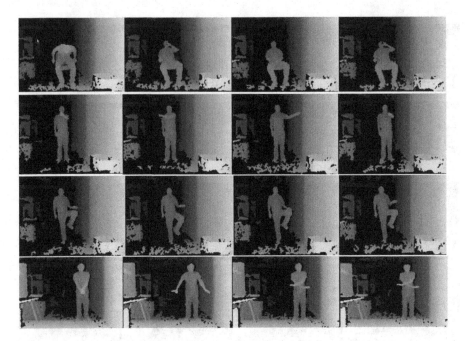

Fig. 3. Depth images at four different positions during the sitting and drinking, four signal, leg bye signal and dead ball signal activities respectively

Support Vector Machines (SVM) is an increasingly becoming a popular tool to solve this kind of classification problem. SVM has produced accurate results in many areas of machine learning including text categorization, gesture recognition and face detection. We are using multi-class SVM classifier which can classify more than two categories of classes. The classification strategy is based on one-against-one approach where each feature set is matched against all samples in the training data. A voting strategy is used for testing where label with maximum positive votes is assigned. Readers interested in learning SVM and other kernel based learning methods are directed towards [3].

Our dataset contains 200 videos in total with 25 videos of each activity. We have trained our SVM based algorithm on 120 videos (15 videos of each activity) while the testing of the algorithm is done on remaining 80 videos (10 videos of each activity) . OpenCV library LibSVM has been used for training and testing of data. We have used linear Gaussian kernel for training the SVM.

6   Experimental Results

It was initially assumed that tracking only two joints or body parts in any activity would be sufficient for high accuracy activity recognition. The results with two joints tracking were satisfactory but not to our expectation. We then added a third joint to be tracked in each activity. The results for both the cases are presented separately.

6.1   Recognition Results While Tracking Two Joints

The two joints undergoing continuous and distinct motion were selected for this experiment. The first two joints in table 2 against each activity were selected for tracking in this experiment. The recognition accuracy turned out to be 92.5% (74 correct recognitions against 80 test videos). The confusion matrix for this experiment is shown in table 3.

Table 3. Confusion matrix for activity recognition while tracking two joints

No.	Activity	1	2	3	4	5	6	7	8
1	Wave	9	0	0	0	0	1	0	0
2	Check watch	0	10	0	0	0	0	0	0
3	Pick from ground	0	0	10	0	0	0	0	0
4	Sit stand	0	0	0	10	0	0	0	0
5	Sit and drink	0	0	0	1	8	0	1	0
6	Four signal	2	0	0	0	0	8	0	0
7	Leg bye signal	0	0	0	0	0	0	10	0
8	Dead ball signal	0	0	1	0	0	0	0	9

Two notable activities that are overlapping and causing a confusion are waving hello and four signal. Both the activities involve similar movements of the right hand with the difference being the position of the hand. Three other activities have one incorrect label prediction each in table 3. These confusions have been avoided with the inclusion of a third joint in tracking algorithm.

6.2   Recognition Results While Tracking Three Joints

A third joint (the third entry in table 2 for each activity) was included in tracking algorithm keeping in mind the activities being confused with each other. For example, the position of right elbow was tracked with reference to neck in four signal and waving activities to further enhance the distinguishing features. With the addition of one more joints for tracking, we were able to obtain 98.75% result with only one video in 80 test videos being confused with another. The confusion matrix for this experiment is given in table 4.

7   Results Analysis and Future Works

The accuracy of results signifies the potential of skeleton tracking for activity recognition. A depth image based skeleton tracker is even better because it makes the image background, lighting conditions etc irrelevant producing robust algorithm. A high accuracy in results can also be attributed to fairly distinct activities in our dataset. Future works should test the algorithm on slightly similar activities (for example head rotation when saying yes versus head rotation when saying no). We are also working on making our dataset robust to include

Table 4. Confusion matrix for activity recognition while tracking three joints

No.	Activity	1	2	3	4	5	6	7	8
1	Wave	10	0	0	0	0	0	0	0
2	Check watch	0	10	0	0	0	0	0	0
3	Pick from ground	0	0	10	0	0	0	0	0
4	Sit stand	0	0	0	10	0	0	0	0
5	Sit and drink	0	0	0	0	9	0	1	0
6	Four signal	0	0	0	0	0	10	0	0
7	Leg bye signal	0	0	0	0	0	0	10	0
8	Dead ball signal	0	0	0	0	0	0	0	10

more activities and more users performing the activities. It can also be a good challenge to include activities involving multiple people (for example two people shaking hands). Another goal would be to integrate object recognition and tracking with skeleton tracker to distinguish between actually drinking something and making drinking like hand movement.

8   Conclusion

We have presented an RGB-D activity dataset and an OpenNI tracker based activity recognition algorithm. The use of skeleton tracker is especially significant because with proper selection of joints to be tracked, we can recognize very complex and long activities. We have tested the algorithm for activities involving movement of arm, hand, head, torso and knees. Feature vectors for each activity have been constructed using relative position of three joints with respect to different reference joints for each activity. The joints undergoing continuous and distinct movement have been selected for feature vector construction in each activity. SVM multi-class classifier has been used for training and testing of data. The experimental results show 98.75% accuracy when tracking three joints in each activity.

References

1. Ahmad, O., Bona, B., Anjum, M.L., Khosa, I.: Using time proportionate intensity images with non-linear classifiers for hand gesture recognition. In: Sakim, H.A.M., Mustaffa, M.T. (eds.) The 8th International Conference on Robotic, Vision, Signal Processing & Power Applications. LNEE, vol. 291, pp. 343–354. Springer, Heidelberg (2013)
2. Choudhury, T., Consolvo, S., Harrison, B., Hightower, J., LaMarca, A., Legrand, L., Rahimi, A., Rea, A., Bordello, G., Hemingway, B., Klasnja, P., Koscher, K., Landay, J., Lester, J., Wyatt, D., Haehnel, D.: The mobile sensing platform: An embedded activity recognition system. IEEE Pervasive Computing 7(2), 32–41 (2008)
3. Cristianini, N., Shawe-Taylor, J.: An Introduction to Support Vector Machines and Other Kernel-based Learning Methods. Cambridge University Press (2000)

4. Ermes, M., Parkka, J., Mantyjarvi, J., Korhonen, I.: Detection of daily activities and sports with wearable sensors in controlled and uncontrolled conditions. IEEE Transactions on Information Technology in Biomedicine 12(1), 20–26 (2008)
5. Hu, J., Boulgouris, N.V.: Fast human activity recognition based on structure and motion. Pattern Recognition Letters 32(14), 1814–1821 (2011)
6. Koppula, H.S., Gupta, R., Saxena, A.: Learning human activities and object affordances from rgb-d videos. The International Journal of Robotics Research 32(8), 951–970 (2013)
7. Naseer, T., Sturm, J., Cremers, D.: Followme: Person following and gesture recognition with a quadrocopter. In: IEEE RSJ International Conference on Intelligent Robots and Systems (IROS), pp. 624–630 (2013)
8. Ni, B., Wang, G., Moulin, P.: Rgbd-hudaact: A color-depth video database for human daily activity recognition. In: Consumer Depth Cameras for Computer Vision Advances in Computer Vision and Pattern Recognition, pp. 193–208 (2013)
9. Parkka, J., Ermes, M., Korpipaa, P., Mantyjarvi, J., Peltola, J., Korhonen, I.: Activity classification using realistic data from wearable sensors. IEEE Transactions on Information Technology in Biomedicine 10(1), 119–128 (2006)
10. Qian, H., Mao, Y., Xiang, W., Wang, Z.: Recognition of human activities using svm multi-class classifier. Pattern Recognition Letters 31(2), 100–111 (2010)
11. Schuldt, C., Laptev, I., Caputo, B.: Recognizing human actions: a local svm approach. In: 17th International Conference on Pattern Recognition (ICPR 2004), pp. 32–36 (2004)
12. Zhu, Y., Fujimura, K.: A bayesian framework for human body pose tracking from depth image sequences. Sensors 10(5), 5280–5293 (2010)

Prime: Towards the Design
of a Small Interactive Office Robot

Dante Arroyo[1], Cesar Lucho[2], Pedro Cisneros[1], and Francisco Cuellar[1]

[1] Department of Engineering
[2] Department of Art
Pontificia Universidad Catolica del Peru
{darroyo,pedro.cisneros}@pucp.edu.pe,
{di.ce.lucho,cuellar.ff}@gmail.com

Abstract. This paper presents the design and implementation of Prime, a small interactive office robot with features to support daily office activities by transporting small desktop supplies, carrying reminder notes and performing other gadget utilities. In order to create an effective inclusion of the robot in this particular workspace, the design of Prime is centered in three important aspects: functionality, aesthetics and interaction. This work is an exploratory research aimed to study the novel inclusion of small service robots in office environments and serve as a research platform to conduct human-robot interaction theories and experiments. The design and implementation of the presented robot results from an interdisciplinary work, including a survey to define Prime's functionality and behavior in response to specific office needs, as well as its design process that comprehends sketching, scale modeling and 3D prototyping.

1 Introduction

In the last years, the field of social and interactive robotics has presented a widespread development comparing to other robotic fields [1, 7, 11]. Within the insertion of robots in society, an emerging category of service robots are showing a more significant presence in daily-life activities performing as social agents [1, 8, 24]. For instance, these robots are able to assume roles, performing as receptionists, assistants, hosts, therapeutic and social companions [2, 7, 13, 17, 26, 27]. Recent design approaches for service robots does not consider functionality as the only priority, whereas interaction and aesthetics are playing a major role [14].

In this context, office oriented robots are considered as agents that exhibit some dynamic behavior and reside within a workplace [24], while performing tasks such as telepresence [18], cleaning [9], and supplies or snack delivering [10, 20]. Among these examples, telepresence robots are mostly commercially available [18], which demonstrates the acceptance of this type of robots. According to our research, there's no extensive published work concerning small office robots, especially those that can be portable and work in a desktop environment [6].

This paper considers the design and implementation of a small interactive office robot, in contrast to more robust and non-portable office robots found in literature.

M. Beetz et al. (Eds.): ICSR 2014, LNAI 8755, pp. 34–43, 2014.

This robot is intended to be used as small assistant agent, supporting different desktop related tasks found in daily office activities. The presented work is an exploratory research which will serve as a framework for future research on human-robot interaction in an office context.

The remaining part of this paper is organized as follows. Section 2 covers the design components of the proposed robot, based on literature and supported by a survey. Section 3 presents the design and implementation of our robot. Finally, the last section presents conclusive remarks and directions of future work.

2   Design Components

This work is a first approach towards the design of a small interactive office robot. Therefore, it is necessary to define the robot's design components. Particularly, a theoretical approach by defining three design guidelines and experimental data collection conducted by a survey are presented.

2.1   Design Guidelines

There are three aspects considered in the theoretical design of the robot: functionality, aesthetics and interaction.

Functionality
Utility or performance along with functionality is one of the main pillars during the design process of everyday products [21]. Robots designed for office environments shouldn't distance from this focus. Previous research has evidenced that perceived usefulness of a robotic service is one of the main facilitators for the user's initial acceptation [2, 28].

Additionally, the daily exposure of an office robot requires mechanisms to ensure a long-term interaction [19], otherwise the user will cease using the robot after the novelty effects of its introduction vanishes [17]. We consider that by ensuring functionality as the main design consideration of an office robot, a long-standing bond with the user will be held.

Aesthetics
Aesthetics, from the product's design perspective is one of the major aspects that influences the response or reaction of people with an object, appliance or system [21], and it is important for determining if the product is rejected or evokes attraction to people [15]. Particularly, visual aesthetics has a symbolic function that influences how a product is comprehended and evaluated [3].

In the context of an office robot, aesthetics is intrinsically linked to the user, serving as a tool for holding the user's attraction to the robot while evoking strong emotions. It is suggested that if aesthetics is considered along with functionality in the entirely design process of the robot, then, it is perceived as being more usable by the target public [14]. For instance, by encouraging the user to ask for the robot's services

[2, 22]. In this way, aesthetics is a catalyzer for establishing user-robot interaction bonds, a desired feature for our continuously exposed robot [17]. This quality is comprised in the concept of aesthetic functionalism [12].

Interaction
Interaction is the design guideline that could differentiate the office robot from any other office machine or supply, because it can generate new user experiences that could attain preference for the robot and achieve a deeper bond with it [22]. For instance, interaction complements the robot's aesthetic functionalism by adding a sort of dynamism to the robot which boosts the user's perception of this object. From this approach, the robot works as an interactive gadget. Additionally, interaction is able to decrease the initial difficulty for the user to identify how to use the robot [23].

Furthermore, the level of interaction determines how a person perceives the robot as a sociable entity, influencing the user's acceptance of the robot [8]. By including a dynamic behavior, it is possible to transcend the robot from being regarded as a "mechanical utility" to a scope in which the user recognizes it as a helpful autonomous entity capable to relate with him [16]. For example, some robots include sophisticated social cues like an expressive head or anthropomorphic limbs to denote an elaborate corporal language [4, 5, 19]. However, a small office robot may exploit simpler social cues based on motion [8], for instance, by naturally wandering throughout its environment while offering its services.

2.2 Survey Analysis

A survey was performed in order to collect information to support the design guidelines and implementation process of the proposed robot from the user's perspective. The examination was taken to 32 office workers, 11 male and 21 female, in order to explore their expectations regarding the inclusion of a robot in their workplace. The participants may be biased towards female workers due to the female gender predominance in this particular surveyed work context.

The survey was structured in two parts. The first part consisted of exploratory questions about the personal opinion of the participants according to visual appearance, functions and behavior of what they considered an office robot. The second part presented the concept of a small interactive desktop robot, and questions about its features were requested.

Results of the exploratory question about the robot's visual appearance showed that 62.5 % of participants preferred an anthropomorphic office robot, 25.0% a zoomorphic appearance and the last 12.5%, a machine-like appearance. Regarding the desired robot's behavior, participants were asked to pick one of the following conducts:

- **Option A:** The robot is placed in a corner or specific spot, waiting for the user to send a command to come and carry out the service it offers. After offering its service, the robot will go back again to its spot and wait for anybody to use it.

- **Option B:** The robot is wandering around the office workplace in a natural way, so the user can approach to it in order to require the service it offers. Eventually, the robot may approach and look for interaction.

The 37.5% of participants chose option A, while a significant 62.5% preferred option B, the more dynamic behavior. Additionally, participants were asked to justify their selection. In the case of participants that selected the second option, they chose a more dynamic behavior, therefore the robot could look for attention, encourage the surroundings people to use it, and show that it is not a mere decoration. Furthermore, explanations related to an interactive behavior of the robot were registered. Participants stated they didn't want a slave or a lazy robot employee, they expected something more natural and easier to relate with, something that could distinguish itself from other electronic devices. On the other hand, participants who chose option A mainly explained that a dynamic behavior in an office robot may represent a physical obstacle for the labor of workers.

In the second part of the survey, the concept of a small office robot for operating in a desktop was introduced. Among the functions defined by the participants, a 34.5% preferred the robot to keep papers and envelops, a 17.2% to carry supplies, and a 51.7% specified gadgets functionalities such as USB storage, music playing, alarm notification and date displaying. Additionally, participants specified the means of interaction with the desktop robot: a 70.6% preferred talking to the robot and a 30.0% interacting by touching it. Participants who chose the first option explained that oral communication was more intuitive. In contrast, the group that selected touching the robot discarded oral interaction as the robot could obey surrounding voices and mentioned that touching was a way of how a person relates with a pet. Finally, participants were asked to list which office supplies they considered important for the desktop office robot to carry. A total of 13 different types of objects could be identified, among them, carrying pencils and highlighter 33.3%, envelops or papers 19.7%, clips 9.1% and reminder notes 7.6%.

The examination results supported two important statements from our design guidelines. First, functionality is the most important consideration for the perception of an office robot, and that this will have a strong impact on accomplishing a long-standing interaction with the robot. Second, the survey showed that participants appreciated a continuous dynamic behavior of the robot, because it corresponds to the busy working context of an office. In this way, motion could be an important social cue. Finally, the survey demonstrates that a desktop robot might not be the type of robot an office worker expects, thus, it is an attractive field for further research.

3   The Desktop Office Robot Implementation

After analyzing the survey results and contrasting them with the design guidelines, an adequate focus for the development of a small office robot can be established. In this context, our proposed robot, Prime, was conceived as a desktop office robot and as a platform to explore and research in human-robot interaction.

Prime's functionalities are basically those related to be used as an additional office utility. Therefore, its main application is to be functional without disturbing the user or being a physical obstacle in its working environment. Additionally, Prime distinguishes from other office tools by creating an appropriate mimicry with the worker's environment such that it will not be regarded as a mere "service supplier", but a more dynamic interactive gadget. For instance, Prime will display a self-explanatory functionality while being an aesthetically pleasant robot. Furthermore, Prime's mimicry with the office environment will be achieved through displayed motion, which, as noted in the survey, will encourage the user to interact with it and obtain its services.

3.1 Design Process

A first consideration for Prime's design process was the definition of its particular physical features. Even though an anthropomorphic appearance was preferred for the robot in the survey, it is impractical for desktop environments due to its limited space, compromising the robot's movement and dimensions. Therefore, we chose the second best option: a zoomorphic appearance. Particularly, Prime resembles an ape. Additionally, previous works suggest that zoomorphic features in robots generate adequate human responses during interaction [13, 25].

Regarding the utility of Prime as an office tool, it was designed to carry small office supplies, such as pens, pencils, highlighters, post-its and clips. Prime's main purpose of carrying the previously mentioned small office supplies spans almost 50% of the expected objects from the survey.

Freehand Sketches

Hand-made sketching was an important stage for defining the morphological considerations of the robot. Prime, addresses the goal of integrating functional and interactive qualities in an object [22], by having a self-explanatory anatomy, which is partly achieved by Prime's ape-like appearance with two relative big limbs, as seen in Fig. 1.a. Additionally, Prime holds a backpack in order to carry the defined small office supplies, a self-explanatory feature which visually communicates the user that "things must be placed here". Notice in Fig. 1.a. that the inclusion of this backpack does not affect Prime's zoomorphic appearance, so the robot's functional and aesthetics design guidelines do not conflict with each other.

Scale Model

The dimensions' definition of the desktop robot is a critical factor to be considered due to its constraint workspace. This urged the need of experimenting with a tangible object before going to a further prototype complementing the 3D modelling software process. As a result, a scale model was constructed using a wire structure covered with modelling clay, as seen in Fig. 1.b. This physical representation contributed not only to define Prime's true dimensions, but to generate additional utilities for Prime, for instance, the fact that post-its could be stick on Prime, serving as a living reminder utility.

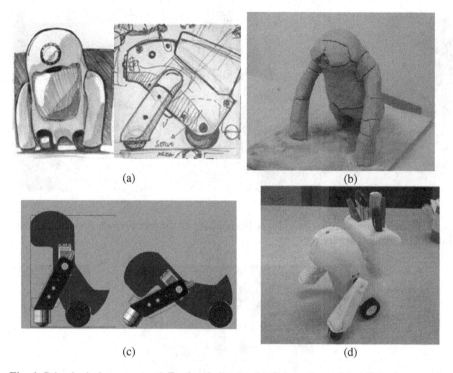

(a)   (b)

(c)   (d)

Fig. 1. Prime's design process a) Freehand sketches b) Clay scale model c) Prime's stretching and spreading movements d) 3D prototype

3.2   Interactive Behavior

Prime can perform animal inspired movements that have demonstrated effective responsiveness in previous research [1, 16, 22], and which have been previously stated as adequate social cues in office workplaces according to the survey results.

First, the robot can autonomously display erratic and random displacements throughout its environment while avoiding obstacles and without disturbing the user or falling from the desk. This allows Prime to be seen as an autonomous attentive agent [16], actively looking for establishing interaction with the user and being busy during its stay in the office. Complementarily, as shown in Fig. 1.c. Prime is able to perform stationary animal-like movement such as stretching and spreading its limbs like an ape. Lastly, another possible benefit is that Prime might add some playfulness to the work environment without being distractive.

Prototype
Prime's 3D printed prototype, as shown in Fig. 1.d., served to explore the user interaction with the robot. The user will need to touch the robot's head as he would be

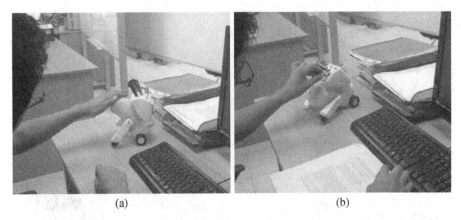

Fig. 2. Prime's dynamic behavior a) Prime being petted by the user b)Prime stretches its body and the user acquires its services

Fig. 3. Prime carrying small office supplies

petting a pet in order to request the robot's service. This action is enough for Prime to interpret that the user's needs its service, and then it will pause and spread its body, so the user can take or place the office supply he wants, as depicted in Fig. 2. According to the design guidelines, this intuitive communication enhances the functionality of the robot. Fig. 3 shows how Prime carries three types of office supplies. The proto-type demonstrated that Prime has the potential of including other gadget functionali-ties demanded by people in the survey, such as USB storage, a display for timing, a cell phone holder, etc.

Prime was implemented with the necessary electronics and mechanical compo-nents in order to accomplish all the requirements defined in the design process. The electronics are detailed in Fig. 4.a., and a more extensive appearance description along with functionality details are shown in Fig.4.b.

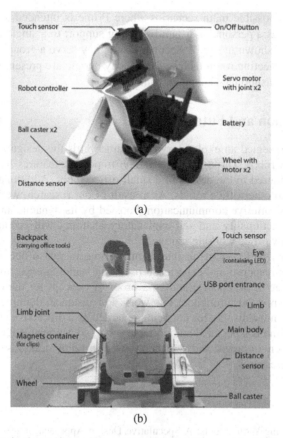

Fig. 4. Prime's electronics and appearance details a) Inner view b) Frontal view

Fig. 5. A group of Prime robots on a meeting room table

There are two possible main scenarios where Prime is intended to be used. First, Prime may serve as a personal assistant by giving support to a single user in its daily office routines, as shown in Fig. 2. Second, Prime may serve a group of people, wandering around in meeting room tables where many people are present, so each person is always sufficient supplied. This is shown in Fig. 5.

4   Conclusion and Future Work

This work has presented an exploratory research towards the design and inclusion of competent office robot. Throughout our study, important features have been recognized to be considered in the design of a small office robot. Additionally, they might be extrapolated to other office robots. As a result of our research, we believe that the presented robot's intuitive communication, boosted by its dynamic animal-like behavior, aesthetic functionality and self-explanatory anatomy will encourage people to use it in their daily office working routine.

Future research will consist on introducing Prime in real office workplaces and testing the user's response and experience in a long-standing experiment, in order to measure its degree of acceptance and be a proof of concept for the different design considerations expressed throughout this work. For instance, defining the amount of dynamism Prime must display for not being regarded as a distractive element. Finally, we know that Prime is a personal and portable robot, so we believe its usage can transcend the office environment and start being used in home desktops: another impact study for further research.

References

1. Auger, J.: Living With Robots: A Speculative Design Approach. International Journal of Human-Robot Interaction 3(1), 20–42 (2014)
2. BenMessaoud, C., Kharrazzi, H., MacDorman, K.F.: Facilitators and barriers to adopting robotic-assisted surgery: Contextualizing the Unified Theory of Acceptance and use of Technology. PLOS One 6(1) (2011)
3. Bloch, P.H., Brunel, F.F., Arnold, T.J.: Individual differences in the centrality of visual product aesthetics: concept and measurement. Journal of Consumer Research 29(4), 551–565 (2003)
4. Breazeal, C.: Emotion and sociable humanoid robots. International Journal on Human-Computer Studies 59(1-2), 119–155 (2003)
5. Bruce, A., Nourbaksh, I., Simmons, R.: The role of expressiveness and attention in human–robot interaction. In: Proceedings of the 2001 AAAI Fall Symposium (2001)
6. Filgueiras, E., Rebelo, F., Moreira da Silva, F.: Human-computer interaction in office work: Evaluation of interaction patterns using office equipment and software during data entry and navigation. In: Robertson, M.M. (ed.) EHAWC 2011 and HCII 2011. LNCS, vol. 6779, pp. 40–48. Springer, Heidelberg (2011)
7. Fong, T., Nourbakhsh, I., Dautenhahn, K.: A survey of socially interactive robots. Robotics and Autonomous System 42(1), 143–166 (2003)
8. Forlizzi, J.: How robotic products become social products: An ethnographic study of cleaning in the home. In: Proceedings of the 2nd ACM/IEEE International Conference on Human-Robot Interaction, pp. 129–136 (2007)

9. Fuji Heavy Industries Ltd., http://www.fhi.co.jp/contents/pdf_56722 .pdf (accessed June 14, 2014)
10. Fujitsu, http://pr.fujitsu.cckom/jp/news/2005/09/13.html (accessed June 14, 2014)
11. Goodrich, M.A., Schultz, A.C.: Human-Robot Interaction: A survey. Foundations and Trends in Human-Computer Interaction 1(3), 203–275 (2007)
12. Hansson, S.O.: Aesthetic Functionalism. Contemporary Aesthetics (2005) (Online)
13. Heerink, M., Albo-Canals, J., Valenti-Soler, M., Martinez-Martin, P., Zondag, J., Smits, C., Anisuzzaman, S.: Exploring requirements and alternative pet robots for robot assisted therapy with older adults with dementia. In: Herrmann, G., Pearson, M.J., Lenz, A., Bremner, P., Spiers, A., Leonards, U. (eds.) ICSR 2013. LNCS, vol. 8239, pp. 104–115. Springer, Heidelberg (2013)
14. Hegel, F.: Effects of a Robot's Aesthetic Design on the Attribution of Social Capabilities. In: The 21st IEEE International Symposium on Robot and Human Interactive Communication, pp. 469–475 (2012)
15. Hekkert, P.: Design aesthetics: principles of pleasure in design. Psychology Science 48(2), 157–172 (2006)
16. Hoffman, G., Wendy, J.: Designing Robots with Movement in Mind. Journal of Human-Robot Interaction 3(1) (2014)
17. Kirby, R., et al.: Designing Robots for Long-Term Social Interaction. In: Proceedings of the IEEE/RSJ International Conference on Intelligent Robots and Systems (2005)
18. Kristoffersson, A., Coradeschi, S., Loutfi, A.: A Review of Mobile Robotic Telepresence. Advances in Human-Computer Interaction (2013)
19. Lee, M.K., et al.: Personalization in HRI: A longitudinal field experiment. In: Proceedings of the 7th ACM/IEEE International Conference on Human-Robot Interaction (2012)
20. Lee, M., et al.: The snackbot: documenting the design of a robot for long-term human-robot interaction. In: Proceedings of the 4th ACM/IEEE International Conference on Human Robot Interaction (2009)
21. Norman, D.: Design of Everyday Things. Basic Books (2013)
22. Osawa, H., et al.: Embodiment of an agent by anthropomorphization of a common object. Web Intelligence and Agent Systems: An International Journal (2012)
23. Overbeeke, C., Djajadiningrat, J., Hummels, C., Wenseen, S.: Beauty in usability: forget about ease of use! In: Green, W.S., Jordan, P.W. (eds.) Pleasure with Products: Beyond Usability (2002)
24. Šabanović, S., Reeder, S.M.: Designing Robots in the Wild: In situ Prototype Evaluation for a Break Management Robot. International Journal of Human-Robot Interaction (2014)
25. Singh, A., Young, J.E.: Animal-Inspired Human-Robot Interaction: A Robotic Tail for Communicting State. In: Proceedings of the 2012 7th ACM/IEEE International Conference on Human-Robot Interaction (2012)
26. Takayuki, K., et al.: An affective guide robot in a shopping mall. In: Proceedings of the 4th ACM/IEEE International Conference on Human-Robot Interaction (2009)
27. Tanaka, F., Cicourel, A., Movellan, J.R.: Socialization between toddlers and robots at an early childhood education center. Proceedings of the National Academy of Sciences (2007)
28. Venkatesh, V., et al.: User acceptance of information technology: Toward a unified view. MIS Quarterly (2003)

Group Comfortability When a Robot Approaches

Adrian Ball[1], David Silvera-Tawil[2], David Rye[1], and Mari Velonaki[2]

[1] Australian Center for Field Robotics, The University of Sydney, Australia
{a.ball,d.rye}@acfr.usyd.edu.au
www.acfr.usyd.edu.au
[2] Creative Robotics Lab, The University of New South Wales, Australia
{d.silverat,mari.velonaki}@unsw.edu.au
www.crl.niea.unsw.edu.au

Abstract. This paper investigates the level of comfort in people with different robot approach paths. While engaged in a shared task, 45 pairs of participants were approached by a robot from eight different directions and asked to rate their level of comfort. Results show that comfortability patterns of individuals in pairs is different to lone individuals when they are approached by a robot. This in turn influences how comfortable a group is with different robot approach paths.

Keywords: Human-robot interaction, comfort, group.

1 Introduction

When robots interact with people in social environments it is important to consider how they can initiate interactions without making people feel uncomfortable. How a robot approaches a person will play a strong part in achieving a 'successful' interaction.

When a robot approaches a single person, it is known [1, 2, 3] that people are most comfortable with approaches from the front—where they can see the robot—and are least comfortable when the robot approaches from behind the person. Approaches from a person's front-right and front-left directions are considered more comfortable than a direct frontal approach [3]. These results hold when the person is sitting or standing in the center of the room or with their back against a wall [4].

Algorithms have been developed to allow a robot to approach individuals at home [5, 6], to maintain social awareness while navigating public places [7, 8] and to approach a pedestrian in a public place [9]. Although these algorithms improve how robots navigate in, and use, social spaces they do not consider how a robot should approach a group of interacting people. By knowing what people in groups find comfortable, social awareness can be incorporated into a robot's path planning algorithms so that the robot will approach a person from a direction that is not likely to cause them discomfort.

Preliminarily research into the comfort levels of groups of people when approached by a robot has been conducted by Karreman et al. [10], who investigated the comfort of a group of two people approached by a robot. The current

M. Beetz et al. (Eds.): ICSR 2014, LNAI 8755, pp. 44–53, 2014.

paper builds on the findings of [10] by also investigating the comfort levels of individuals in the pair, and extending the experiment to 45 pairs of participants.

When interacting with each other, people form a shared interaction region and face this region [11]. The relative positions of people in the group will often lead to multiple 'front' regions that define frontal approaches to individuals and multiple 'rear' regions that are usually avoided when approaching. When a robot is to approach a group of interacting people it is not obvious which approach path would be most comfortable for the group as a whole.

Note that in all the cited works the notion of a person's 'comfort' is consistent with a natural language understanding of mental comfort as tranquil enjoyment and contentedness; as freedom from unease, anxiety and fear, and is typically assessed simply by asking a person "how comfortable" they are. The same approach is adopted here.

This paper presents the results of an experiment designed to investigate the levels of comfort in a group of two people seated in various configurations when approached from different directions by a robot. The experiment allowed the comfort level of the pair, and the influence of the presence of a second person on the comfort of an individual, to be measured. Two hypotheses were tested: (H1) A group of two people is more comfortable with robot approach directions from a common 'front' direction and less comfortable with approaches from a common 'rear' direction; and (H2) The presence of a second person does not influence the level of comfort of an individual approached by a robot. Hypothesis H2 is derived from the construction of (H1). If it is possible to estimate the comfort levels of groups interacting with robots from the comfort levels of lone individuals interacting with robots, then the presence of other people (H2) cannot influence an individual's comfort levels with different robot approach paths.

2 Experiment Design

For each experimental trial, two participants were seated in low armchairs adjacent to a small square table in the center of the room. The participants were asked to work on a cooperative task for the duration of the experiment. A robot periodically approached and interrupted the participants, asking each to rate their level of comfort with that particular approach direction. Once the robot had approached the group from eight different directions, the experiment concluded with a post-experiment questionnaire. Further details are given below.

2.1 Seating Configuration

Kendon describes [11] how groups of people use physical space while interacting. Three spatial regions are defined in Kendon's formulation: 'o-space', 'p-space' and 'r-space'. The o-space is a transactional space shared between interactants and maintained for the duration of the interaction. The central o-space is surrounded by the p-space; an agent must occupy the p-space to be considered part of the interaction. The nearby area outside the p-space is the r-space. The r-space encapsulates both the p-space and o-space and is the portion of the rest of the world that is monitored by the interactants.

The experiment used the three maximally different ways that two people working on a common task can be seated. These configurations are: opposite each other; in an 'L-shape' and side-by-side, referred to here as Configuration A, B and C respectively (Figure 1).

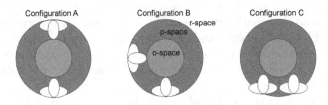

Fig. 1. Seating configurations of two people

2.2   Group Activity

Participants were asked to complete a task to provide a cognitive load that would distract them from the presence and movement of the robot, minimizing participant anticipation of the robot approaches. A jigsaw puzzle was chosen as the task as it is easy to understand, time consuming to finish and doesn't involve taking turns. Tasks that are performed in turns have an increased chance of participants being less focused on the task when awaiting their turn. A three-dimensional puzzle was chosen to increase task novelty.

2.3   Experimental Space

It is desirable that the experimental space is symmetrical to remove spatial bias due to asymmetric placement of participants in the room. It should also have multiple exits so there is always an exit available to a participant avoiding confrontation with the robot. The room used in this work was square, with six-metre sides and with exits on three of the four walls. Although the exit locations were not completely symmetrical there were exits readily available to participants. Figure 2 shows the arrangement of the experimental space.

2.4   Robot Approach Directions

During each experimental trial the robot continuously circled the seated pair of participants and then approached once from each of the eight directions shown in Figure 2. The approaches were made in random order. Participant familiarity with the robot was expected to increase as they observed it moving around the room, potentially influencing their comfort level during the experiment. Randomizing the order of approach direction across all participants will remove any bias due to increasing familiarity with the robot during each experimental trial.

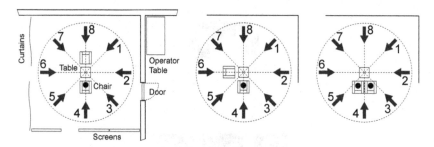

Fig. 2. Experimental space with chairs arranged in Configurations A, B and C. The dots represent reference locations referred to in Section 3.5.

The robot directly approached the center of the table which—as the focus of the group task—was assumed to be the center of the o-space. In every seating configuration there were approach directions where the robot could not reach the p-space surrounding the table by approaching in a straight line; for example, when the robot approached from behind a participant. In such situations the robot approached the p-space as closely as possible without a collision. The robot departed from each encounter along the approach path.

2.5   Robot Design

An Adept Pioneer 3 DX robot was used as the motion platform in this work. The motion platform was augmented with an aluminium frame that supported an Asus Xtion Pro Live RGB-D sensor and a speaker. A laptop computer was placed on the base of the aluminium frame. The robot can be seen in Figure 3. It was intended that the robot be mechanical in appearance to facilitate comparison of results with other research using similar robots.

The robot was controlled using the Wizard of Oz methodology. This decision allows for the robot to be operated in an ordinary room with only an overhead camera to assist the operator with robot movement. Should an unexpected situation arise, a Wizard of Oz methodology allows for safe control of the robot.

2.6   Conduct of the Experiment

Each pair of participants was brought into the room and seated in one of the three configurations. The robot was then wheeled into the room and placed in a corner of the space. The experiment was described to the participants; they were not told that the robot was being controlled remotely. Once the participants understood the experiment and the task, the experimenter left the room and the experiment began. On each approach, when the robot reached the p-space surrounding the table it stopped and prompted the participants via an audio message to answer the next question on the questionnaire. All questions were identical, and asked "Please rate your comfort level regarding the robot's most recent approach path", to be answered on a five point Likert scale. Following a

Fig. 3. Robot used for the experiment

short pause, the robot departed along its approach path. In the time between encounters, the robot travelled counter-clockwise around the periphery of the room. This movement was intended to reduce the predictability of when and from where the robot would next approach the participants. Once the group had been approached from all directions, the robot was steered to its initial location in the corner and the experimenter returned to the room with a post-experiment questionnaire. This questionnaire incorporated two commonly used tools: the NASA-TLX [12] and the Godspeed [13] questionnaires, together with questions on participant demographics and comfortability.

The NASA-TLX questionnaire measures a user's perceptions of the mental, physical and temporal demands required to perform a task. This questionnaire was included to determine whether participants were focused on the jigsaw puzzle task rather than the presence and movement of the robot. The Godspeed questionnaire was included to determine how participants perceived the robot. There are five sub-categories that form this questionnaire: anthropomorphism, animacy, likeability, perceived intelligence and perceived safety of the robot.

3   Results

3.1   Participants

Fifteen trials were conducted with participants seated in each of the three configurations. Thirty two of the 90 participants were male and 58 were female. The mean age of the group was 24.4 years old, with a standard deviation of 10.2 years, a minimum age of 18 and a maximum age of 73 years old. Most participants were university students; all were naive to the experiment. Although the variance was raised by the participation of four persons older that 60, no age-dependent effects were observed in the data. Although the variance was raised by the participation of four persons older that 60, no age-dependent effects were observed in the data.

3.2   Perceptions of the Robot

Responses to the three relevant questions in the Godspeed questionnaire, Table 1, show that the robot was perceived as being of mechanical appearance. For example, 74 of 90 (82%) participants gave a score ≤ 2 on a Likert scale of machine-like (1) to human-like (5).

Table 1. Results from the Godspeed questionnaire showing number of participants against scores rounded to the nearest integer

Score	1	2	3	4	5	
Machine-like	27	47	15	1	0	Human-like
Artificial	28	36	21	5	0	Lifelike
Mechanical	30	43	16	1	0	Organic

3.3   Perceptions of the Task

Table 2 shows how the mental demand and effort required to complete the 3-D jigsaw puzzle were rated. The majority of participants scored the mental demand and effort required for the task as 2 or 3, suggesting that moderate mental demand and effort were required to progress towards completing the puzzle.

Table 2. Results from the NASA-TLX questionnaire showing number of participants against scores rounded to the nearest integer

Score	1	2	3	4	5	
Low Mental Demand	16	29	21	17	7	High Mental Demand
Low Effort	12	27	24	25	2	High Effort

3.4   Group Comfort with Direction of Robot Approach

In each experimental trial the pair of participants was approached by the robot from all eight directions in a random order. Each approach direction was assigned a pair comfort score calculated as the sum of the pair's individual comfort scores for that direction. The eight pair scores were then ranked in descending order. Ranks were used in place of scores to remove individual participant bias by effectively using a measure of *relative* comfort rather than an absolute comfort level. Table 3 shows the mean rank of all pair comfort scores for each approach direction for the three seating configurations.

Table 3. Means and standard deviations (in parentheses) of *group* rankings for each robot approach direction and for the three seating configurations, across all pairs of participants

Direction	Config. A	Config. B	Config. C
1	4.4 (2.3)	2.7 (2.1)	3.2 (2.3)
2	4.6 (2.5)	3.9 (2.6)	4.9 (2.3)
3	3.1 (2.4)	3.5 (2.0)	4.3 (2.1)
4	4.9 (2.5)	4.3 (2.9)	6.2 (2.6)
5	3.6 (1.3)	4.9 (2.2)	4.4 (2.3)
6	4.2 (2.6)	5.2 (2.7)	3.6 (1.9)
7	4.5 (2.5)	3.4 (2.1)	2.7 (2.4)
8	5.3 (2.6)	4.5 (2.3)	3.5 (1.7)

A Kruskal-Wallis non-parametric one-way analysis of variance (KW-ANOVA) test was used to determine if there were statistically significant differences in group comfort levels with different robot approach directions. Where significant differences were found, multiple comparisons were made using the Mann-Whitney U test to determine which pairs of directions were significantly different. The p values from this set of comparisons were ranked, and compared with Q values calculated using the False Discovery Rate (FDR) control method with $q = 0.05$ [14]. The FDR method was preferred over the use of the—more conservative—Bonferroni correction factor as it leads to fewer Type I errors.

In Configuration A there was no significant difference in group comfort levels between any of the approach directions ($\chi^2(7,112) = 8.64$, p $= 0.28$, $\eta^2 = 0.07$). In Configuration B, there was also no significant difference in group comfort levels ($\chi^2(7,112) = 11.94$, p $= 0.10$, $\eta^2 = 0.10$). In Configuration C there was a highly significant difference in group comfort levels ($\chi^2(7,112) = 22.16$, p $<$ 0.01, $\eta^2 = 0.19$). Multiple comparison testing showed that the group comfort ranking for direction 4 was different to directions 6, 7, 8 and 1. The preferred directions can be seen in Table 3; approach directions with ranks nearest to one are most comfortable.

Analysis under the assumption of normally-distributed sample populations showed that at a significance level $\alpha = 0.05$, a statistical power $(1 - \beta) = 0.80$ and a sample size of 15 the smallest difference in rank that was statistically detectable was approximately 1.8. All results reported here and in the following section have mean differences greater than this value.

These results reject the first hypothesis. When the seating configuration had no common 'front' or 'rear' direction (Configuration A), there was no statistically significant difference in comfort level with different robot approach directions. In Configuration B, the common 'front' direction was not statistically more comfortable than the common 'rear' direction. In Configuration C, the participants shared a common immediate 'rear' direction that the robot could approach from. Approaches from all 'front' directions were found to be more comfortable than from this shared rear direction.

3.5   Individual Comfort with Direction of Robot Approach

The previous table summarizes the comfort level of the pair, making no distinction between the two individuals. By analyzing individual preferences it is possible to see how the presence of a second person influences the comfort level of an individual when the pair are approached by a robot. Table 4 shows the mean rank of each approach direction for each of the five different relative seating positions of an individual. The robot approach directions are numbered relative to the positions marked with dots in Figure 2.

Table 4. Means and standard deviations (in parentheses) of *individual* rankings for each robot approach direction for the three seating configurations, across all pairs of participants. The labels 'Left' and 'Right' identify where the person of interest was sitting in the pair.

Dir	Config. A	Config. B (Left)	Config. B (Right)	Config. C (Left)	Config. C (Right)
1	3.1 (1.9)	2.5 (1.9)	2.3 (1.7)	2.9 (2.1)	3.7 (2.1)
2	4.1 (2.5)	3.1 (2.4)	3.7 (2.6)	4.4 (2.3)	5.1 (2.3)
3	4.2 (2.5)	4.9 (1.9)	4.1 (2.5)	3.3 (2.7)	4.3 (2.5)
4	5.9 (2.8)	5.5 (2.7)	4.5 (3.1)	4.8 (3.1)	6.1 (2.4)
5	4.4 (2.4)	4.5 (2.6)	4.4 (2.7)	4.7 (2.5)	3.3 (2.5)
6	4.0 (2.7)	3.3 (2.2)	3.8 (2.6)	3.5 (1.7)	2.9 (2.0)
7	3.0 (2.3)	2.7 (2.1)	2.9 (1.7)	3.1 (2.7)	2.6 (2.4)
8	2.7 (1.9)	3.6 (2.6)	4.2 (2.8)	2.7 (2.3)	3.3 (2.0)

Since the relative position of the second person in Configuration A is identical for each participant, twice as much data is available for this configuration. Performing a KW-ANOVA test showed that there was a highly significant difference between individual participant comfort levels with different robot approach directions ($\chi^2(7,232) = 30.20$, $p < 0.01$, $\eta = 0.13$). Multiple comparison testing using the previously described procedure showed that direction 4 was ranked differently from all other directions. In addition, approach direction 8 was found to be more comfortable than direction 5.

For the person sitting on the left in Configuration B, a KW-ANOVA test showed a highly significant difference in individual comfort levels ($\chi^2(7,112) = 20.80$, $p < 0.01$, $\eta = 0.18$). The multiple comparison test showed that the distribution of rankings for direction 4 was different to that of directions 7, 1 and 2. Direction 3 was also different to both directions 1 and 7. The KW-ANOVA test for people sitting on the right in Configuration B found no significant difference in comfort levels ($\chi^2(7,112) = 8.33$, $p = 0.30$, $\eta = 0.07$).

For Configuration C with the person sitting on the left, the KW-ANOVA test showed that there was no significant difference in individual comfort levels ($\chi^2(7,112) = 11.28$, $p = 0.13$, $\eta = 0.10$). The test for the person sitting on the right in Configuration C showed that there were highly significant differences in comfort levels ($\chi^2(7,112) = 23.88$, $p < 0.01$, $\eta = 0.20$). The multiple comparison tests found that direction 4 was different to directions 1, 5, 6, 7 and 8. Direction 2 was also different to direction 6.

These results collectively show that hypothesis H2 is false; the presence and location of a second person does influence the comfort level of an individual approached by a robot. The patterns of participant comfortability also differ from prior results [1, 2, 3] where a lone individual was approached by a robot. It is interesting to note that there is a left-right asymmetry in the results between Configurations B and C.

4   Discussion

When the different robot approach directions are compared to each other, pairs of people are least comfortable when they are approached from directions where the robot cannot be seen by either individual. This agrees with previous results for lone individuals approached by a robot. The comfort levels of individuals within the group are influenced by the presence and location of another person. Most notably, if the second person can see the robot approach directions to the 'rear' of the first person, then the levels of comfort felt by the first person are increased for these directions.

There is a curious asymmetry present in the findings. When seated in Configuration B (L-shaped), individuals seated on the left of the pair showed highly significant comfort preferences for robot approach directions while individuals seated on the right had no preference. These results were reversed when participants were seated side-by-side in Configuration C. We are not able to explain this asymmetry. A deeper investigation of the psychology of group interactions may shed some light on these results.

5   Conclusion

This paper describes an experiment that measured the comfort levels of seated pairs of people engaged in a shared task when approached by a robot. It was found that the presence and location of a second person influenced how comfortable someone was with different robot approach paths. The comfort patterns of individuals within the pairs were also shown to differ from prior results for those of lone individuals approached by a robot.

References

1 Dautenhahn, K., Walters, M., Woods, S., Koay, K.L., Nehaniv, C.L., Sisbot, A., Alami, R., Siméon, T.: How May I Serve You?: A Robot Companion Approaching a Seated Person in a Helping Context. In: 1st ACM SIGCHI/SIGART Conf. Human-Robot Interaction, pp. 172–179. ACM (2006)
2 Walters, M.L., Dautenhahn, K., Woods, S.N., Koay, K.L., Te Boekhorst, R., Lee, D.: Exploratory Studies on Social Spaces Between Humans and a Mechanical-Looking Robot. Connection Science 18(4), 429–439 (2006)

3 Walters, M.L., Dautenhahn, K., Woods, S.N., Koay, K.L.: Robotic Etiquette: Results from User Studies Involving a Fetch and Carry Task. In: 2nd ACM/IEEE Int. Conf. Human-Robot Interaction, pp. 317–324. IEEE (2007)

4 Walters, M.L., Koay, K.L., Woods, S.N., Syrdal, D.S., Dautenhahn, K.: Robot to Human Approaches: Preliminary Results on Comfortable Distances and Preferences. In: AAAI Spring Symposium: Multidisciplinary Collaboration for Socially Assistive Robotics, p. 103 (2007)

5 Sisbot, E.A., Alami, R., Siméon, T., Dautenhahn, K., Walters, M., Woods, S.: Navigation in the Presence of Humans. In: 5th IEEE-RAS Int. Conf. Humanoid Robots, pp. 181–188. IEEE (2005)

6 Kessler, J., Scheidig, A., Gross, H.-M.: Approaching a Person in a Socially Acceptable Manner Using Expanding Random Trees. In: Proc. ECMR, pp. 95–100 (2011)

7 Qian, K., Ma, X., Dai, X., Fang, F.: Robotic Etiquette: Socially Acceptable Navigation of Service Robots with Human Motion Pattern Learning and Prediction. J. Bionic Engineering 7(2), 150–160 (2010)

8 Luber, M., Spinello, L., Silva, J., Arras, K.O.: Socially-Aware Robot Navigation: A Learning Approach. In: IEEE/RSJ Int. Conf. Intelligent Robots and Systems, pp. 902–907. IEEE (2012)

9 Satake, S., Kanda, T., Glas, D.F., Imai, M., Ishiguro, H., Hagita, N.: A Robot That Approaches Pedestrians. IEEE Trans. Robotics 29(2), 508–524 (2013)

10 Karreman, D., Utama, L., Joosse, M., Lohse, M., van Dijk, B., Evers, V.: Robot Etiquette: How to Approach a Pair of People? In: ACM/IEEE Int. Conf. Human-Robot Interaction, pp. 196–197. ACM (2014)

11 Kendon, A.: Spacing and Orientation in Co-present Interaction. In: Esposito, A., Campbell, N., Vogel, C., Hussain, A., Nijholt, A. (eds.) COST 2102 Int. Training School 2009. LNCS, vol. 5967, pp. 1–15. Springer, Heidelberg (2010)

12 Hart, S.G., Staveland, L.E.: Development of NASA-TLX (Task Load Index): Results of Empirical and Theoretical Research. Advances in Psychology 52, 139–183 (1988)

13 Bartneck, C., Kulić, D., Croft, E., Zoghbi, S.: Measurement Instruments for the Anthropomorphism, Animacy, Likeability, Perceived Intelligence, and Perceived Safety of Robots. Int. J. Social Robotics 1(1), 71–81 (2009)

14 Benjamini, Y., Hochberg, Y.: Controlling the False Discovery Rate: A Practical and Powerful Approach to Multiple Testing. Journal of the Royal Statistical Society. Series B (Methodological) 57, 289–300 (1995)

Human Robot Interaction and Fiction: A Contradiction

Eduardo Benitez Sandoval[1], Omar Mubin[2], and Mohammad Obaid[3]

[1]HIT Lab NZ, University of Canterbury, New Zealand
[2]School of Computing, Engineering and Mathematics and the MARCS Institute, University of Western Sydney, Australia
[3]t2i Lab, Chalmers University of Technology, Gothenburg, Sweden

Abstract. In this position paper a perspective on how movies and science fiction are currently shaping the design of robots is presented. This analysis includes both behaviour and embodiment in robots. We discuss popular movies that involve robots as characters in their storyline, and how people's beliefs and expectations are affected by what they see in robot movies. A mismatch or contradiction emerges in what the robots of today can accomplish and what the movies portray. In order to overcome this mismatch we present design implications that may be of benefit to HRI designers.

Keywords: Fiction robots, Research robots, Commercial robots.

1 Introduction

One of the major aspects of research in Human Robot Interaction (HRI) is to design social robots that look and behave as humans anticipate and desire, ultimately allowing easier and more seamless integration into society. In order to achieve this, researchers in HRI conduct lab-based experiments to determine user needs, expectations and requirements. In our research, we reflect on what we have learned as HRI researchers and robot designers, and look to what we can predict for the future from non-research domains. The primary category employed for this purpose in this paper is robots as seen in science fiction movies. Prior work in HRI has mostly concentrated on reviewing research work or empirical research in the laboratory. However, more often that not, results are presented only with a subset of users and the general public is not fully exposed to or aware of these robots. In addition, real life evaluations for HRI and away from the lab are already occurring [26,30], and in the related domain of Human Computer Interaction (HCI) there is movement to ascertain the future trend of interfaces based on fiction and movies [28,20]. We can also see examples of fictional material being used as a pedagogical instrument in Computer Science Education [11]. Therefore, in order to ascertain the public sentiment and perception of robots it may well be worthwhile to study the more public and accessible media where humans are exposed to robots, such as robots found in movies and fiction. There are several overview articles that base their results and conclusions on research

M. Beetz et al. (Eds.): ICSR 2014, LNAI 8755, pp. 54–63, 2014.
© Springer International Publishing Switzerland 2014

work in HRI and on what makes effective HRI, however little is determined using robots that humans see in movies. In our opinion this is an under-treated area and therefore the niche of our research.

1.1 Motivation and Related Work

In this position paper we restrict our analysis of non-research domain areas to social robots, namely robots that socially engage with humans and display social characteristics. A more formal definition of social robots can be found in [18]. Other robots, such as industrial robots, do not interact with humans as intensely and are therefore out of the scope of this paper.

As stated, prior HRI research overviews have been based on extrapolations of lab-based results on human-robot behavioural studies and design. In particular, research has minutely discussed what physical features a robot should have based on the robots function [29]. An example which discusses the design of social robots in line with anthropomorphism is outlined in [17]. In addition, several researchers have presented position papers in the field of social robots in specific domains such as education [24]. However, it would be naive to perceive complex concepts such as anthropomorphism from the perspective of research projects only; there are deeper philosophical underlying issues which can be explored from our real life interactions with robots, and not just by experimental user-centred HRI design [27]. These can include robots which humans visualise in various media, such as caricatures, figurines, toys and domestic robots that assist in chores at home. Little is known about how these beliefs and perceptions connect to findings reported in HRI literature. Can we use such beliefs to inform the design of future HRI research? Why is that the Aibo robot which reported such successful sales initially [25] is now almost obsolete? Why is it that other social robots like Nao have limited their sales to the research market? These and similar questions need addressing.

We are not the first to approach the concept of analysing HRI from the perspective of non-research based themes. We find inklings of certain topics such as Culture, Media, Fiction, Religion, and Ethics [8] in HRI literature. However, most of this work is focused on a single theme and does not present a holistic picture in terms of human expectations and preferences. In addition, concrete design implications and linkages to research based results in HRI are not sufficiently dealt with. For instance, the study of Bartneck et. al. on LEGO figurines [10] concludes that facial expressions of such figurines are becoming less happy and examines how this has affected the sale of such figurines. However, in this study the effect of the faces on the users/children themselves is not discussed and neither is any subjective feedback or review evaluated. We can be positive about the transition of beliefs and perceptions about robots that humans have in fiction and in the media. Prior work has suggested [9] and shown [15] that portrayal of robots in the media can both negatively and positively affect human perceptions when they interact with real robots. In order to conduct a deeper analysis on the impact of media on how humans perceive robots we firstly determined the most popular movies emerging from four databases, and used their popularity rankings and revenue generated. We then discuss the contradiction that emerges from what the human sees the robot doing in the movie and what a robot can do in the real world.

2   Robots in Media and Popular Culture

Fiction in books and movies has been an important inspiration for many robots in the real world. It is likely that a high percentage of this fiction has been inspired, consciously or unconsciously, by models of human-human interaction. Although there is great diversity of robots appearing in movies and books, there are two main characteristics in these fantasy robots that are highly social and at that same time functional in nature: (a) These robots tend to show human characteristics such as anthropomorphic bodies, communication skills and even emotional capabilities, (b) they are capable of generating interaction with humans through diverse mechanisms such as non-verbal and verbal communication cues (for example, body language, voice, gestures, gaze or facial expressions), and they use this interaction to accomplish certain tasks.

Figure 1 shows several robots extracted from popular movies and cartoons, where it is evident that most of the presented robots share anthropomorphic characteristics. Several of the presented robots can be easily identified by humans as they may be familiar, beloved characters appearing in movies, series and cartoons. Definitely robots, androids and cyborgs are popular characters in modern culture. In order to evaluate the impact of such characters we have attempted to classify and rank films that involve robots based on their popularity and viewing numbers. We chose movies as our selected medium due to the fact that the film industry is capable of enormous distribution all around the globe. After exhibition in theatres, movies are further distributed in physical and digital formats and repeated constantly on TV.

Allmovie.com [6] reports 95 movies with the word "robot" in the title. We can extend the search using the terms "robots and androids" and the catalogue shows 400 titles. We find movies dating from the production of "Robot vs the Aztec Mummy" in 1958, to "Robot and Frank" in 2012. The number of titles and their diversity give us an idea of the details of the movies that have been produced; their budgets and artistic intentions. According to boxofficemojo.com [5], it is evident that the film industry has benefited greatly, financially, in movies that have robots as the main characters; 47 movies classified as robot/android/cyborg movies have generated a revenue of $3,482,508,447 USD since 1984 (average income of $74,095,924 USD), and this excludes movies with robots acting as secondary characters, such as "Star Wars", which themselves have generated enormous revenue.

In order to observe which movies have higher impact on the movie goer or viewer we extracted movie rankings using the search terms robot, android and humanoid on the following four movie databases: imdb.com, about.com, denofgeek.com and totalfilm.com. We would like to clarify that these databases do not specify the methods used for their rankings. Our final numbers excluded

Fig. 1. Robots in popular culture. Most of them display certain human characteristics (adopted from [7]).

any rankings that mentioned they were made based on a single personal preference, and only the top ten movies across each database were retained.

We found the movie "The Terminator" was the only movie that appeared in the top ten of the four databases. "Terminator 2", "Wall-E", "The Iron Giant", and "Star Wars" appeared in three of the top ten rankings and "Forbidden Planet", "Transformers", and "Metropolis" appeared in two of the top ten positions. The rest of the movies appeared just once in every ranking. See Table 1.

We ran a Spearman's rank correlation in order to find the level of agreement between the four different databases, however we did not find any significant correlation amongst the four rankings. See Table 2.

The non-significant correlations are possibly due to the lack of clarity and transparency in the methodology used for the design of the rankings. Apparently rankings and film critiques tend to be very subjective, furthermore the rules around allocating awards to movies are usually secretive. In other words, there is no homogeneity in their evaluation, which complicates the process of evaluating their impact on popular culture.

In our search of popular robot movies, we noticed that there is a predominant influence of Hollywood productions, which may be due to the fact that Hollywood directors have the resources to create attractive movies with state-of-the-art special effects, and can propagate a proper marketing strategy to generate the interest of the public and distribute their movies to a very large public.

Table 1. The Top-ten rankings of robot movies from four databases. The number indicates the position in the ranking. * indicates that this case was ranked two times in 4th and 10th positions. The ranking considered two characters for each ranking.

Movie	about.com	imdb.com	denogeek.com	totalfilm.com
Star Wars	1		4*	1
Wall-E	2	8		6
A.I.: Artificial Intelligence	3			
The Terminator	4	1	9	5
Robocop	5			3
Short Circuit	6			
Forbidden Planet	7			4
Star Trek: Generations	8			
The Iron Giant	9	6	3	
I, Robot	10			
Terminator 2: Judgment Day		2	2	2
Transformers		3	6	
Terminator 3: Rise Of The Machines		4		
Terminator Salvation		5		
Screamers		7		
Virus		9		
Evolver		10		
2001 A Space Odyssey			1	
Blade Runner			5	
Metropolis			7	10
Edward Scissorhands			8	
The Black Hole				
Blade Runner				7
The Day The Earth Stood Still				8
Alien				9

Other movies that include robotic characters that are not part of these rankings show a strong influence coming from research directions in the HRI field. These movies show clear inspiration from the psychological, sociological, behavioural and technological aspects of HRI literature. It is not clear if the writers are inspired by the HRI literature or if their design of a fantasy world emerges from scientific consultants. We can find recent examples like Robot and Frank (2012), Black Mirror (2013) episode "Be right back", and Surrogates (2009), that explore the interaction between robot agents and humans. Other movies that explore the interaction between humans and other kinds of agents are Her (2013), SimOne (2002) and Black Mirror (2013) episode "The Waldo moment". All of these can be studied for their insights into the science of reciprocity, persuasion, anthropomorphism and other theories in the fields of HRI, HCI and Human Agent Interaction.

An important part of the drama storyline in all fiction movies with robot characters is that humans react towards the robots in such a way that that the robots play an integral role in the plot development. In the past, the public would only tend to think of robots in terms of fantasy, yet although real science

Table 2. Correlations among the different rankings. None is significant p <0.05.

Spearman's rho	about.com	imdb.com	denogeek.com
imdbcom	-.500		
denogeekcom	-.500	-.400	
totalfilmcom	.100	.500	.600

and technology are perhaps lagging behind what is portrayed in movies the role of the robot is consistent across the real world and the movie screen.

Generally, humans tend to perceive objects as anthropomorphic. This conception exists in most cultures, for instance, with the idea of anthropomorphic Gods. This conception within cultures can be extended to the representation of robots, cyborgs and other technological agents in the media. Science has played an important role in the creation of fiction by stimulating the imagination of artists and providing information to be used for creative work. Scientific explanations try to explain the world and justify incredible and inspiring fiction stories. Frankenstein by Mary Shelley is one of the first examples of modern, fantasy-based creatures whose origin lies in an acceptable scientific explanation. Similarly, the novels and short stories of Asimov, Clarke, Dick and other science fiction writers have also had an impact on popular culture with their stories about HRI. Therefore it does not seem out of the ordinary that robots appearing in movies allow human viewers to easily associate with them and anthropomorphise them. Generally, science fiction movies involving robots are popular if the robot appears to be lifelike and engaging in normal conversations with a human, rather than machine-like. Results from research show a similar inclination towards interaction with anthropomorphic robots, where it has been shown that users would prefer to interact with a robot that is more human-like [16].

The perception of robots could be biased by fiction deriving from books and movies rather than scientific research. B. Sandoval et. al. [12][13] showed that the perceptions of robots in children are strongly biased toward mechanical representations of robots they are familiar with when they are asked to draw a robot. In several cases children tried to sketch robots that they had watched in movies or on TV. However, deeper studies are required to determine the influence of media in the design of robots. In general, we can say that there is an interaction between the development of science in HRI and robots presented in media, however, it is a challenge to delimit how media and HRI interact amongst themselves to inspire each other.

3   Discussion and Design Implications

We find a contradiction with the expectations created by the media, the real capabilities of the robots and the needs of people. In general, when people interact with humanoid social robots, demonstrated by HRI researchers, they are mesmerized about how it works and what are the possible uses. A usual conversation after the demonstration is the comparison between real robots and fiction robots (as most people know only about the robots they have seen in movies), and we find their expectation of robots to be similar to their favorite fiction robot. Typical questions that come from the audience are on the useful functionality of robots, such as whether the robot can wash their dishes, babysit children, or be their driver. However, the limitations of the robot's capabilities are quite disappointing to the general audience (we are unable to freely talk with a robot even in this day and age [23]). Recently Aldebaran robotics released a social robot known as Pepper, and they specifically say "At the risk of disappointing you, he doesn't clean, doesn't cook and doesn't have super powers... Pepper is a social robot able to converse with you, recognize and react to your emotions, move and live autonomously." Pepper is not capable of doing exactly what people want from a robot in their daily life. It is obvious that there are differences between what people expect (which emerges from what they see in movies) and their experience with experimental robots, which are incapable of delivering most of the easiest human activities. This mismatch is well grounded in extensive literature on the uncanny valley [22].

The design of humanoid robots is at times inspired by fictional robots; intentionally or unintentionally, scientists try to explore all possibilities to design robots and acquire as much knowledge and inspiration as possible from fiction in their experiments. This approach can ultimately result in humans having unrealised expectations (created by fiction) as social and humanoid robots (i.e. the technology in them) are still limited in how they can help humans. For example, the state of art androids and geminoids do not have enough skills to walk properly, grab objects or have a basic conversation with a human. Other robots are capable of walking in a controlled manner, have a command and control-based interaction and even jump or do a few funny tricks, such as ASIMO from Honda [1]; however, such robots are not affordable for most of the people. We strongly believe that robot designers should match the ideas coming from fiction with the actual necessities of the people in order to have commercial robots capable of doing chores and tasks for the costumers. Therefore, do we as users really require robots in our daily lives that appear like robots from the movies? For example, generally, the robots in movies are highly anthropomorphic and life-sized.

There are many aspects of design involved in this contradiction of expectation and need. Some HRI researchers design and improve humanoid robots with helpful functionality with an aim to deploy these in our private space such as homes, schools and offices. Can we expect the design traditions and conventions of robots in movies and fiction being easily applied to our domestic environments? Is the approach of extracting design guidelines from movies and fiction to design real-world robots practical and recommended? It would mean that HRI at the home could be invasive; as robots would be big, fully autonomous and occupy space, with the capability to talk and permeate our personal space, and they would use a lot of resources at home. Furthermore, HRI research has shown that even at a proximity of 3m, users start to feel uncomfortable interacting with a human scaled robot [21], which begs the question as to what robotic creatures will ultimately live in our homes and how will they look and behave.

In this context and looking at non-robotic objects, people invest in appliances such as washing machines and dryers that are big and expensive because they are useful, but very importantly they do not consume us or our private space. The general location for such appliances is in a space that does not interfere with daily human activities, and we would not generally see a washing machine in the middle of a room because of its fashionable characteristics. Other appliances are less invasive and their appearance in our living spaces can be considered as a form of a display. An example of an appliance that people feel proud to own is the robotic vacuum cleaner Roomba [2] and the service that these give justifies the investment in them. Even though such robots are not fully anthropomorphic [17], users still tend to attribute social interactions towards them [19]. In addition, although a Roomba is not precisely cheap, it is so far the best selling robot that is not a toy. A-priori, in our opinion the requirements for future robots are: a) the user can engage emotionally with them, b) they do not consume a lot of resources, and c) the justification in the investment made in the robot is correlated with its functionality.

People such as Colin Angle claim strongly that the robotic costumer industry has been stunted because research is too focused on biped robots[4]. However, our thought is that common users want a mixture of the subtle interactive abilities similar to fictional robots with the useful capabilities that real robots like robot Roomba offers and the good design of other robots such as NAO [3] or AIBO [14] deliver.

4   Conclusion and Future Work

We have mentioned that robots in the media have an effect on peoples expectations of the capabilities of robots in the real world. We have found that there is a conflict between the expectations of the users (that are primarily shaped by movies and fiction), the goals of HRI research, and the needs of the users. Consequently, we propose that robots should have certain characteristics in order to become socially acceptable. Robot designers should match the expectations created by media with commercial possibilities. We need to push for common goals in industry and academic research in order to invest resources into robots that are capable of delivering tasks at homes and offices regardless of their embodiment and shape. We also argue that certain degrees of anthropomorphism and a size that is comparable with current robots can justify having a robot at home.

In the current work we have given an overview of robots in movies to address their significance and influence on popular culture as emerging from the movies rankings. We have touched upon the drawbacks of our methodology; which is that the ranking method of each database can be expected to differ. Therefore we aim to extend our research in a number of ways to overcome this bias. Firstly, we aim to study other mediathat can have a strong impact on how we perceive robots and their design, for example printed and written literature. Similarly, user perception of robots appearing in movies could be extracted from public media such as message boards where people post reviews and comments about movies. We expect to run such a data extraction/mining phase, followed by a quantitative research method such as content analysis. In addition, we aim to develop a questionnaire in which we will collect data related to user perception of specific robots appearing in movies and science fiction. Consequently we may be able to negate the inconsistencies that arise from considering only data from movie rankings.

References

1. Honda ASIMO, http://asimo.honda.com/
2. iRobot: iRobot roomba 760, http://www.irobot.co.nz/Roomba760
3. Nao robot, http://www.aldebaran.com/en/humanoid-robot/nao-robot
4. The problem with asimo: Tabloid humanoid is holding back bots,
   http://www.popularmechanics.com/technology/engineering/robots/4264593
5. Revenue of Cyborg/Android and robot movies,
   http://www.boxofficemojo.com/genres/chart/?id=cyborg.htm
6. robot movies, http://www.allmovie.com/search/movies/Robot/all/70
7. ACrezo: Geeks are sexy. tech, science & social news,
   http://www.geeksaresexy.net/2011/08/19/wheres-wall-e-pic/u1gu1/
8. Anya, O., Tawfik, H., Nagar, A., Westaby, C.: An ethics-informed approach to the development of social robotics. From Critique to Action: The Practical Ethics of the Organizational World 231(253), 23 (2011)
9. Bartneck, C.: From fiction to science–a cultural reflection of social robots. In: Proceedings of the CHI 2004 Workshop on Shaping Human-Robot Interaction, pp. 1–4 (2004)
10. Bartneck, C., Obaid, M., Zawieska, K.: Agents with faces-what can we learn from lego minifigures? In: Proceedings of the 1st International Conference on Human-Agent Interaction, Sapporo (2013)
11. Bates, R., Goldsmith, J., Berne, R., Summet, V., Veilleux, N.: Science fiction in computer science education. In: Proceedings of the 43rd ACM Technical Symposium on Computer Science Education, pp. 161–162. ACM (2012)
12. Benítez Sandoval, E., Penaloza, C.: Children's knowledge and expectations about robots: A survey for future user-centered design of social robots. In: Proceedings of the Seventh Annual ACM/IEEE International Conference on Human-Robot Interaction, HRI 2012, pp. 107–108. ACM, New York (2012)
13. Benítez Sandoval, E., Reyes Castillo, M., Rey Galindo, J.A.: Perceptions and knowledge about robots in children of 11 years old in méxico city. In: Proceedings of the 6th International Conference on Human-Robot Interaction, HRI 2011, pp. 113–114. ACM, New York (2011)
14. Borl, J.: Sony puts aibo to sleep - CNET news,
    http://news.cnet.com/Sony-puts-Aibo-to-sleep/2100-1041_3-6031649.html
15. Bruckenberger, U., Weiss, A., Mirnig, N., Strasser, E., Stadler, S., Tscheligi, M.: The good, the bad, the weird: Audience evaluation of a "Real" robot in relation to science fiction and mass media. In: Herrmann, G., Pearson, M.J., Lenz, A., Bremner, P., Spiers, A., Leonards, U. (eds.) ICSR 2013. LNCS, vol. 8239, pp. 301–310. Springer, Heidelberg (2013)

16. Dautenhahn, K., Woods, S., Kaouri, C., Walters, M.L., Koay, K.L., Werry, I.: What is a robot companion-friend, assistant or butler? In: International Conference on Intelligent Robots and Systems, pp. 1192–1197. IEEE (2005)
17. Fink, J.: Anthropomorphism and human likeness in the design of robots and human-robot interaction. In: Ge, S.S., Khatib, O., Cabibihan, J.-J., Simmons, R., Williams, M.-A. (eds.) ICSR 2012. LNCS, vol. 7621, pp. 199–208. Springer, Heidelberg (2012)
18. Fong, T., Nourbakhsh, I., Dautenhahn, K.: A survey of socially interactive robots. Robotics and Autonomous Systems 42(3), 143–166 (2003)
19. Forlizzi, J.: How robotic products become social products: an ethnographic study of cleaning in the home. In: Proceedings of the ACM/IEEE International Conference on Human-Robot Interaction, pp. 129–136. ACM (2007)
20. Iio, J., Iizuka, S., Matsubara, H.: The database on near-future technologies for user interface design from sciFi movies. In: Marcus, A. (ed.) DUXU 2014, Part I. LNCS, vol. 8517, pp. 572–579. Springer, Heidelberg (2014)
21. Koay, K.L., Dautenhahn, K., Woods, S., Walters, M.L.: Empirical results from using a comfort level device in human-robot interaction studies. In: Proceedings of the 1st Conference on Human-Robot Interaction, pp. 194–201. ACM (2006)
22. Mitchell, W.J., Szerszen, K.A., Lu, A.S., Schermerhorn, P.W., Scheutz, M., Mac-Dorman, K.F.: A mismatch in the human realism of face and voice produces an uncanny valley. i-Perception 2(1), 10 (2011)
23. Mubin, O., Bartneck, C., Feijs, L., Hooft van Huysduynen, H., Hu, J., Muelver, J.: Improving speech recognition with the robot interaction language. Disruptive Science and Technology 1(2), 79–88 (2012)
24. Mubin, O., Stevens, C.J., Shahid, S., Al Mahmud, A., Dong, J.J.: A review of the applicability of robots in education. Journal of Technology in Education and Learning 1 (2013)
25. Pransky, J.: Aibo–the no. 1 selling service robot. Industrial Robot: An International Journal 28(1), 24–26 (2001)
26. Sabanovic, S., Michalowski, M.P., Simmons, R.: Robots in the wild: Observing human-robot social interaction outside the lab. In: 9th IEEE International Workshop on Advanced Motion Control, pp. 596–601. IEEE (2006)
27. Salvini, P., Laschi, C., Dario, P.: Design for acceptability: improving robots coexistence in human society. International Journal of Social Robotics 2(4), 451–460 (2010)
28. Schmitz, M., Endres, C., Butz, A.: A survey of human-computer interaction design in science fiction movies. In: Proceedings of the 2nd International Conference on Intelligent Technologies for Interactive Entertainment, p. 7 (2008)
29. Shin, E., Kwak, S.S., Kim, M.S.: A study on the elements of body feature based on the classification of social robots. In: The 17th IEEE International Symposium on Robot and Human Interactive Communication, RO-MAN 2008, pp. 514–519. IEEE (2008)
30. Sung, J., Christensen, H.I., Grinter, R.E.: Robots in the wild: understanding long-term use. In: 2009 4th ACM/IEEE International Conference on Human-Robot Interaction (HRI), pp. 45–52. IEEE (2009)

Robots in Older People's Homes to Improve Medication Adherence and Quality of Life: A Randomised Cross-Over Trial

Elizabeth Broadbent, Kathy Peri, Ngaire Kerse, Chandimal Jayawardena,
IHan Kuo, Chandan Datta, and Bruce MacDonald

The University of Auckland, Auckland, New Zealand
{e.broadbent,k.peri,n.kerse,ikuo005,b.macdonald}@auckland.ac.nz,
cjayawardena@unitec.ac.nz,
work.chandan@gmail.com

Abstract. Healthcare robots are being developed to help older people maintain independence. This randomised cross-over trial aimed to investigate whether healthcare robots were acceptable and feasible and whether the robots could impact quality of life, depression and medication adherence. 29 older adults living in independent units within a retirement village were given robots in their homes for 6 weeks and had a non-robot 6-week control period, in a randomised order. The robots reminded people to take medication, provided memory games, entertainment, skype calls, and blood pressure measurement. The robots were found to be acceptable and feasible, and many participants described them as useful and as friends although not all comments were positive. There were relatively few problems with robot functions. The participants' perceptions of the robots' agency reduced over time. The robots had no significant impact on adherence, depression or quality of life. While the robots were feasible and acceptable, improvements in their reliability and functionality may increase their efficacy.

Keywords: robots, quality of life, medication, adherence, blood pressure, companion, acceptance.

1 Introduction

1.1 Eldercare Robots

Research teams across the world are developing robots that can provide services to older people to help them cope with age-related declines in physical health and cognitive abilities [1-3]. These robots include companion type robots, such as Paro, that are designed to provide companionship and reduce agitation in patients with dementia [4]. Other robots have been designed to help people with more practical tasks, such as physical assistance [5], health-care related tasks [6] and rehabilitation [7,8].

User trials of such robots to date have often been lab-based, observational, and used wizard of oz scenarios [1]. Few studies have been conducted in real-world settings and even fewer have been randomised controlled trials. Results to date have

M. Beetz et al. (Eds.): ICSR 2014, LNAI 8755, pp. 64–73, 2014.

been promising, with two RCTs showing that companion robots can reduce loneliness in a rest-home/hospital aged care facility [9, 10].

Recent advances in technology mean that service robots are reaching a stage where they too can be tested more autonomously in real-world settings. This paper reports the results of a randomised cross-over trial of two types of autonomous service robots in an aged care facility over three months.

1.2   Background to This Study

This research relates to the multidisciplinary, cross-faculty, international healthcare robotics project, jointly funded by New Zealand's Science and Innovation Group within the Ministry of Business, Innovation and Employment (MBIE), and South Korea's Ministry of Knowledge Economy (MKE). The long term goal of the project was to develop an affordable healthcare robot for use in aged communities. The University of Auckland researchers come from engineering, computer science, health informatics, health psychology, general practice, gerontology nursing, integrated care, and geriatrics.

This project started with a questionnaire and focus-group study of staff, residents and relatives preferences for robots within an aged care facility [11]. This indicated several key roles for robots in the centre: falls detection and calling for help, detection of wandering, reminders for schedules & medication, and vital signs assessment. We also added an entertainment and socialization function as this was identified as a need by several senior staff. These applications were developed by Auckland UniServices Ltd and the University of Auckland. Two robots from Yujin Robot Co., Ltd were programmed with these functions (Cafero robot and iRobiQ). These robots were tested in several small studies at the retirement village, and were shown to be acceptable to trial participants [12-15].

This paper reports on a larger controlled trial held at the retirement village with several robots of each kind [16]. The aim of the trial was to investigate whether personal service type robots in the homes (independent units) of older people at the retirement village were feasible and acceptable and to provide pilot information regarding their ability to improve quality of life, reduce depression, and improve medication adherence compared to a control group. We also investigated how participants' attitudes towards the robots changed over the course of the trial using both questionnaires and open-ended interviews.

2   Method

2.1   Trial Design

Repeated measures randomised controlled cross-over trial.

2.2   Setting and Participants

The study was held at a large retirement village in Auckland, New Zealand, with a range of services, from independent living apartments to hospital facilities. The village has over 700 residents, who are mostly aged over 65 years. Recruitment for the study began in September 2011. One hundred and sixteen residents in the independent living apartments at the village were invited to take part in the study through letterbox flyers and advertised talks and demonstrations held at the village centre in November 2011. Thirty residents gave written informed consent to take part in the trial. Ethics Approval was obtained from the University of Auckland, and permission was gained from the CEO of the Selwyn Foundation. One person withdrew prior to randomization leaving 29 participants who were randomized..

2.3   Procedure

The study was divided into two 6-week periods with an washout period of 18 days between them. The robots were placed in the village from the 1st November 2011 until the 21st December 2011. They were removed over the Christmas holiday period. The robots were repositioned from the 8th January 2012 to the end of March 2012.

After baseline measures were taken, participants were randomized to receive the robot during the first 6 week period or the second using computer generated random numbers. Primary outcome measures were taken by blinded interviewers. Interviews about attitudes to robots were not blinded.

2.4   Interventions

During the intervention period the robots were installed into the residents' apartments by the researchers, usually in the dining room. IrobiQ (small robot) was installed on a table top to make it an appropriate height for interactions, while Cafero (taller robot) was freestanding. While both robots have navigation systems and wheels and can be mobile, they were both stationary for this project in order to reduce risks of injury. Both robots run on batteries and have a charging station.

The robot applications were idividualised to the participants needs and communicated over wireless network services to a server which held participants' profiles and measurements, medication prescriptions, and logs of robot activity. The participants were shown through all of the modules available on the robots. Each participant was also provided with a user manual, which informed him/her how to perform basic troubleshooting steps. All residents were provided with a phone number to call if they had any questions or problems.

What Did the Robots Do? Two types of autonomous robots were used in the study, both manufactured by Yujin Robot, Korea (http://yujinrobot.com/eng/). The first robot, iRobiQ, is 45 by 32 by 32 cm and weighs 7kg. It has a 7 inch touch screen, microphone, camera, speakers, a face capable of expression through Led lights, IR obstacle sensors, and runs on windows XP (see Figure 1). It could take blood pressure and pulse oximetry, had music videos and quotes, and a medication management program. The robot's head could swivel and tilt in response to sound, its arms could raise, and its base could swivel. The robot 'danced' when it played music by raising its arms, swivelling its base and head, and showing face lights. The robot displayed a menu screen, and the user could touch the function they wanted to use at any time. Once selected, each function ran autonomously. Seven iRobiQ were used in this trial.

The medication management system was programmed by the health provider with the participants' usual medications. The robot autonomously raised it arms, sounded a bell, and said '[participant name], it is time for your medication' each time medications were due. The robot asked a series of questions which the participant answered via the touch screen, and guided the participant through taking each medication. The robot asked the participant to confirm if they had taken each medication, if they felt unwell or had any side effects. The system had been developed and successfully piloted [15]. If a resident's medication was missed, the robot sent an alert to a duty cell phone. On receiving the alert a nurse familiar with the resident's medication regime would make a clinical decision to take any action regarding the noncompliance. This action was generally taken immediately but in some circumstances contacting the resident was left until the following day. The risk assessment of non-compliance primarily related to the type of medication missed. If the alert also included a message to contact the resident as they were feeling unwell, then an immediate phone call was made to ascertain the urgency of the clinical situation and appropriate action was taken following this conversation. If the resident was unable to be contacted by phone, a research assistant made a visit to the resident's apartment.

The second robot, Cafero, is approximately four feet tall and has a touchscreen, microphone, camera, speakers, IR obstacle sensors, (see Figure 1). It also took blood pressure and pulse oximetry and had music videos and quotes, but not the medication management program. Instead Cafero had a simple Skype calling function, had a commercially available program that provides cognitive exercises, Dakim Brain Fitness (http://www.dakim.com/), a website showing information about the village, and a calendar reminder system. Cafero also had data exchange with Lifetime Health Diary™ (an online platform created by the Lifetime Health Diary Ltd, New Zealand. It integrates clinical data and background health information about the patient). The robot displayed a menu screen and the user could touch the function they wanted to use at any time. Once selected, each function ran autonomously.

Fig. 1. IrobiQ (left) and Cafero (right) robots used in the study

2.5   Measures

Measures were taken and interviews conducted by trained interviewers using standardized techniques at baseline, after the first 6-week period, and after the second 6-week period. At baseline, information was collected about age, gender, ethnicity and education. The 10-item Abbreviated Mental Test Score was administered to assess cognitive impairment and scores below 7 indicate impairment [17].

The *primary outcomes measures* were health related quality of life, depression and adherence. Health related quality of life was measured using the SF-12 [18]. This 12-item measure contains a physical and mental health component, and has been validated by Quality Metric Incorporated. Higher scores indicate better health related quality of life. Depression was measured with the Geriatric Depression Scale (GDS-15) [19], comprising 15 yes or no items which assess depressive symptoms over the last week. Self-reported adherence was measured using the Medication Adherence Report Scale (MARS)[20], in which higher scores indicate higher levels of adherence.

To assess the *acceptability* and feasibility of the robots, a combination of open-ended interviews, questionnaires, a diary and data logs from the robots were used. The interviewers asked participants about how they felt about the robot, how often they used the robot, how the robot affected their lives, in what ways the robot was or was not useful, what they thought of its appearance, and what they would change about it, and how their visitors reacted. The questionnaires included the Robot Attitudes Scale (RAS) [10], which assesses what people think about robots on 10 items (e.g. friendly, useful, trustworthy), and the Mind Perception Questionnaire [21]. This psychometrically validated scale assesses perceptions of the robot's ability to experience things (E.g. feel pain, and pleasure) and have agency (E.g. have thought, and memory). We asked participants to keep a diary of adverse/positive events (E.g. grandchildren played with robot and had fun; robot would not turn on). The data logs recorded when the robot was used, for how long, and which applications were used, as well as the data from the applications such as blood pressure readings. For reasons of limited space, this paper reports the questionnaire and interview results only.

2.6   Data Analysis

Descriptive statistics are used to summarize the characteristics of the sample. Primary outcomes were compared between the period with the robot and the period without for all participants. To analyse the effects of the robot on depression, QOL, and adherence, three mixed ANOVA were conducted (with the repeated measures variable being the scores at each timepoint, and the group factor the order of the robot or no robot phases). We were looking for a significant group by time interaction to indicate that the robot made a difference to outcomes. Attitudes towards the robots and perceptions of mind were analysed using repeated measures t-tests.

3   Results

The mean age of the sample was 85.23 years (SD 5.14, range 72 to 94). There were 14 males and 15 females. Fourteen participants had three years of secondary school education or less, 3 had four or five years of secondary education and 12 had higher education (technical, trade or university degrees). Their self-rated computer experience averaged 3.17 (SD 2.17) of a maximum of 8. The mean score on the Abbreviated Mental Test was 9.24 (SD 0.99). One participant scored 6, which is suggestive of cognitive impairment; this person was married to and living with another participant.

There were no significant differences between groups at baseline in depression, cognition, adherence, or physical QOL scores. There was a baseline difference in mental QOL scores, t(23) = 3.64, p=.001). Table 1 shows these scores.

To analyse the effects of the robot on depression, QOL, and adherence, mixed ANOVA were conducted (with the repeated measures variable the time-point (Dec/Feb) and the group factor the order of the robot or no robot phases). A significant group by time interaction would indicate that the robot made a difference to outcomes. There were no significant group by time effects (p>.05). Analysing the data by paired samples t-tests for robot versus no robot periods showed the same results.

Table 1. Residents' depression, physical quality of life, mental quality of life, and adherence scores between groups and across time

	Baseline Robot 1st	Dec Robot 1st	Feb Robot 1st	Baseline Robot 2nd	Dec Robot 2nd	Feb Robot 2nd
Depression	1.32(1.13)	1.82(2.04)	1.82(2.48)	2.44(2.34)	2.10(1.74)	2.33(1.82)
SF12 PCS	36.68(11.59)	36.85(10.88)	40.79(12.37)	42.71(10.72)	40.28(10.57)	39.80(13.23)
SF12 MCS	59.21(5.07)	54.76(1.90)	52.55(2.34)	49.54(8.86)	56.71(1.81)	55.86(2.23)
MARS	23.16 (2.13)	23.00 (3.03)	22.17 (5.53)	24.33 (0.82)	24.83 (0.41)	24.67 (0.52)

Note - SF12 PCS SF-12 physical health component summary score, SF12 MCS SF-12 mental health component summary score, MARS – Medication Adherence Report Scale. MARS is only analysed for those with the medication management robot (iRobiQ). Follow up means (SE) are controlling for baseline for SF12 MCS.

Residents' attitudes towards robots and mind perception scores at baseline and after using the robot are displayed in Table 2. There was a significant decrease in how much agency people perceived the robot to have after using it for the trial.

There were 75 calls made to the phone line, of which 68 pertained to errors on the robots (such as a frozen screen). The majority of these problems could be fixed by the research assistant rebooting/restarting the robot. Seven calls concerned medication management: a set-up issue (1), not reminding (2), missed medication (2), medication reminding at 4am (1), medication change possibly coming up (1). Medication issues were discussed with the nurse on the research team on three occasions, and the participant was called and reminded on one occasion, the medication reminding program was checked on three occasions.

Table 2. Attitudes towards robots and robot mind perception scores at baseline and after using the robot

	Baseline	Follow-up	t	p
Robot Attitude Scale	56.62 (8.43)	53.60 (9.02)	1.67	.10
Robot agency	21.22 (10.42)	16.87 (8.62)	2.95	.007
Robot experience	12.27 (7.22)	12.08 (7.81)	3.27	.90

3.1   Interviews about the Robots (N=26)

When asked how they felt about the robot, 17 responses were positive and included: being a friend e.g. "I felt like I had a friend in it..."; feeling "OK" or "comfortable" with it; and finding it interesting. There were three negative responses, e.g. "It was a chore. It would be less of a chore if it worked properly", and one mixed response "Quite comfortable except 3-4 times in the night when it woke us up".

They were asked how people reacted to the robots: 22 described positive reactions, e.g. "Amazed and delighted". There were two mixed responses. Five people said that no-one else used it, 8 said 1 or 2 people, 5 said 3-5 and 3 said more than 7 others.

There was a range of how often people reported using the robots. 17 people reported using it at least once a day everyday primarily for medication reminders, 2 people said every second day, and 6 people reported not using it much, and one not at all.

Ten reported that the robot had no effect on their lives. Seven reported positive effects, 3 neutral and 4 negative. The positive effects included companionship (2), relief from going to the medical centre (1), photos bringing back memories (1), adding interest (1), entertainment increasing happiness (1), and reminders (1). Negative effects included: "boring and frustrating", "the camera... was an invasion of privacy...it was quite disturbing when it got the medications wrong".

Participants reported the comments that other people made about the robot. 14 reported positive comments, 5 reported negative comments and 3 reported mixed comments. Positive comments included interest, cuteness, amusement, privilege, "wow that's great", "quite amazed", "good for keeping brain alert, they wanted to take their BP and the brain test". Negative comments included "it was a pain in the backside", "husband does not like it", "grandson was disappointed". Mixed comments included "Good idea but needs more developing", and "some people passed it off as something to learn about and enjoy. Some people ignored it. Some thought it was stupid."

Seven people reported that the robot was not useful. Seven reported that the medication reminding was useful, 6 reported entertainment was useful (2 entertainment, 1 athletic pictures, 3 music, 1 brain fitness), one reported the reminders, 2 reported the Skype function, 3 blood pressure, one the date and time, one that it was useful as a night light, "keeping wife happy", "Felt like someone was taking care of me".

Suggested changes were: continuous music (2), relaxation music (1), better music (1), more music (1), do the washing up (2), vacuuming (2), make easier to turn off (4), fewer leads (1), able to pick up emotions (1), hold a conversation (2), more internet (1), exercise program (1), turn down noise/brightness (1), program it myself (1), make voice less condescending (1), did not like the metallic squeaky voice (1), did not like how it said "have a nice day", increase its memory; 7 said no changes.

The participants were asked how they would feel about having a similar robot all the time. Nineteen said that they would not want it – reasons included: being out (1), don't need it (1), not enough room (4), don't need medication reminder as can remember (3), does not do anything for me (1), needs to be built up (1), not enough time to use it (1), hard to move/switch off/needs more programs. Three people said they would like to have one, and two that they wouldn't mind, and one that it would be like a television or radio that they would switch on/off.

Sixteen people reported that the size/shape/appearance were OK or good, two said the robot was too big, two wanted the screen bigger/clearer. One liked the robot because it looked like a teletubby. Three said it was cute and one that it was attractive; one liked the eyes and grin. One said it was unusual, and one that it looked like a robot. One wanted to be able to turn it off and put it away.

Other comments included: would like it to have an event reminder, video Skype, radio stations, the robot was friendly and comfortable, too many researchers and forms to fill out, the robot would be useful for others but not for me, it should show the medications, if it worked properly it would be useful, not good for someone who is really sick because not interested, blood pressure was unreliable; marvellous bit of technology, enjoyed the study, the study was fun, useful and interesting.

4   Discussion

This report describes a trial of healthcare assistant robots that were deployed at a retirement village for a three month period. The robots provided a number of services and communicated with a server over wireless links. The trial shows that it is possible to deploy robots in such an environment, and that people can use the robots. There were no benefits or harms to QOL, depression or adherence with medications.

Together with our previous studies the results of this larger deployment of robots show that some older people are able to interact with robots and may accept robots, and that it is feasible to deploy robots in a retirement village setting. There were many positive and some negative reactions expressed about robots. No other studies, to our knowledge, show improvement in QOL, medication adherence or depression due to robots in homes. It may be that a much larger trial would show significance; quality of life changes are difficult to show and require large numbers of interactions.

There were a number of challenges conducting this trial. First, while the number of robots was larger than previous trials, the power of the study to find significant effects was small. Second, care had to be taken that the robots did not cause participants to make mistakes in adherence. For safety, we chose participants who usually managed their own medications, and a nurse and physician remotely monitored responses.

There was a significant decrease in perceived agency of the robots by older people in independent living after interacting with the robot. This may reflect an adjustment from unrealistic expectations about robots to a more realistic position, an effect we have previously observed [13].

People's comments about the robots were mixed. Many negative comments were of a relatively minor nature, and might be addressed by improving the robots' software, providing more customization of services to the particular needs of the users, and more business work flow focus on the purpose of robots in particular scenarios.

Overall in the participant responses there is a background theme that reflects the lack of a clear value proposition for the robots' activities in the trial, including direct comments about what the eventual purpose of the robots would be, and indirectly some ambivalence to the robots. The robots were deployed doing various tasks and people were asked to respond about how they considered the robots. However the robots were not deployed in the operational activities of the retirement village, and there was no articulation or expectation that robots were required to achieve any operational objectives; the main goal was to see whether adding robots to the environment would alter quality of life or have any risks. Operational deployment is the next phase for the Healthbots project, where the efficacy, costs and benefits of the robots and applications will be evaluated for specifically designed activities in an operational scenario where the robots are filling a clear operational role in healthcare.

Acknowledgements. This work was jointly supported by the Robot Pilot Project program of the Korea Ministry of Knowledge and Economy (MKE) and Korea Institute for Robot Industry Advancement (KIRIA) and the New Zealand Ministry of Business, Innovation & Employment (IIOF 13635). We thank the Electronics and Telecommunications Research Institute (ETRI) for their valuable collaboration, the ED consortium of South Korean companies ED Corporation, Isan Solutions Corporation, and Yujin Robot Co Ltd.

References

1. Bemelmans, R., Gelderblom, G.J., Jonker, P., de Witte, L.: Socially assistive robots in elderly care: A systematic review into effects and effectiveness. J. Am. Med. Dir. Assoc. 13, 114–120 (2012)
2. Broekens, J., Heerink, M., Rosendal, H.: Assistive social robots in elderly care: A review. Gerontechnology 8, 94–103 (2009)
3. Broadbent, E., Stafford, R., MacDonald, B.: Acceptance of healthcare robots for the older population: Review and future directions. Int. J. Soc. Robot. 1, 319–330 (2009)
4. Wada, K., Shibata, T., Saito, T., Tanie, K.: Effects of robot-assisted activity for elderly people and nurses at a day service center. Proc. IEEE 92, 1780–1788 (2004)
5. Dario, P., Guglielmelli, E., Laschi, C., Teti, G.: MOVAID: a personal robot in everyday life of disabled and elderly people. Technology and Disability 10, 77–93 (1999)
6. Noury, N.: AILISA: experimental platforms to evaluate remote care and assistive technologies in gerontology. In: Proc. of 7th International Workshop on Enterprise networking and Computing in Healthcare Industry, HEALTHCOM 2005, pp. 67–72 (2005)
7. Fasola, J., Mataric, M.A.: Socially assistive robot exercise coach for the elderly. Journal of Human-Robot Interaction 2, 3–32 (2013)
8. Mehrholz, J., Hadrich, A., Platz, T., Kugler, J., Pohl, M.: Electromechanical and robot-assisted arm training for improving generic activities of daily living, arm function, and arm muscle strength after stroke. Cochrane Database of Systematic Reviews 6, CD006876 (2012)

9. Robinson, H., MacDonald, B.A., Kerse, N., Broadbent, E.: The psychosocial effects of a companion robot: A randomized controlled trial. Journal of the American Medical Directors Association 14, 661–667 (2013)

10. Banks, M.R., Willoughby, L.M., Banks, W.A.: Animal-assisted therapy and loneliness in nursing homes: use of robotic versus living dogs. J. Am. Med. Dir. Assoc. 9, 173–177 (2008)

11. Broadbent, E., Tamagawa, R., Patience, A., Knock, B., Kerse, N., Day, K., MacDonald, B.A.: Attitudes towards health care robots in a retirement village. Australasian Journal on Ageing 31, 115–120 (2012)

12. Stafford, R.Q., Broadbent, E., Jayawardena, C., Unger, U., Kuo, I.H., Igic, A., Wong, R., Kerse, N., Watson, C., MacDonald, B.A.: Improved robot attitudes and emotions at a retirement home after meeting a robot. In: IEEE International Symposium on Robot and Human Interactive Communication, pp. 82–87 (2010)

13. Stafford, R.Q., MacDonald, B.A., Jayawardena, C., Wegner, D.M., Broadbent, E.: Does the Robot Have a Mind? Mind Perception and Attitudes Towards Robots Predict Use of an Eldercare Robot. International Journal of Social Robotics 6, 17–32 (2014)

14. Datta, C., Tiwari, P., Yang, H.Y., Kuo, I., Broadbent, E., MacDonald, B.: An interactive robot for reminding medication to older people. In: Proceedings of the 9th International Conference on Ubiquitous Robots and Ambient Intelligence (URAI 2012), Deajeon, Republic of Korea (2012)

15. Datta, C., Tiwari, P., Hong Yul, Y., Broadbent, E., MacDonald, B.A.: Utilizing a closed loop medication management workflow through an engaging interactive robot for older people. In: IEEE 14th International Conference on Paper Presented at: e-Health Networking, Applications and Services, Healthcom (2012)

16. Jayawardena, C., Kuo, I., Datta, C., Stafford, R.Q., Broadbent, E., MacDonald, B.A.: Design, implementation and field tests of a socially assistive robot for the elderly: HealthBot Version 2. In: RAS/EMBS International Conference on Biomedical Robotics and Biomechatronics, pp. 1837–1842 (2012)

17. Hodkinson, H.M.: Evaluation of a Mental Test Score for Assessment of Mental Impairment in the Elderly. Age Ageing 1, 233–238 (1972)

18. Ware, J.E., Kosinski, M., Keller, S.D.: A 12-item short-form health survey: Construction of scales and preliminary tests of reliability and validity. Med. Care 34, 220–233 (1996)

19. Yesavage, J.A., Brink, T.L., Rose, T.L., Lum, O., Huang, V., Adey, M., Leirer, V.O.: Development and validation of a geriatric depression screening scale: A preliminary report. J. Psychiatr. Res. 17, 37–49 (1983)

20. Horne, R., Weinman, J.: Patients' beliefs about prescribed medicines and their role in adherence to treatment in chronic physical illness. J. Psychosom. Res. 47, 555–567 (1999)

21. Gray, H.M., Gray, K., Wegner, D.M.: Dimensions of mind perception. Science 315, 619 (2007)

Real-Time Gender Based Behavior System for Human-Robot Interaction

Pierluigi Carcagnì[1], Dario Cazzato[1], Marco Del Coco[1], Marco Leo[1], Giovanni Pioggia[2], and Cosimo Distante[1]

[1] National Research Council of Italy, Institute of Optics, Arnesano (LE), Italy
pierluigi.carcagni@ino.it
http://www.ino.it/en/
[2] Institute of Clinical Physiology of CNR, Pervasive Healthcare Center, Messina, Italy

Abstract. This work introduces a real-time system able to lead humanoid robot behavior depending on the gender of the interacting person. It exploits Aldebaran NAO humanoid robot view capabilities by applying a gender prediction algorithm based on the face analysis. The system can also manage multiple persons at the same time, recognizing if the group is composed by men, women or is a mixed one and, in the latter case, to know the exact number of males and females, customizing its response in each case. The system can allow for applications of human-robot interaction requiring an high level of realism, like rehabilitation or artificial intelligence.

Keywords: human-robot interaction, artificial intelligence, gender recognition.

1 Introduction

Each human-human communication is based on a form of interaction that involves faces. In the light of this, for the design of a human-computer interaction, it is natural to expect to find faces playing an essential role. In fact, there has been considerable technical progress within artificial intelligence in the field of computer vision to open the possibility of positioning faces at a very significant place within human-machine interaction [14]. In the field of artificial intelligence and human-robot interaction (HRI), even gender recognition can significantly improve the overall user experience quality, giving to the person the opportunity to interact with an entity that can change its behavior depending on the sex of the user that is interacting with it. Beyond realism and variance of the interaction, a gender recognition system able to work in real-time could lead to several applications in the field of socially assistive robotics, like people in rehabilitation or autistic children, considering their well-known interest on computers and electronic devices [19].

Since its importance, this topic has been well investigated in the last decades by computer vision and machines learning scientists. As a preliminary step, especially in order to create a fully automatic face analysis system, facial images of men and women must be extracted. The well-known Viola-Jones [26] algorithm introduces a robust cascade detector (based on AdaBoost [9] and Haar features) for the face recognition in image, and is actually considered as a state-of-art approach.

M. Beetz et al. (Eds.): ICSR 2014, LNAI 8755, pp. 74–83, 2014.
© Springer International Publishing Switzerland 2014

Gender recognition can be viewed as a two-class classification problem, and methods can be roughly divided in feature-based and appearance-based. Mäkinen and Raisamo [18] and Sakarkaya et al. [22] introduced two wide interesting surveys that exhaustively cover the topic.

The very first results were simultaneously shown in [7] and [10], in 1990. A following study, that investigated the use of geometrical features in order to achieve gender recognition, was performed by Brunelli and Poggio [4](1995), while Abdi at al. [11], in the same year, applied pixel based methods and used a radial-basis function (RBF) network. Lyons at al. used Gabor wavelets with PCA and Linear discriminant analysis (LDA) [17]. In 2002, Sun at al. showed the importance of features selection for generic algorithms [24] first and, successively, tested the efficiency of Local Binary Pattern (LBP) for gender classification [23]. Seetci at al. applied Active Apparence Models (AAM) to this scope [21], with the support of an SVM classifier. Recently, Ihsan et al. showed the performance of a spatial Weber Local Descriptor (SWLD) [25].

The problem of gender estimation, together with all the other information extractable from facial images, as a way to be considered in the design of HRI applications has been taken into account already in [27], but gender has been considered only for the design of humanoid faces, and not as a possibility of improving social interaction thanks to the possibility to perform a recognition task on the user's face. Recently, in [13], performances comparison of gender and age group recognition to carry out robot's application service for HRI has been proposed, but with the usage of audio information only. The work of [16] addresses the same problem, but using a RGB-D device and basing its processing on the body shape.

Although several works on the topic of gender recognition have been proposed over the years, in both academia and industry, it seems that very few applications of it in the field of human-robot interaction have been taken into account. Moreover, the only work of this kind in the state of the art does not explore 2D visual information. To overcome to these limitations, in this work, a real-time system that, processes data coming from a camera on board the robot is automatically able to provide more situation awareness if the person in front of it is a male or a female, is proposed. The system can also manage multiple persons at the same time, recognizing if the group is composed by men, women or is a mixed one and, in the latter case, to know the exact number of males and females, customizing its response in each case. The manuscript is organized as follows: in section 2, our system is presented. After introducing the overall scheme, we will focus on the used gender estimation algorithm. Section 3 shows experimental results. Finally, obtained results and future developments are discussed in section 4.

2   NAO Gender Based Behavior System

In Fig. 1 a scheme of the proposed system is shown. It is composed by two main units: the first unit is the Aldebaran NAO humanoid robot, while the second one is a Remote Computational Unit (RCU) aimed to perform all the computational tasks. RCU and NAO are connected by a local network, as shown in Fig. 1. This architecture allows to satisfy the fundamental requirement to work in real-time, avoiding an overload on the low computational power of the robot CPU (an ATOM Z530).

Video frames, coming from the camera mounted on the top of the head of the robot, are taken by means of the API (Application Programming Interface) provided with the NAO Software Development Kit. Captured video frames are sent to the Gender Prediction Module (GPM) subsystem in order to detect the presence of a human being and predict his/her gender. Gender predictions are then sent to the Behavior Decision Module (BDM) that sends a message to the robot in order to activate gender-specific behaviors.

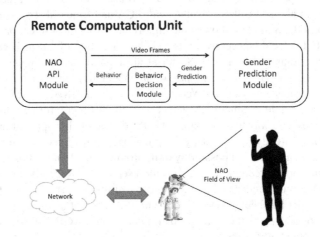

Fig. 1. A scheme of the proposed gender based behavior system

Communication between NAO and the RCU has been achieved using the NAOqi framework, that allows homogeneous communication between different modules (motion, audio, video), homogeneous programming and homogeneous information sharing. After connecting to the robot using an IP address and a port, it is then possible to call all the NAO's API methods as with a local method. For further informations, refer to the official documentation [1].

2.1 Gender Prediction Module

The system core is the *gender prediction module*. It uses the raw video frames as input to detect the presence of a human being and predict his/her gender. As illustrated in Fig.2, the first step is to recognize the presence of a face (consequently a human being) in the scene and to extract the normalized face to analyze. To this end, a *face detection and normalization* process is done by means of the procedure proposed by Castrillon et al. in [5] and the `processed face` is obtained. Moreover, this procedure allows to detect and to track multiple faces in the scene assigning them unique IDs, allowing for particularizing a behavior only one time for a specific person. Once the normalized face image is available, a *features extraction* phase is performed. In particular, we chose to work with Histogram of Oriented Gradients (HOG) that shows, since previous tests, better performance against other low complex features. The procedure aimed to the features extraction is well discussed in section 2.1. HOG `features` `data vectors` are than projected in a low-dimensional subspace through the *subspace projection* block. Subspace projection makes use of a precomputed *projection model* trained over the features extracted from a dataset of thousands of faces. Successively, the `reduced features data vector` is given to the *SVM prediction* block, that gives as output the gender prediction. As well as the subspace projection, the *SVM prediction* needs a model trained over the reduced features data vector of the same faces dataset. *Subspace projection* and *SVM prediction* blocks are detailed respectively in section 2.1. Predicted genders are stored, frame by frame, in a *predicted gender buffer* of length N_{maj} using a FIFO logic. Finally, the *majority filter* compute the gender class with the greater number of occurrences and give in output the `filtered predicted gender class`.

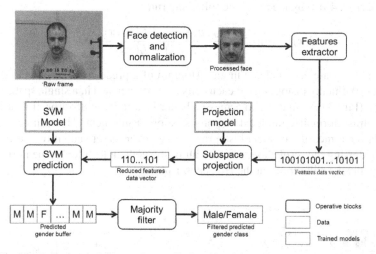

Fig. 2. The block diagram of the gender prediction algorithm: the raw frames are processed in order to obtain a reliable gender-prediction of the people in the scene

Face Detection and Normalization. The detection and normalization of the face in the scene are mainly preprocessing operations whose main steps are illustrated in Fig. 3. It is necessary to guarantee, to the successive operative blocks, a standard face image pose. Castrillon et al. in their face detection and normalization processes [5] perform the sequence of these two operation exploiting persistence face information among successive frames. The current frame is gray-scale converted and then the well known Viola-Jones face detector is applied. Successively, an eye detection is done to locate the eye pairs in the image and rotate and scale the face with the aim to obtain standard face image with eyes pair located in the same position. Down-line the process the result is a normalized 65×59 pixel gray-scale face image.

HOG - Features Extraction. HOG is a well known feature descriptor based on the accumulation of gradient directions over the pixel of a small spatial region referred as a "cell", and in the consequent construction of a 1D histogram. Even thought HOG

Fig. 3. The face detection and normalization step: the face is cropped and aligned in order to guarantee a standard pose to the *features extraction* step

has many precursors, it has been used in its mature form in Scale Invariant Features Transformation [15] and widely analyzed in human detection by Dalal and Triggs [8]. This method is based on evaluating well-normalized local histograms of image gradient orientations in a dense grid. Let L be the image to analyze. The image is divided in cells (Fig. 4 (a)) of size $N \times N$ pixels and the orientation θ of each pixel $x = (x_x, x_y)$ is computed (Fig. 4 (b)) by means of the following rule:

$$\theta(x) = \tan^{-1} \frac{L(x_x, x_y + 1) - L(x_x, x_y - 1)}{L(x_x + 1, x_y) - L(x_x - 1, x_y)} \tag{1}$$

The orientations are accumulated in an histogram of a predetermined number of bins (Fig. 4 (c-d)). Finally histograms of each cells are concatenated in a single spatial HOG histogram (Fig. 4 (e)). In order to achieve a better invariance to disturbs, it is also useful to contrast-normalize the local responses before using them. This can be done by accumulating a measure of local histogram energy over larger spatial regions, named blocks, and using the results to normalize all of the cells in the block. The normalized descriptor blocks will represent the HOG descriptors.

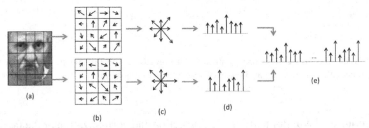

Fig. 4. HOG features extraction: the image is spatially divided in cells and the pixel orientation of each pixel in a cell is computed. Successively orientations histograms are computed and concatenated depending on the cell-space image division.

Subspace Projection. The number of used features for face description is highly influenced in computational complexity and accuracy of classification. Indeed, a reduced number of features allows SVM to use easier functions and to perform better division of clusters. Anyway, the reduction of original features space is a non trivial step.

Principal component analysis (PCA) is a widely used approach for subspace reduction. It chooses a dimensionality reducing linear projection that maximizes the scatter of all projected samples. Simply speaking, the more informative subspace direction are selected for the subspace reduction. The number of components should be selected as the one able to preserve the desired total variance of data.

On the other hand, Linear Discriminant Analysis (LDA) [3] is a class specific method that tries to shape the scatter in order to make it more reliable for classification. This method selects the projection matrix in such a way that the ratio of the between-class scatter and the within-class scatter is maximized. Moreover, in LDA analysis the number of non-zero generalized eigenvalue, and so the upper-bound in eigenvectors numbers, is $c - 1$, where c represents the number of class.

SVM Prediction. Support Vector Machines (SVM)s are techniques aimed to data classification. A classification task uses a training set to generate the model used for the prediction. The training set is usually made up by many instances each of which contains a *class label* and several *features*. The prediction step uses just the features set and the trained model to predict a class for the current instances. As well as for the subspace, projection either the *SVM* accuracy need to be tested over a set of instances different by the training one. At this purpose, in Section 3, a k-fold validation approach has been applied over data.

3   Experimental Results

The *gender prediction module* accuracy evaluation has been realized with a k-fold test over the whole model estimation and prediction process.

We employed a fusion of two of the most representative datasets in face classification problems (on the following referred as "Fusion"): the Morph [2] and the Feret [20] datasets. Both datasets consist of face images of people of different gender, ethnicity and age and are equipped with a complete CVS file with gender, race and other information. Anyway, due to face recognition errors the real number of tested faces is of 55915 male and 9246 females. Even a balanced subset has been taken into account for evaluation. The procedure, showed in Fig.5, consists of two step: a model estimation and a prediction estimation. The whole face-images dataset is randomly split in k subfolds. For each of the k validation steps, $k - 1$ sub-fold for the training and 1 sub-fold for the prediction/validation process have been used. We performed face detection and normalization over each image and successively, the `features data vector` is extracted. The set of `features data vector` is then used to train, in sequence, the subspace reduction algorithm and the SVM prediction one. When both models are available, the one-out fold is tested over them.

The process is repeated over each of 5 to one-out sub-fold combination and the accuracy results is averaged.

For HOG operator, the *VLFeat library*[1] has been used using standard parameters as in [8] with a feature vector length of 2016 elements. Both PCA and LDA subspace projection reduce the features vector dimension. In our case, a number of 100 component for the PCA was taken into account in order to preserve the 95% of the total variance of data. On the other hand, the LDA approach is characterized by a dimension of the projection space that is fixed to 1 (i.e. the number of classes minus one).

The SVM classification problem has been treated by means of the publicly available LIBSVM library [6]. More precisely, we used a radial basis function (RBF) that, in the

---

[1] http:www.vlfeat.org

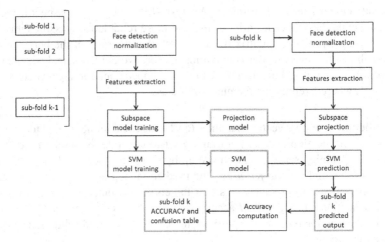

Fig. 5. Test procedure for accuracy estimation: the procedure is done k times in order to obtain the best estimation of total accuracy and confusion table

opinion of the authors of as well as in our experience, seems to be the most reasonable choice [12]. Usually a grid search for penalty parameters C and the others RBF parameters could be desirable. Anyway, our tests does not arise any significant difference in the results as the parameters change. More specifically, we set $C = 1$ and $\gamma = 1/N_f$ where N_f is the number of features.

We obtained a total accuracy of 86.5% and 88.6% for PCA and LDA respectively for the unbalanced dataset, while balanced dataset showed an accuracy of 89.7% with PCA and 80.5% with LDA. Confusion tables are presented in Tables 1 and 2, where M_T, F_T, M_P and F_P represent respectively the true male and female subjects and the predicted ones, TA is the total accuracy and the superscript B stands for *balanced*. All the results are quite close, anyway the HOG+PCA on the balanced dataset gives the best performances both in terms of total accuracy and gap among the two genders.

The whole architecture (presented in section 2) has been tested on a real scenario where people directly interacted with the robot. No constraint in the appearance nor in the background were given to the participants. Each person, one at time, entered in the field of view of the NAO robot. When the face was detected, depending on the gender of the person, the robot acted in a different way. For our purpose, i.e. in order to show the possibility to develop a complete different behavior depending on the user (even originating different learning scheme, since it would be based on the same input), the robot acted in the following way: in the presence of a woman in the scene, it bowed down, while in the presence of a man, the robot greets with his right hand. Fig. 6 illustrate the NAO point of view and the recognize step (a,b) and the consequent action depending on the male (c) or female (d) interacting subject. Even a sentence to be pronounced from the robot has been customized depending on the sex. In the presence of a mixed group (without overlapping of the face area), the robot can say the exact number of men and woman in the scene. Errors are completely related to the errors in the gender prediction algorithm. Moreover, since given a person each prediction is independent from the

Gender confusion tables: each table presents the results for the each specific descriptor/projection pair for both balanced and unbalanced data-set configuration. M_T: Male true; F_T: Female true; M_P: Male prediction; F_P: Female prediction; M_P^B: Male prediction using balanced data-set; F_P^B: Female prediction using balanced data-set; TA: Total accuracy; TA^B: Total accuracy using balanced data-set.

Table 1. HOG + PCA	M_T	F_T
M_P	97.8%	2.2%
F_P	25.8%	74.2%
M_P^B	87.3%	12.7%
F_P^B	7.8%	92.2%
TA	86%	
TA^B	89.7%	

Table 2. HOG + LDA	M_T	F_T
M_P	98.7%	1.3%
F_P	21.5%	78.5%
M_P^B	82.1%	17.9%
F_P^B	21.2%	78.8%
TA	88.6%	
TA^B	80.5%	

(a) NAO gender recognition step (male). (b) NAO gender recognition step (female). (c) Male behavior after recognition. (d) Female behavior after recognition.

Fig. 6. A test of the interaction between the NAO and humans being. The NAO recognizes the gender of the interacting subject (a,b) and reacts with a customized behavior (it bows down for woman and greets with its right hand for male).

possible presence of other faces in the same image, it was possible to estimate the error of the system evaluating the interaction with the robot of one person at time. With our real scenario, we tested the algorithm on 20 persons, 10 males and 10 females, and 3 errors have been reported. Therefore, the estimated error was of 15%. The system was able to detect and classify faces at a distance in the range of [20, 300] cm.

About computational remarks, the system was tested on a local network in order to avoid latency errors in the evaluation of the frame rate. The RCU was a CPU i7@3.20GHz with a RAM of 16 GB DDR3. Images were processed as a resolution of 640×480. In these conditions, our system was able to work at a frame rate of 13 fps. This is a very encouraging result since it allowed to use the predicted gender buffer in order to strengthen the prediction.

4 Conclusions

With this work, a real-time system able to process data coming from a camera installed into an Aldebaran NAO humanoid robot in order to define, depending on the gender of the person, its behavior, has been proposed. Multiple persons in the scene at the same time are also managed. The system can allow for applications of human-robot interaction requiring an high level of realism, like rehabilitation or artificial intelligence. A simple customized behavior has been implemented in order to show the possibility to use the system as a starting point for developing a more complex artificial intelligence for the robot, with a more advanced behavior and different tasks. Moreover, other information can be integrated, like an estimation of race and/or age of the users, augmenting the level of the interaction. Additionally, in the case of false prediction, it could be possible to integrate a technique based on gesture recognition in order to, with a pre-specified gesture, teach the robot the right gender of the user, that will store the information. Finally, a user study to investigate whether and how gender-based interaction scheme can improve HRI could be conducted. An evaluation of these developments will be the subject of future works.

References

1. https://community.aldebaran-robotics.com/doc/1-14/index.html
2. Morph-noncommercial face dataset, http://www.faceaginggroup.com/morph/
3. Belhumeur, P., Hespanha, J., Kriegman, D.: Eigenfaces vs. fisherfaces: recognition using class specific linear projection. IEEE Transactions on Pattern Analysis and Machine Intelligence 19(7), 711–720 (1997)
4. Brunelli, R., Poggio, T.: Hyberbf networks for gender classification (1995)
5. Castrillón, M., Déniz, O., Guerra, C., Hernández, M.: Encara2: Real-time detection of multiple faces at different resolutions in video streams. Journal of Visual Communication and Image Representation 18(2), 130–140 (2007)
6. Chang, C.C., Lin, C.J.: LIBSVM: A library for support vector machines. ACM Transactions on Intelligent Systems and Technology 2, 27:1–27:27 (2011), software available at http://www.csie.ntu.edu.tw/~cjlin/libsvm
7. Cottrell, G.W., Metcalfe, J.: Empath: Face, emotion, and gender recognition using holons. In: Advances in Neural Information Processing Systems, pp. 564–571 (1990)
8. Dalal, N., Triggs, B.: Histograms of oriented gradients for human detection. In: IEEE Computer Society Conference on Computer Vision and Pattern Recognition, CVPR 2005, vol. 1, pp. 886–893 (2005)
9. Freund, Y., Schapire, R.E.: A decision-theoretic generalization of on-line learning and an application to boosting. J. Comput. Syst. Sci. 55(1), 119–139 (1997)
10. Golomb, B.A., Lawrence, D.T., Sejnowski, T.J.: Sexnet: A neural network identifies sex from human faces. In: NIPS, pp. 572–579 (1990)
11. Abdi, H., Valentin, D., Edelman, B., O'Toole, A.J.: More about the difference between men and women: evidence from linear neural networks and the principal-component approach. Neural Comput. 7(6), 1160–1164 (1995)
12. Hsu, C.W., Chang, C.C., Lin, C.J., et al.: A practical guide to support vector classification (2003)
13. Lee, M.W., Kwak, K.C.: Performance comparison of gender and age group recognition for human-robot interaction. International Journal of Advanced Computer Science & Applications 3(12) (2012)
14. Lisetti, C.L., Schiano, D.J.: Automatic facial expression interpretation: Where human-computer interaction, artificial intelligence and cognitive science intersect. Pragmatics & Cognition 8(1), 185–235 (2000)

15. Lowe, D.G.: Distinctive image features from scale-invariant keypoints. Int. J. Comput. Vision 60(2), 91–110 (2004)
16. Luo, R.C., Wu, X.: Real-time gender recognition based on 3d human body shape for human-robot interaction. In: Proceedings of the 2014 ACM/IEEE International Conference on Human-Robot Interaction, pp. 236–237. ACM (2014)
17. Lyons, M.J., Budynek, J., Plante, A., Akamatsu, S.: Classifying facial attributes using a 2-d gabor wavelet representation and discriminant analysis. In: Proceedings of the Fourth IEEE International Conference on Automatic Face and Gesture Recognition, pp. 202–207 (2000)
18. Mäkinen, E., Raisamo, R.: An experimental comparison of gender classification methods. Pattern Recognition Letters 29(10), 1544–1556 (2008), http://www.sciencedirect.com/science/article/pii/S0167865508001116
19. Moore, D.: Computers and people with autism. Asperger Syndrome, 20–21 (1998)
20. Phillips, P., Moon, H., Rizvi, S., Rauss, P.: The feret evaluation methodology for face-recognition algorithms. IEEE Transactions on Pattern Analysis and Machine Intelligence 22(10), 1090–1104 (2000)
21. Saatci, Y., Town, C.: Cascaded classification of gender and facial expression using active appearance models. In: 7th International Conference on Automatic Face and Gesture Recognition, FGR 2006, pp. 393–398 (April 2006)
22. Sakarkaya, M., Yanbol, F., Kurt, Z.: Comparison of several classification algorithms for gender recognition from face images. In: 2012 IEEE 16th International Conference on Intelligent Engineering Systems (INES), pp. 97–101 (June 2012)
23. Sun, N., Zheng, W., Sun, C., Zou, C.-r., Zhao, L.: Gender classification based on boosting local binary pattern. In: Wang, J., Yi, Z., Żurada, J.M., Lu, B.-L., Yin, H. (eds.) ISNN 2006. LNCS, vol. 3972, pp. 194–201. Springer, Heidelberg (2006)
24. Sun, Z., Bebis, G., Yuan, X., Louis, S.J.: Genetic feature subset selection for gender classification: A comparison study. In: IEEE Workshop on Applications of Computer Vision, pp. 165–170 (2002)
25. Ullah, I., Hussain, M., Muhammad, G., Aboalsamh, H., Bebis, G., Mirza, A.: Gender recognition from face images with local wld descriptor. In: 2012 19th International Conference on Systems, Signals and Image Processing (IWSSIP), pp. 417–420 (April 2012)
26. Viola, P., Jones, M.: Rapid object detection using a boosted cascade of simple features. In: Proceedings of the 2001 IEEE Computer Society Conference on Computer Vision and Pattern Recognition, CVPR 2001, vol. 1, pp. I–511. IEEE (2001)
27. Walker, J.H., Sproull, L., Subramani, R.: Using a human face in an interface. In: Proceedings of the SIGCHI Conference on Human Factors in Computing Systems, pp. 85–91. ACM (1994)

Activity Recognition for Natural Human Robot Interaction

Addwiteey Chrungoo[1], S.S. Manimaran, and Balaraman Ravindran[2]

[1] School of Engineering and Applied Science, University of Pennsylvania,
Philadelphia PA 19104, USA
[2] Department of Computer Science,
Indian Institute of Technology Madras, Chennai, India

Abstract. The ability to recognize human activities is necessary to facilitate natural interaction between humans and robots. While humans can distinguish between communicative actions and activities of daily living, robots cannot draw such inferences effectively. To allow intuitive human robot interaction, we propose the use of human-like *stylized gestures* as communicative actions and contrast them from conventional activities of daily living. We present a simple yet effective approach of modelling pose trajectories using directions traversed by human joints over the duration of an activity and represent the action as a histogram of direction vectors. The descriptor benefits from being computationally efficient as well as scale and speed invariant. In our evaluation, the descriptor returned state of the art classification accuracies using off the shelf classification algorithms on multiple datasets.

1 Introduction

As robots are employed to perform wide range of tasks, especially in human environments, the need to facilitate natural interaction between humans and robots is becoming more pertinent. In many roles, such as, indoor personal-assistants, robots must be able to infer human activities and decipher whether or not a human needs assistance. For e.g., if a robot could recognize whether a person is drinking water, it could offer to pour more and react appropriately based on the person's response. In such scenarios, in addition to recognizing the drinking activity, the robot needs to be capable of recognizing communicative actions, so as to infer whether it should pour more or stop. This is similar in principle to *how humans assist others*, i.e., either they assist if assistance is sought or they foresee the need for assistance based on perception and acquired knowledge. Though past works [10] have focussed on estimating human intent to take such decisions, this work is motivated by the need for interaction between the robot and human as a factor in deciding on an appropriate behaviour. Incorporating such natural interactions is not easy when robots work in highly cluttered environments where people carry out activities in different ways leading to high variability [14,7]. However, to best support humans, assistive robots need to behave interactively like humans, making it imperative to correctly understand the human actions involved.

M. Beetz et al. (Eds.): ICSR 2014, LNAI 8755, pp. 84–94, 2014.

As a result, we are particularly interested in developing a concise representation for a wide variety of actions; both communicative and conventional activities of daily living. We propose the use of human-like *stylized gestures* as communicative actions and contrast them from conventional activities of daily living. *Stylized gestures* are symbolic representations of activities and are widely used by humans across cultures to communicate with each other when verbal communication is not possible.We hypothesize that such actions have distinct motion intrinsics as compared to conventional activities of daily living and can hence be used effectively to communicate with robots in the absence of verbal means.

Before we can begin to develop a system for activity recognition, we need an efficient representation mechanism for human motion.

In this work we introduce a novel activity descriptor: Histogram of Direction vectors (HODV) that transforms 3D spatio-temporal joint movements into unique directions; an approach that proves to be highly discriminative for activity recognition. As shown in Figure 1, we represent skeletal joint movements over time in a compact and efficient way that models pose trajectories in terms of directions traversed by human joints over the duration of an activity. The issue we address in this paper is as follows: Learn to recognise various human actions given a direction-vector histogram representation using three dimensional joint locations as raw data. Further,

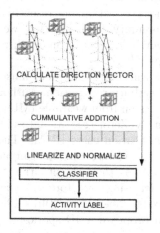

Fig. 1. The general framework of the proposed approach

learn to distinguish communicative actions to instruct a robot from conventional activities of daily living and obtain a descriptive labelling of the same. We show that our proposed approach is efficient in distinguishing Communicative and Non Communicative activities in our novel RGBD dataset and also performs equally well on two public datasets: Cornell Activity Dataset (CAD -60) and UT-Kinect Dataset using off the shelf classification algorithms.

1.1   Contributions and Outline

The contributions of this work are are as follows: Firstly, we introduce the problem of communicative vs non-communicative actions. Secondly, we propose a novel and computationally efficient activity descriptor based on pose trajectories. We provide analysis of our algorithm on two public datasets and demonstrate how the algorithm could be used for both Communicative/Interactive and Non-Communicative/Non-Interactive activity recognition. We will also release an annotated RGBD Human Robot Interaction dataset consisting of 18 unique activities including 10 *stylized gestures* as well as 8 conventional activities of daily living (within the same dataset) along with full source code of our algorithm.

The rest of the paper is organized as follows. Section 2 presents a brief literature review. Section 3 explains our dataset, while section 4 and 5 describe our algorithm and experimental results in detail respectively. We conclude the paper in section 6 and also present directions for future work.

2   Related Work

Human activity recognition has been widely studied by computer vision researchers for over two decades. The field, owing to its ability to augment human robot interaction, has recently started receiving a lot of attention in the robotics community. In this section, we restrict ourselves largely to research relevant to robotics, and for an in-depth review of the field, one can refer to recent survey papers [2].

Earlier works focussed on using IMU data and hidden Markov models(HMMs) for activity recognition. Authors in [18] proposed a model based on multi sensor fusion from wearable IMUs. They first classified activities into three groups, namely: Zero, Transitional and Strong displacement activities, followed by a finer classification using HMMs. Their approach was however restricted to very few activity classes and was computationally expensive. Mansur et al.[8] also used HMMs as their classification framework and developed a novel physics based model using joint torques as features; claimed to be more discriminative compared to kinematic features [12]. Zhang et al.[17] followed a vision based approach and proposed a 4D spatio-temporal feature that combined both intensity and depth information by concatenating depth and intensity gradients within a 4D hyper-cuboid. Their method was however dependant on the size of the hyper-cuboid and could not deal with scale variations. Sung et al.[12] combined human pose and motion, as well as image and point-cloud information in their model. They designed a hierarchical maximum entropy Markov model, which considered activities as a superset of sub-activities.

While most of these works focussed on generating different features, work on improving robot perception, including recognizing objects and tracking objects [4] led to the incorporation of domain knowledge [13] within recognition frameworks. Authors in [5] proposed a joint framework for activity recognition combining intention, activity and motion within a single framework. Further, [7,10] incorporated affordances to anticipate activities and plan ahead for reactive responses. Pieropan et al.[9] on the other hand introduced the idea of learning from human demonstration and stressed the importance of modelling interaction of objects with hands such that robots observing humans could learn the role of an object in an activity and classify it accordingly.

While past works excluded the possibility of interaction with the agent, this work aims to understand activities when interaction between robots and humans is possible and realistic, especially, in terms of the human providing possible instructions to a robot while also performing conventional activities of daily living. The focus of our work is to utilize distinctions in motion to differentiate between communicative/instructive actions and conventional activities of daily living. Having said this, we do not see motion information alone as a replacement, but as a complement to existing sensory modalities, to be fused for particularly robust activity recognition over wide ranges of conditions.

3 Our Dataset

Recent advances in pose estimation [11] and cheap availability of RGBD cameras, has lead to many RGBD activity datasets [12,14]. However, since none of the datasets involved communicative/interactive activities alongside conventional activities of daily living, we collected a new RGBD dataset involving interactive as well as non interactive actions. Specifically, our interactive actions were between a robot and a human; where the human interacts with the robot using *stylized gestures*; an approach commonly used by humans for human-human interaction.

The activities were captured using a kinect camera mounted on a customized pioneer P3Dx mobile robot platform. The robot was placed in an environment wherein appearance changed from time to time, i.e., the background and objects in the scene varied. In addition, the activities were captured at various times of the day leading to varied lighting conditions. A total of 5 participants were asked to perform 18 different activities, including 10 Communicative/Interactive activities and 8 Non-Interactive activities, each performed a total of three times with slight changes in viewpoint from the other instances. *'Catching the Robots attention', 'Pointing in a direction', 'Asking to stop', 'Expressing dissent', 'Chopping', 'Cleaning', 'Repeating', 'Beckoning', 'Asking to get phone' and 'facepalm'* were the 10 Robot-Interactive activities. In Robot-Interactive activities like *'Facepalm'*, the human brings his/her hand up to his head, similarly, the activity *'chopping'* involved a human repeatedly hitting one of his hands with the other hand, creating a stylized chopping action and so on. The non interactive activities were more conventional activities of daily living like *'Drinking something', 'Wearing a backpack', 'Relaxing', 'Cutting', 'Feeling hot', 'Washing face' 'Looking at time'* and *'Talking on cellphone'*.

We stress that our dataset is different from publicly available datasets as we represent a new mix of activities, more aligned with how humans would perform these in real life. In addition, the dataset involves wide variability in how the activities were performed by different people as subjects used both left and right hands along with variable time durations. For e.g., in the *'Drinking something'* activity, some subjects took longer to drink water and brought the glass to their mouth couple of times, while others took the glass to their mouth just once. The wide variety and variability makes recognition challenging. We have made the data available at: `http://rise.cse.iitm.ac.in/activity-recognition/`

4 Action Representation

Activities usually consist of sequences of sub-activities and can be fundamentally described using two aspects: a) Motor Trajectory and b) Activity context. For eg., in a drinking activity, a subject picks a glass or a cup, brings it closer to his/her mouth and returns it. While there are numerous possibilities behind the context of the activity, as a glass could contain juice while a cup could contain coffee, thereby giving more meaning to the activity 'drinking' and answering a question: *What is probably being drunk?* The motor trajectory followed by most people for a generic drinking activity would predominantly be similar. We aim to exploit this similarity and introduce a *local motion* based action representation called *Histogram of Direction Vectors*, defined as the distribution of directions taken by each skeleton joint during all skeleton pose transitions during an activity.

The intuition behind the descriptor is that directions have a clear physical significance and capturing motion intrinsics as a function of direction should be discriminative across classes. We describe the 3D trajectory of each joint separately and construct the final descriptor by concatenating the direction vector histogram of each joint.

4.1   Direction Vectors from Skeletons

The algorithm takes RGBD images as input and uses the primesense skeleton tracker [1] to extract skeleton joints at each frame. For each joint i, P_f^i represents the 3D cartesian position of joint i at time frame f. The joint locations are then normalized by transforming the origin to the human torso, thereby making them invariant to human translation. Direction vectors are then calculated for each joint i by computing the difference between joint coordinates of frame f and frame $f + \tau$, where τ is a fixed time duration (e.g., 0.1 seconds) in terms of frame counts. Mathematically, direction vectors are estimated for each joint at every frame as:

$$d_f^i = \left[P_f^i - P_{f+\tau}^i\right], \forall f \in [1, 2, \ldots, f_{max} - \tau] \tag{1}$$

The next section explains the construction of our action descriptor, Histogram of direction vectors, and the final descriptor used to classify activities.

4.2   Histogram of Direction Vectors

At each frame f, the local region around a joint i is partitioned into a 3D spatial grid. We chose 27 primary directions in the 3D space and represented the direction taken by a joint by the nearest primary direction in that grid. The grid entries represent real world directions such as, up, down, up-left, down-right and so on; resulting in a total of 27 directions. The direction vector corresponding to a joint i is mapped onto the index of one of 27 directions, by estimating the 3D euclidean distance between grid coordinates σ_q and the direction vector d_f^i; with a vector being allotted a particular direction index q corresponding to the minimum distance. The goal is to find the specific direction index q^* that represents the direction which is at minimum euclidean distance from the direction vector.

$$q^* = \operatorname{argmin}\|d_f^i - \sigma_q\| \quad \forall q \in [1, 2, \ldots, 27] \tag{2}$$

where σ_q is the coordinate of grid index q.

Let Q^f denote the vector of directions, with Q_q^f denoting the entries of vector Q^f at index q. The grid index q^* is then used to update vector Q^f. To attain the total number of times a particular direction was taken during an activity, we perform cumulative addition of vector Q^f at each frame as shown in equation 4 where h^* is a vector revealing the number of times each direction was taken by a joint during the course of an activity.

The vector h^* is then normalized to compute the feature vector h_i for joint i. Normalizing the vector h^* gives us a histogram h_i, representing the probability of occurrence of each direction for a particular joint i, during the course of an activity. Further, each histogram h_i is concatenated to gen-

$$Q_q^f = \begin{cases} 1 & \text{if } q = q^* \\ 0 & \text{otherwise} \end{cases} \tag{3}$$

$$h^* = \sum_f Q^f \tag{4}$$

$$h_i = \frac{h^*}{\|h^*\|_1} \tag{5}$$

erate the final feature vector $H = [h_1, h_2, \ldots, h_i]$; namely the Histogram of direction Vectors.

5   Experimental Results

In this section we present detailed analysis of our experiments. In addition to our dataset, we test our algorithm on two public datasets: The Cornell activity dataset (CAD-60) [12] and the UTKinect-Action Dataset [14]. Our results reveal that the proposed approach performs comparable to the state of the art approaches, which in general, are computationally expensive and involve complicated modelling. We show how our algorithm, despite being very simple, returns better results; while being computationally inexpensive as well as lower in dimensionality. We use an SVM (LIBSVM) as our classification algorithm along with histogram intersection as the kernel choice. We optimize the cost parameter using cross validation.

5.1   Our Dataset

On our dataset, we ran experiments using three different settings. In the first, we classified actions into their respective categories using the entire dataset. In the second setting, we manually separated the activities into Communicative/Interactive activities and Non-Interactive activities and ran our classification algorithm on the two groups independently. In the third setting, we trained a two class classifier and labelled the activities as belonging to either of the two groups. All experiments were performed using 5 fold cross subject cross validation, such that, at a time, all instances of one subject were used for testing and the instances from the other subjects were used for training. None of the instances used for training were ever present in the test set at the same time.

It was our observation that not all joints contributed towards an activity. This lead to many joints being binned into the grid representing *no movement*, leading to reduced accuracy. To counter this phenomenon, we masked the feature vector i.e., made the contribution of the corresponding *no movement* bin zero and renormalized. Feature masking resulted in in-

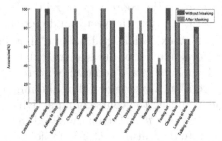

Fig. 2. Comparison on accuracies with and without feature masking

creased accuracy in not only our dataset (Figure 2) but also the CAD 60 and UTKinect Action Datasets.

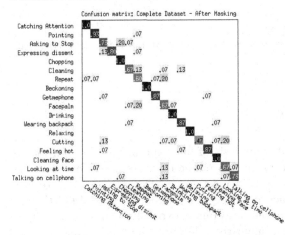

Fig. 3. Confusion matrix of entire dataset using Feature Masking

Figure 3 shows the confusion matrix of our first experimental setting. Most activities are classified with good accuracy apart from *Repeat* and *Facepalm*, mostly because of the similar motion trajectories. Also, as visible in Figure 2 activities such as *Asking to stop, Repeat, Drinking, Wearing backpack* and *Cleaning face* were better classified after feature masking. The average accuracy attained without feature masking was 80%, while with feature masking the average accuracy improved to 82.59%.

Figure 4 shows the confusion matrix of our second experimental setting. The average classification accuracy for Interactive actions was 84%, while for Non Interactive actions, the average accuracy was 86.67%. Like in the previous setup, the algorithm was able to accurately classify actions which had distinct motion trajectories but gets confused with actions with very similar motion like *Repeat* and *Facepalm*.

In the third experimental setup, we classified an activity into either of the two groups. The algorithm achieved an average classification accuracy of 89.26%. Interactive actions were classified with an accuracy of 92.67% while Non Interactive activities were classified correctly with an accuracy of 85%. This classification paradigm could be essential for the development of hierarchical models where the first level could be an Interactive Vs Non-Interactive classification, followed by a finer categorization into an exact activity.

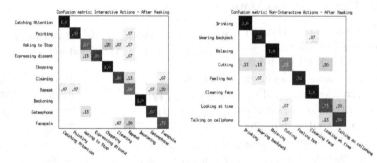

Fig. 4. Left: Confusion matrix of Interactive/Communicative actions after Feature Masking. Right: Confusion matrix of Non-Interactive actions after Feature Masking

Our algorithm is able to distinguish between Interactive and Non Interactive activities with good accuracy. It works well even when subjects take different time duration to complete an activity. Further, since we follow a histogram based representation, classification is invariant to the number of times an action is performed within an activity. For. e.g., a circle could be made once or five times. As long as the feature vector is normalized and if an action is symmetric (activities involving mirror directions eg: waving), the number of times the action is performed or the starting point of the activity would not hamper classification. The descriptor also benefits from being computationally efficient as the only calculations involved for each joints are:

- Calculation of direction vectors, which can be performed in constant time.
- Updating appropriate Histogram bins which is linear in the number of frames and can be performed real-time as and when new frames are captured.

This makes HODV an efficient, yet effective feature vector for classifying human activities.

5.2   Cornell Activity Dataset (CAD 60)

The dataset comprises of 60 RGBD video sequences of humans performing 12 unique activities of daily living. The activities have been recorded in five different environments: Office, Kitchen, Bedroom, Bathroom, and Living room; generating a total of 12 unique activities performed by four different people: two males and two females. We used the same experimental setup (4 fold cross-subject cross validation) and compare precision-recall values for the 'New Person' setting as described in [12]. Table 1 shows a comparison of our algorithm with other state of the art approaches. All of the algorithms mentioned in table 1 use visual features in addition to skeleton data. This work is largely restricted to the use of skeleton data for classification. Hence it would be fair to compare with an approach that uses just skeleton data. The precision recall scores in [12] without visual features is 67.20 and 50.20 respectively. Considering that we use only skeleton data, our approach still outperforms other algorithms.

5.3 UTKinect Action Dataset

The UTKinect Action Dataset [14] presents RGBD video sequences and skeleton information of humans performing various activities from different views. 10 subjects perform 10 different activities namely: *walk, sit down, stand up, pick up, carry, throw, push, pull, wave hands* and *clap hands*. Each subject performs an activity twice.

Table 1. Comparison of our algorithm with other approaches on the CAD 60 dataset

Method	Precision	Recall
Sung et. al[12]	67.90	55.50
Yang, Tian[15]	71.90	66.60
Ni. et al[3]	75.90	69.50
Gupta et. al[6]	78.10	75.40
Koppula et. al[7]	80.80	71.40
Zhang, Tian[16]	86.00	84.00
Our Descriptor	71.76	70.23
Our Descriptor + Masking	83.77	82.06

There are a total of 200 instances of different activities in this dataset. Since each skeleton is described by 20 joints, our feature vector is of dimensions 20 × 27, i.e., a total of 540 features were used for classification in this dataset. For this dataset, we compare our approach with the state of the art methodology called histogram of 3D skeleton joint positions (HOJ3D)[14] using Leave one Sequence out Cross validation (LOOCV) and cross subject validation as defined previously in this paper. This dataset has activities which look very similar e.g., Sit down and Stand Up. Our high accuracies reveal the superiority of our algorithm in distinguishing such actions, which despite looking similar, have distinct trajectory directions, aptly captured by our approach. The overall accuracies attained on the dataset are shown in Table 2. Clearly, our approach generates better accuracy as compared to the Histogram of 3D joints algorithm under the LOOCV setting. The performs drops a bit under the cross subject crossvalidation scheme. Authors in [14] do not report cross subject results.

6 Conclusion

This paper presented the problem of Communicative vs Non-Communicative actions and human activity recognition in general. We proposed a novel and computationally efficient activity descriptor, Histogram of Direction Vectors, which aptly

Table 2. Comparison of our algorithm with HOJ3D on the UT-Kinect dataset

Method	Accuracy
HOJ3D [14] (LOOCV)	90.92
Ours (Cross Subject)	84.42%
Ours (LOOCV)	87.44%
Ours + Masking (Cross Subject)	89.45%
Ours + Masking (LOOCV)	91.96%

captured motion intrinsics and returned good accuracies on our new RGBD dataset. The descriptor proved beneficial in distinguishing between Interactive/Communicative and Non-Interactive activities. Further, results on two public datasets depict its potential in conventional activity recognition frameworks. As part of future work, we would like to combine the descriptor with visual features to cater to cases where the motion trajectories are very similar.

References

1. Nite Skeleton Tracking, `http://wiki.ros.org/nite` (accessed: July 30, 2014)
2. Zhu, C., Sheng, W.: Human daily activity recognition in robot-assisted living using multi-sensor fusion. In: IEEE International Conference on Robotics and Automation, ICRA 2009, pp. 2154–2159 (May 2009), doi:10.1109/ROBOT.2009.5152756
3. Ni, B., Moulin, P., Yan, S.: Order-preserving sparse coding for sequence classification. In: Fitzgibbon, A., Lazebnik, S., Perona, P., Sato, Y., Schmid, C. (eds.) ECCV 2012, Part II. LNCS, vol. 7573, pp. 173–187. Springer, Heidelberg (2012)
4. Collet, A., Martinez, M., Srinivasa, S.S.: The moped framework: Object recognition and pose estimation for manipulation. The International Journal of Robotics Research (2011)
5. Gehrig, D., Krauthausen, P., Rybok, L., Kuehne, H., Hanebeck, U., Schultz, T., Stiefelhagen, R.: Combined intention, activity, and motion recognition for a humanoid household robot. In: 2011 IEEE/RSJ International Conference on Intelligent Robots and Systems (IROS), pp. 4819–4825 (September 2011)
6. Gupta, R., Chia, A.Y.S., Rajan, D.: Human activities recognition using depth images. In: Proceedings of the 21st ACM International Conference on Multimedia, MM 2013, pp. 283–292. ACM, New York (2013), `http://doi.acm.org/10.1145/2502081.2502099`
7. Koppula, H., Gupta, R., Saxena, A.: Learning human activities and object affordances from rgb-d videos. IJRR 32(8), 951–970 (2013)
8. Mansur, A., Makihara, Y., Yagi, Y.: Action recognition using dynamics features. In: 2011 IEEE International Conference on Robotics and Automation (ICRA), pp. 4020–4025 (May 2011)
9. Pieropan, A., Ek, C., Kjellstrom, H.: Functional object descriptors for human activity modeling. In: 2013 IEEE International Conference on Robotics and Automation (ICRA), pp. 1282–1289 (May 2013)
10. Saxena, A.: Anticipating human activities using object affordances for reactive robotic response. In: RSS (2013)
11. Shotton, J., Fitzgibbon, A., Cook, M., Sharp, T., Finocchio, M., Moore, R., Kipman, A., Blake, A.: Real-time human pose recognition in parts from single depth images. In: CVPR (March 2011)
12. Sung, J., Ponce, C., Selman, B., Saxena, A.: Unstructured human activity detection from rgbd images. In: 2012 IEEE International Conference on Robotics and Automation (ICRA), pp. 842–849 (May 2012)
13. Teo, C., Yang, Y., Daume, H., Fermuller, C., Aloimonos, Y.: Towards a watson that sees: Language-guided action recognition for robots. In: 2012 IEEE International Conference on Robotics and Automation (ICRA), pp. 374–381 (May 2012)
14. Xia, L., Chen, C., Aggarwal, J.: View invariant human action recognition using histograms of 3d joints. In: 2012 IEEE Computer Society Conference on Computer Vision and Pattern Recognition Workshops (CVPRW), pp. 20–27. IEEE (2012)
15. Yang, X., Tian, Y.: Effective 3d action recognition using eigenjoints. J. Vis. Comun. Image Represent. 25(1), 2–11 (2014), `http://dx.doi.org/10.1016/j.jvcir.2013.03.001`

16. Zhang, C., Tian, Y.: Rgb-d camera-based daily living activity recognition. Journal of Computer Vision and Image Processing 2(4) (December 2012)
17. Zhang, H., Parker, L.: 4-dimensional local spatio-temporal features for human activity recognition. In: 2011 IEEE/RSJ International Conference on Intelligent Robots and Systems (IROS), pp. 2044–2049 (September 2011)
18. Zhu, C., Sheng, W.: Human daily activity recognition in robot-assisted living using multi-sensor fusion. In: IEEE International Conference on Robotics and Automation, ICRA 2009, pp. 2154–2159 (May 2009)

Intuitive Robot Control
with a Projected Touch Interface

Lennart Claassen, Simon Aden,
Johannes Gaa, Jens Kotlarski, and Tobias Ortmaier

Institute of Mechatronic Systems,
Leibniz Universität Hannover,
Appelstraße 11a, 30167 Hanover, Germany
{lennartc,simon.aden}@stud.uni-hannover.de,
{johannes.gaa,jens.kotlarski,tobias.ortmaier}@imes.uni-hannover.de
www.imes.uni-hannover.de

Abstract. This work proposes an intuitive and adaptive interface for human machine interaction that can be used under various environmental conditions. A camera-projector-system is added to a robot manipulator allowing for a flexible determination of a suitable surface to project a graphical user interface on. The interface may then be used to select different autonomous tasks to be carried out by the robot. In combination with an implemented person tracking algorithm our approach offers an intuitive robot control, especially for repetitive tasks as they occur inside domestic or working environments.

Keywords: Robots, Intuitive Control, Projection, Touch Interface.

1 Introduction

Today's working environments of industrial production sites are still characterized by a combination of static machines and human workers. Due to safety reasons, areas where robots operate are often shielded from human access. This applies both to stationary and to mobile robotic systems. It has been shown that the joint actions of humans and robotic systems can lead to more flexibility and new possibilities [1]. In the field of service robotics the interaction of human and robotic systems became a main exploratory focus, i.e. see [2].

Previous approaches of integrating projectors into the human machine interaction have either used static robot configurations, thereby limiting the range of possible projection surfaces [3,4], or used the projectors as hand held devices to control robotic movement [5]. The objective of the presented work is the creation of an innovative user interface which allows for simple accessibility and easy operation of a robot. The application is especially designed for recurring tasks characteristic for domestic and working environments such as collecting and delivering materials or products. By projecting a graphical user interface using the robots manipulator the need for additional input equipment such as computers, mobile devices or other control panels would become redundant. Through the

M. Beetz et al. (Eds.): ICSR 2014, LNAI 8755, pp. 95–104, 2014.

flexibility offered by the robots manipulator any suitable surface in the reach of the robot may be used as a projection area for the user interface. The projected user interface then allows for selection and start of different implemented tasks. This way the proposed interface is most effective when combined with autonomous systems, which are able to carry out several tasks by their own.

An important aspect in the interaction with a mobile robot is to tell the robot where to move. In a well defined environment this can be done by choosing one of several predefined locations via the graphical interface. The additional option of manual movement control however allows more flexibility and lets the user e.g. teach new locations or direct the robot to a desired target area. For this purpose we implemented a track and follow algorithm that lets the robot track the user and follows him to any location. Combined with the freely selectable projection area, full advantage can be taken of the robots mobility.

The interface projection system and the user input detection are described in section 2. Section 3 presents the tracking system and user following algorithm. The user input detection is evaluated in section 4 followed by conclusions and possible future enhancements in section 5.

2   The Interface Projection System

To realize an intuitive control of a robot the cooperation of different components is required. These include the distortion free projection of a user interface onto a given area and the detection of the user input.

The projection system consists of two devices: a small laser projector used to project the graphical user interface and a 2D video camera to detect suitable projection areas and to capture the selection made by the user. Therefore, the first requirement for the camera-projector-system is a high degree of correlation between the camera field of view (FOV) and the projector FOV. This was achieved by creating a mounting which allowed a fixed arrangement of the two devices on the robots manipulator as shown in figure 2. The second requirement for the system is to allow for perspective transformations between the camera image frame and the projector image frame in order to transform the detected user input into the scope of the projected interface. To identify intrinsic and extrinsic parameters of the camera and the projector a calibration of the system as proposed by Raskar and Beardsley in [6] can be performed. However, since this approach makes use of external sensors a different calibration method was implemented which is based on correspondences between 3D points in the camera coordinate frame and 2D points in the projector image frame.

2.1   Camera Projector Calibration

The chosen calibration method after Leung et al. [7] is based on detected correspondences between the homogeneous 2D points on the projector image plane $P_P = [u_P, v_P, 1]$ and the homogeneous 3D points inside the camera coordinate frame $P_C = [x_C, y_C, z_C, 1]$. This approach is valid because the projection model of a projector is basically the same as the model of a camera. The only difference lies in the projection direction: a camera projects 3D points into a 2D plane while a projector creates a 2D image at the intersection with the 3D points of the image plane. For this reason any known 3D point can also be projected onto the projector image plane if the projector is treated as a camera.

After identification of the intrinsic camera parameters using a camera calibration based on Zhang's method [8] the 3D point coordinates can be determined in the camera coordinate system. Afterwards a transformation of the points into the projector coordinate system would be possible given the relative rotation and translation between camera and projector. However, if this transformation is combined with the unknown intrinsic parameters of the projector to form the projection matrix M_P, the relationship between the points can directly be expressed as:

$$
\begin{bmatrix} u_P \\ v_P \\ 1 \end{bmatrix} = \underbrace{\begin{bmatrix} m_{11} & m_{12} & m_{13} & m_{14} \\ m_{21} & m_{22} & m_{23} & m_{24} \\ m_{31} & m_{32} & m_{33} & m_{34} \end{bmatrix}}_{M_P} \begin{bmatrix} x_C \\ y_C \\ z_C \\ 1 \end{bmatrix} . \tag{1}
$$

Division of lines one and two and lines one and three of the equation system in (1) leads to the equations:

$$
v_P(x_C m_{11} + y_C m_{12} + z_C m_{13} + m_{14}) - u_P(x_C m_{21} + y_C m_{22} + z_C m_{23} + m_{24}) = 0 \; , \tag{2}
$$

$$
u_P(x_C m_{31} + y_C m_{32} + z_C m_{33} + m_{34}) = (x_C m_{11} + y_C m_{12} + z_C m_{13} + m_{14}) \; . \tag{3}
$$

By dividing the n detected correspondences into two subsets of $n_1 \geq 8$ and $n_2 \geq 4$ the equations (2) and (3) can be solved using Singular Value Decomposition to give an estimation of the projection matrix M_P.

The executed calibration now enables the transformation of any given point in 3D camera coordinates to the projector image plane.

2.2   Projection Plane Detection

After successfully calibrating the camera-projector-system the detection of a suitable projection area in the camera coordinate system is required. An area is considered suitable for projection if it is a flat plane in which a rectangular shape of at least ten centimeters in width and six centimeters in height may be fitted. The plane detection can be achieved using either the camera directly or using the forward kinematics of the robot.

Plane Detection Using the Camera. Since we are using a 2D camera the detection of a plane in 3D coordinates is only possible with at least some prior knowledge about the projection area. Using given information, such as dimension, shape or color of the plane, different image processing techniques e.g. Harris corner detection [9] or Hough transformation [10] may be applied to extract the corner points which then can be used to determine if the area is suitable for projection of the user interface.

Plane Detection Using the Forward Kinematics. If the pose of the camera has been integrated into the robot model e.g. using hand-eye calibration the forward kinematics of the robot may be used to determine a suitable projection area. In our implementation the forward kinematics is used to determine the camera pose in relation to a given plane in the robot environment, e.g. the floor plane the robot is moving on. The algorithm then searches for the largest possible projection area to fit the FOV of the projector starting from the intersection of the central projection ray and iteratively incrementing the projection area until the limitation of the projector FOV is reached.

2.3   Image Projection

After the determination of a suitable projection area and transformation of the plane into the projector image frame, the projection image has to be transformed to fit the projection area in order to be displayed to the user without any perspective distortion. This is achieved by calculating the homography matrix and applying perspective transformation to the output image as in the example shown in figure 1.

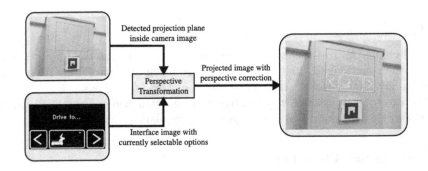

Fig. 1. Perspective transformation of the output image

2.4   Projection Area Alignment

As an example application we implemented the detection and usage of different projection stations. The stations are equipped with labels containing augmented reality (AR) code markers as well as suitable projection areas. By detecting the AR code a coordinate system can be determined inside the camera frame for every marker. Using coordinate transformations it is then possible to align the robot and the projection system to the marker and thereby to the projection surface.

Figure 2 shows the transformation used to align the robot to the projection station. First the detection of the AR marker gives the transformation $^{AR}T_C$ from the Coordinate System of the AR marker $(CS)_{AR}$ to the coordinate system of the camera $(CS)_C$. Using the given transformation $^{C}T_R$ between the camera and the robot coordinate system $(CS)_R$ from the hand-eye calibration the transformation between the AR marker and the robot can be determined as:

$$^{AR}T_R = {}^{AR}T_C \, {}^{C}T_R \ . \tag{4}$$

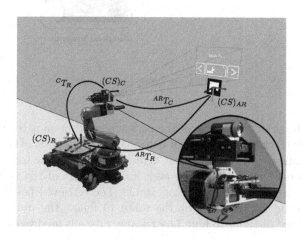

Fig. 2. Coordinate Transformation at a Projection Station

2.5   User Input Detection

As described, the detection of the user input is achieved using the 2D camera of the camera-projector-system. First, a perspective transformation of the projection area into the camera image plane is carried out, to restrict the processed section of the camera image to the projection area. It is then possible to divide the input image into sections that relate to different areas of the projected user interface. The user input e.g. touching of the projected interface buttons is then detected using the implementation of the Gaussian mixture model for background subtraction described by Zivkovic [11]. The complete process of the user input detection is shown in figure 3. By implementing an additional color filter the possible input devices may be restricted and noise in the input image can be reduced to enhance the robustness of the input detection.

3   The User Tracking System

For the task of following the user the robot is equipped with a depth camera to detect and subsequently track the person standing in front of the robot. Taking

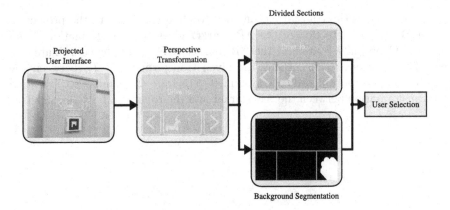

Fig. 3. User Input Detection

advantage of the prior knowledge of the person's position on activation we used an approach for tracking that does not depend on the detection of specific human features. This allows for a more robust detection and tracking from any camera angle and a person can be detected even if only a part of the body is visible or if the person's silhouette is unrecognizable. The retrieved position from the tracking algorithm is then used by the robot to follow the user by trying to maintain a defined distance to him. Laser scanner data is used to avoid obstacles along the way.

3.1   Tracking Algorithm

The algorithm uses the centroid of the tracked object in the previous frame as a seed point for a region growing algorithm that segments the tracked object in the current frame of the depth image. In order to verify that the correct object was found the geometric extent of the object is calculated using a principle component analysis (PCA) of the segmented point cloud. If the change of geometry is greater than a defined threshold the object is rejected and the tracker is reinitialized.

Transformation. The depth values are needed in a coordinate frame orthogonal to the tracked object. If the sensor is mounted with an angle the point cloud must be transformed to a suitable frame. The depth sensor provides a point cloud $P \in \mathbb{R}^{I \times J \times 3}$ where I and J are the height and width of the depth image and $p(i,j) \in \mathbb{R}^3$ with $i \in [0, I[$ and $j \in [0, J[$ is one Cartesian point. With the rotational matrix R the points are transformed to a coordinate frame orthogonal to the object being tracked. The transformed point cloud P_t consists of the points

$$p_t(i,j) = [x_{ij}, y_{ij}, z_{ij}]^T = R\, p(i,j)\ . \tag{5}$$

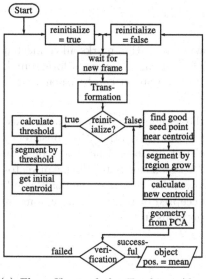

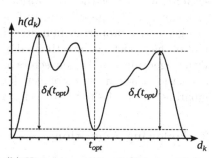

(a) Flow Chart of the Tracking Algorithm

(b) Histogram of Depth Values with Optimal Threshold

Fig. 4. Flow Chart and Depth Value Histogram of the Tracking Algorithm

Initialization. At initialization it is assumed that the object to be tracked is the dominant object in the foreground of the scene. An initial mask for the object can then be obtained by applying a threshold to the depth image.

In the histogram $h(d_k)$ of the K discretized depth values d_k the optimal threshold t_{opt} can be calculated using the following left and right distances in the histogram (also see figure 4 (b))

$$\delta_l(d_k) = \max_{i=0,\ldots,k-1} \left(h(d_i) - h(d_k) \right) ,$$ (6)

$$\delta_r(d_k) = \max_{i=k+1,\ldots,K-1} \left(h(d_i) - h(d_k) \right) .$$ (7)

The optimal threshold maximizes the sum of both distances.

$$t_{opt} = \arg\max_{k=0,\ldots K-1} \left(\delta_l(d_k) + \delta_r(d_k) \right) .$$ (8)

The centroid of all pixels with a depth value $z_{ij} < t_{opt}$ is used to find the initial seed point.

Segmentation. First a good seed point $s = [i_s, j_s]$ has to be found in the neighborhood N of the given centroid $c = [i_c, j_c]$. If the depth value of the centroid from the last frame is d_{last} then the seed point is chosen as the point with the closest depth value to d_{last}.

$$s = \arg\min_{[i,j]\in N} (|d(i,j) - d_{\text{last}}|) \ . \tag{9}$$

Starting from this seed point a region growing algorithm marks all connected pixels that have a depth value within a given tolerance range as foreground. Given the foreground pixels the new centroid of the object can be calculated.

Verification. Under the assumption that the geometric extent of the tracked object can not change drastically from one frame to another the consistency of the geometric properties indicates if the object has been lost. The geometric extent of the object can be estimated by a PCA which calculates the mean vector and the eigenvalues and eigenvectors of the covariance matrix of the point cloud. The decision if the detected object shall be accepted or rejected can be done by comparing the results to the ones from the previous frame.

3.2   Following Algorithm

Starting with the position of the tracked person a target point is set on the intersection point of the direct line between the robot and the person and a circle around the person's position (see figure 5 (a)). The radius of the circle defines the distance at which the robot tries to follow. The translational velocity vector is set towards the target point with an absolute value proportional to the distance. The heading of the robot is controlled towards the tracked person to ensure that the person is always within the field of view of the depth sensor.

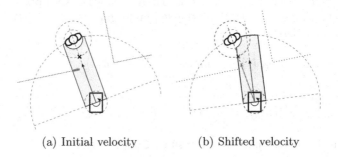

(a) Initial velocity   (b) Shifted velocity

Fig. 5. The observed corridor is limited by a maximum distance from the robot, a maximum distance to the velocity vector, a maximum angle to the velocity vector and a radius around the target. All laser scan points within this corridor are regarded as obstacles and have to be avoided.

To avoid obstacles a corridor surrounding the velocity vector is observed. If laser scan points are detected within this corridor the velocity vector is shifted until the corridor is free or until an abortion criteria is met (see figure 5 (b)).

4   Results

To determine the applicability and robustness of the projected interface in combination with the user input detection the user interface was projected onto a plane as described in section 2.2. The interface was divided into six areas which had to be selected by the user to generate different commands. Overall $n = 1080$ user inputs given by hand were evaluated and used to determine the influence of the relative position between the projection system and the projection plane as shown in figure 6.

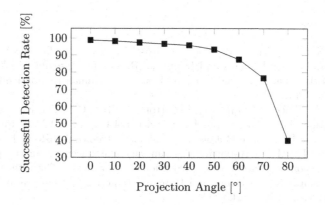

Fig. 6. Successful User Input Detection Rate Depending on the Projection Angle

As it can be seen, the average successful detection rate drops significantly if the projection angle becomes too steep and exceeds 50 degrees. Since the input was generated using hands, part of the increasing error may result from the movement of the hands over the desired interface area before actually touching it. The likelihood of generating a false input this way increases with the projection angle since the space above certain interface areas may occlude other areas. On the other hand the very high rate of successful detections remains relatively constant up to an angle of 30 degrees which is well suitable for most application cases.

5   Conclusion

In our approach we implemented an innovative interface for human-machine interaction. Using a camera-projector-system a graphical user interface is projected onto a suitable surface. The detected input enables the user to control a robot without the requirement of a special input device. The addition of the camera-projector-system to the robots manipulator allows for a high flexibility in the determination of a suitable projection area. In combination with our human tracking algorithm it allows for intuitive control of the robot in various environments.

The developed components are able to operate independently from each other allowing for transfer to any other mobile or stationary robotic system.

Increased robustness, though, especially concerning the user input detection may be achieved by making use of more advanced equipment and computing resources. Possible enhancements of the proposed system include the usage of a projecting device with increased brightness and the replacement of the 2D camera of the camera-projector-system with an RGB-D camera to further improve the determination of plane projection surfaces as well as the precision of the user input detection.

References

1. Lenz, C., Nair, S., Rickert, M., Knoll, A., Rösel, W., Gast, J., Bannat, A., Wallhoff, F.: Joint-Action for Humans and Industrial Robots for Assembly Tasks. In: The 17th IEEE International Conference on Robot and Human Interactive Communication, Munich (2008)
2. De Luca, A., Flacco, F.: Integrated control for pHRI: Collision avoidance, detection, reaction and collaboration. In: The 4th IEEE RAS & EMBS International Conference on Biomedical Robotics and Biomechatronics (BioRob), Rome (2012)
3. Matsumaru, K.: Mobile Robot with Preliminary-announcement and Display Function of Forthcoming Motion using Projection Equipment. In: The 15th IEEE International Symposium on Robot and Human Interactive Communication, pp. 442–450. IEEE Press, New York (2006)
4. Park, J., Kim, G.J.: Robots with Projectors: An Alternative to Anthropomorphic HRI. In: Proceedings of the 4th ACM/IEEE International Conference on Human Robot Interaction, pp. 221–222. ACM, New York (2009)
5. Hosoi, K., Dao, V.N., Mori, A., Sugimoto, M.: VisiCon: A Robot Control Interface for Visualizing Manipulation Using a Handheld Projector. In: Proceedings of the International Conference on Advances in Computer Entertainment Technology, pp. 99–106. ACM, New York (2007)
6. Raskar, R., Beardsley, P.A.: A self-correcting projector. In: Proceedings of the 2001 IEEE Computer Society Conference on Computer Vision and Pattern Recognition, vol. 2, pp. 504–508. IEEE Press, New York (2001)
7. Leung, M.C., Lee, K.-K., Wong, K.H., Chang, M.-Y.: A Projector-based Movable Hand-held Display System. In: IEEE Conference on Computer Vision and Pattern Recognition, pp. 1109–1114. IEEE Press, New York (2009)
8. Zhang, Z.: A Flexible New Technique for Camera Calibration. IEEE Transactions on Pattern Analysis and Machine Intelligence 22, 1330–1334 (2000)
9. Harris, C., Stephens, M.: A Combined Corner and Edge Detector. In: Proceedings of the 4th Alvey Vision Conference, Manchester, pp. 147–151 (1988)
10. Duda, R.O., Hart, P.E.: Use of the Hough Transformation to Detect Lines and Curves in Pictures. Communications of the ACM 15, 11–15 (1972)
11. Zivkovic, Z.: Improved Adaptive Gaussian Mixture Model for Background Subtraction. In: Proceedings of the 17th International Conference on Pattern Recognition, vol. 2, pp. 28–31. IEEE Press, New York (2004)

A Novel Collaboratively Designed Robot to Assist Carers

Lakshitha Dantanarayana, Ravindra Ranasinghe,
Antony Tran, Dikai Liu, and Gamini Dissanayake

Centre for Autonomous Systems, University of Technology, Sydney
15, Boradway, Ultimo NSW 2007, Australia
{Lakshitha.Dantanarayana,Ravindra.Ranasinghe,Antony.Tran,Dikai.Liu,
Gamini.Dissanayake}@uts.edu.au
http://www.cas.uts.edu.au

Abstract. This paper presents a co-design process and an assisted navigation strategy that enables a novel assistive robot, Smart Hoist, to aid carers transferring non-ambulatory residents. Smart Hoist was co-designed with residents and carers at IRT Woonona residential care facility to ensure that the device can coexist in the facility, while providing assistance to carers with the primary aim of reducing lower back injuries, and improving the safety of carers and patients during transfers.

The Smart Hoist is equipped with simple interfaces to capture user intention in order to provide assisted manoeuvring. Using the RGB-D sensor attached to the device, we propose a method of generating a repulsive force that can be combined with the motion controller's output to allow for intuitive manoeuvring of the Smart Hoist, while negotiating with the environment.

Extensive user trials were conducted on the premises of IRT Woonona residential care facility and feedback from end users confirm its intended purpose of intuitive behaviour, improved performance and ease of use.

Keywords: Assistive Robots, Aged Residential Care, Patient Hoist, Human Robot Interaction, Navigation Assistance.

1 Introduction

Assistive robots [1–4] are devices that work collaboratively with a range of human users; as assistants, tools and as companions. These machines are expected to be able to perceive the user's behaviour and needs, communicate in a human-centred manner, and respond safely and efficiently to directions. Although machines for assisting users in performing difficult tasks have already been adopted in many industry sectors, the potential of assistive robotics in aged care has only gained attention in the last few decades.

Many assistive robots have emerged in recent years such as smart wheelchairs [5], smart walkers [6], and telepresence robots [7]. These devices assist people in their daily living activities whether they are disabled or senior citizens, enhancing their quality of life.

M. Beetz et al. (Eds.): ICSR 2014, LNAI 8755, pp. 105–114, 2014.
© Springer International Publishing Switzerland 2014

Due to the increasing demand for aged care services and the continuing decline in the relative availability of informal carers, the scarce trained aged care workforce often find themselves overworked to meet community expectations. This combined with the high rate of work-related musculoskeletal injuries amongst carers [8–11] and injuries to non-ambulatory residents occurring during transfer (eg. bed to chair, chair to toilet and bath), gives rise to significant costs and health & safety risks. Therefore it is equally important to assist these stretched carers in order to improve safety and the quality of care services.

Motivated by this real need, the research work presented in this paper is based on the Smart Hoist, a modified conventional patient lifter(standard hoist) with the primary aim of reducing lower back injuries in carers, and improving the safety of carers and patients during transfer in aged care facilities. The main focus of this paper is to present the collaborative design methodology used in developing a novel motorised patient lifting device and the human robot interaction approaches used in controlling it. The paper also highlights the navigation assistance methodologies incorporated in the Smart Hoist to further improve the carer's experience.

Obstacle detection and avoidance has always been an integral part any mobile robot system [12–14]. This becomes more relevant in cluttered and confined spaces. The main hurdle in implementing obstacle avoidance is the disparity in the user intention and the robot's movement. Either the user's commands override the system, or the robot takes control of the vehicle.

2   Collaborative Design Process of the Smart Hoist

The Smart Hoist device is targeted at a group of professionals specialised in aged care, performing the specific tasks of resident lifting and transferring. Therefore the active participation of the carers was crucial even during the early stages of the design process. The ultimate goal was to build a device the carers would be comfortable working with. This was one of the key pillars of the overall approach in the design of the Smart Hoist.

The co-design approach enabled the involvement of future users who are familiar with the routine activities in an aged care facility in the design process, which empowered the carers to make high level design decisions. The study was conducted at the IRT Woonona aged care facility. Selection of the project participants was done during the preliminary meetings with the help of IRT management. A group of keen participants were chosen from the carers to participate in subsequent co-design workshops.

The process of designing, developing and commissioning a single Smart Hoist was performed over a period of 18 months commencing in December, 2012. At the beginning of the co-design process, several knowledge-building meetings were conducted to gain insight into resident lifting and transporting. During these meetings a productive working relationship between the project participants and the members of the research team was also formed.

Subsequent co-design workshops involving 4-5 carers and one UTS member who specialised in co-design were conducted as model building exercises [15].

Insight acquired during the knowledge-building meetings was quite resourceful in forming the mockup models during these workshops.

The objectives of these co-design workshops were to design the Smart Hoist's user interface, external housing structure, key functionalities of the device and the batteries. Key design considerations that were focused on during the co-design workshops include driving confidence, comfort & ease of use, and safety & efficiency. Some key outcomes from the discussions were:

- Assisted manoeuvring especially when loaded with a patient
- Weight measurement and Body Mass Index(BMI) calculation of the patient
- Ability to monitor the environment (esp. under furniture, beds)
- Rear view mirror to monitor the environment behind the carer

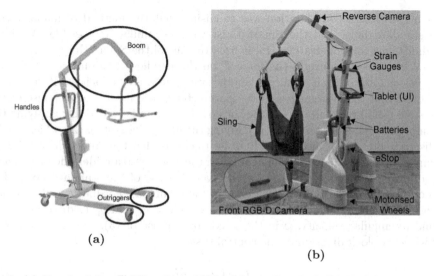

Fig. 1. (a) Standard Joey™ Lifter from AIS healthcare Pty. Ltd. (b) UTS-IRT Smart Hoist

The UTS-IRT Smart Hoist incorporates and builds upon the standard Joey™ Lifter from AIS healthcare Pty. Ltd. As part of the transformation the Joey™ Lifter has undergone a series of modifications which were completed with extreme care to avoid compromising its structural integrity, comparison between the two can be seen in Fig. 1. Modifications include:

- The linear actuators controlling the boom and the outriggers have been retrofitted with encoders to determine their location
- Strain gauges have been placed on the boom, which combined with the boom angle allow for patient weight calculations.
- Strain gauges have also been placed onto the handlebars of the Smart Hoist in order to detect the forces the carer applies to the hoist

- Rear caster wheels replaced with the Revolution 2™ assembly from 221 Robotic Systems
- A High Definition RGB camera placed at the top of the boom which provides the carer with information about the environment behind them
- A RGB-D sensor positioned at the bottom of the hoist facing forward. The camera provides the carer with information which is usually obstructed by a hanging patient, additionally it provides vision of objects at the ground level (eg. underneath beds, furniture)
- A Google®Nexus 7 provides a user interface to display system status(battery charge, time, EStop status, etc), camera views, and weight information.

A more detailed description of the parts and components can be found in [16].

3   User Intention Recognition

A major design consideration was to ensure that the method of manoeuvring a standard hoist and the Smart Hoist were as similar as possible. A simple admittance control strategy [16] is used to control the Smart Hoist.

When the carer exerts the force F on the handles of the hoist, a collective opposing force of Cv is applied when the system moves at velocity v. If we assume the mass of the system m and the parameter C are fixed, the response of the system would be identical whether or not the system is loaded. By applying the simple motion equation we get (1), giving the first order system (2), which can be discretized to (3) at instance k. In order to ensure that the Smart Hoist behaved similarly to a standard mechanical hoist, the research team identified the major motion patterns [16] required for the everyday use of the standard hoist. The motion logic [16] evaluates the trends in the strain gauge values and determines the Smart Hoist's motion and its linear velocities v_x and v_y in directions x, y, and its angular velocity, ω in the z axis by the use of (3). Fig. 2 represents a high level block diagram of the control system.

$$F - Cv = m\dot{v} \tag{1}$$

$$H = v/F = \frac{1/C}{(m/C)s + 1} \tag{2}$$

$$v_k = \frac{F_k + m.v_{k-1}}{m + C} \tag{3}$$

4   Navigation Assistance Based on Environmental Data

During the co-design workshops it was mentioned by carers that, in their busy daily schedules they find it difficult to navigate through narrow passageways and door frames with a hoist, especially when loaded. Therefore the Smart Hoist is designed to provide assistance when performing tight manoeuvres to minimise the effort required by the carer to navigate through these tight spaces.

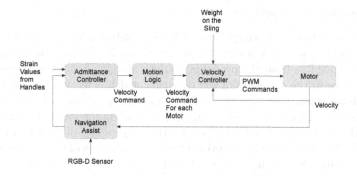

Fig. 2. Block diagram of the Controller

4.1   Sensing the Environment

The Smart Hoist makes use of the point cloud data generated from the RGB-D sensor located at its base. The Asus®Xtion PRO Live is capable of publishing point-clouds at a frequency of $30Hz$. The point clouds are first processed to filter spurious noise and a crop filter is then applied to remove the floor and hoist sling which is normally in the sensor's field of view. Since the RGB-D sensor is mounted at a fixed elevation, this process is fairly trivial. The point cloud is then segmented using a nearest neighbour method to identify and remove the outriggers. It is then projected to the ground plane to generate a 2D birds-eye view image.

The image is then used to extract the Unsigned Euclidean Distance Transform (DT). For a binary image with the set of occupied pixels V, the formed DT image in which each pixel value (x) indicates the minimum distance from that point to the closest occupied pixel$(v \in V)$ is given by (4).

$$DT(x) = \min_{v \in V} |x - \mathbf{v}| \qquad (4)$$

This is a linear time $O(n)$ computation and requires just two passes over the image [17]. Fig. 3 represents outputs of each stage of this process.

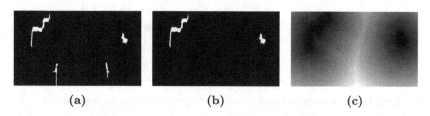

| (a) | (b) | (c) |

Fig. 3. (a) Ground plane projection of the point-cloud, (b) Environment Map after removal of outriggers, (c) DT image of the environment map

4.2   Navigation Assistance

With the DT image, it is simple to obtain the distance from the edge of each outrigger to the closest obstacle. Assistance is provided when an outrigger reports a distance less than the predefined safety margin. A sideway (y direction) repulsive force in the direction of the outrigger which is least susceptible to collision is introduced. This repulsive force is applied in par with the strain gauge inputs, and as a factor of the forward (x direction) input force. The new input force to the system in (3), F is given by the empirically determined equation seen in (5). The parameter P is derived from the closeness of the outrigger given by the DT value above. The constant K scales the output of P so that the wheel angle α is between $0° - 60°$ which is an empirically determined safe operating angle for small confined regions.

$$F_y = F_{y, \text{ handles}} + (K * P) * F_{x, \text{ handles}}$$
$$F_x = F_{x, \text{ handles}} \tag{5}$$
$$\text{wheel angle, } \alpha = tan^{-1}\left(\frac{F_y}{F_x}\right)$$

4.3   Evaluation

To measure the level of assistance required to minimise collision when passing through the narrow doorway shown in Fig. 4, two experiments were conducted. The Smart hoist is initially placed at the start position. In the first experiment, the Smart Hoist is pushed through the doorway and in the second experiment the Smart Hoist is driven using a constant virtually simulated force input of $15N$ to the handles. The Fig. 5 shows the actual input forces and the forces generated by the assistance strategy for the hoist's outriggers to avoid collision with obstacles.

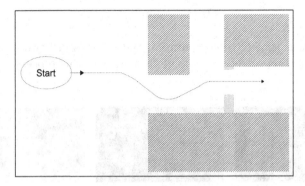

Fig. 4. The narrow doorway used for the experiments in Section 4.3

As seen in the time-force plots, the navigation assistance algorithm automatically generates the repulsive forces in real-time that are necessary to avoid

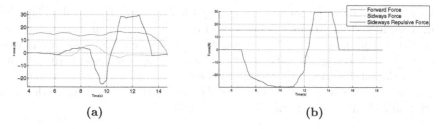

Fig. 5. Push using (a) exerting forces on the handles, (b) simulated force on the handles

collision so that the Smart Hoist can navigate through the doorway without colliding. However, the two scenarios cannot be directly compared as it is impossible to introduce a forward only force on the handles when pushed manually. Because these forces are proportional to the carer's input force there would never be a repulsive spring action that could negatively affect the carer's experience.

5 User Trials

The design of the Smart Hoist was a reiterative process, which involved a series of demonstrations and user trials at IRT Woonona care facility. The user trials described below were conducted prior to the implementation of the navigation assistance strategy to obtain initial user feedback. (Fig. 6).

The first prototype of the Smart Hoist was constructed to meet the design specifications that were laid down during the earlier co-design workshops. A user trial with this prototype was conducted early December, 2013 in order to gauge the carer's first impressions of the Smart Hoist. 15 volunteers were introduced to the use of the Smart Hoist and were asked to perform basic manoeuvres. Feedback received from carers during the early design stages of the project and the user trial was extremely valuable in the development of the Smart Hoist. Table 1 lists a summary of important comments from that trial. The Smart Hoist underwent many hardware and software changes based on the feedback.

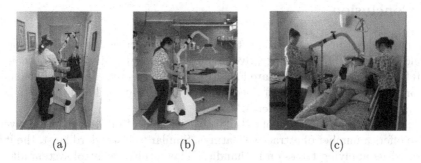

Fig. 6. User trials conducted at IRT Woonoona, Australia

The second trial was conducted in late March 2014 with approximately 50 carers. The carers participated in an interactive training workshop, after which they were asked to perform a complete patient transfer from bed to bathroom in a simulated environment. The exercise included complex manoeuvres such as lifting a patient from a bed, navigating through corridors and around tight corners and lowering the patient into a chair. The aim of this experiment was to assess the intuitiveness and responsiveness of the Smart Hoist in comparison to a standard hoist in a routine exercise. The preliminary outcomes of the second trial upheld the results from the first user trial.

Table 1. Summary of carers' comments from the first user trial

Evaluation category	Evaluation Criteria	Score out of 10	Additional Comments
Confidence	Driving & Turning	7	− "Need to be slow when moving sideways and turning under load" − "Too slow to change to sideways mode"
Comfort & Ease of use	Handles & Grips	6	− "Needs too much force"
	Screen	8	− "Bigger icons and text" − "High contrast and brightness" − "Include descriptions for the icons" − "Arrow to indicate the wheel direction"
	Batteries	9	− "Long but OK" − "No heavier please" − "Must charge fast"
	Cameras	9	− "Need a higher field of view" − "Can I rely on it? Will it give me a false sense of security?"
Safety & Efficiency	Overall	10	− "I was never worried about my feet"

6   Conclusion

This paper describes a navigation strategy for a novel assistive robot developed by the University of Technology, Sydney, working collaboratively with the staff at IRT Woonona Residential Care Facility. The aim was to reduce the likelihood of workplace injuries being sustained by care workers in aged and disabled care sectors when transferring residents.

Smart Hoist is an extension of a standard hoist, apart from being motor driven, it also offers a number of attractive features. Similar to a standard hoist, the it is operated by applying forces on its handles. This intuitive control system allows carers to seamlessly migrate to the Smart Hoist without an added learning curve that is usually associated with most assistive robotic devices. It also senses its environment using an RGB-D sensor in order to provide navigation assistance to the carer in confined spaces.

Further evaluation of the benefits of the Smart Hoist, using Electromyographic (EMG) readings of the major muscles involved in manoeuvring the hoist is planned to be conducted in a forthcoming extended user trial. A thorough post-deployment evaluation and comprehensive comparison with the second user trial outcomes will also be a part of this exercise. Future work also includes further extending and improving the navigation assistance algorithm.

Acknowledgments. This work was supported in part by IRT Research Foundation, Australia and the Centre for Autonomous Systems, University of Technology, Sydney (UTS), Australia.

We thank the management, the staff and the residents of IRT Woonoona care facility, Australia for their involvement in the design development, implementation and evaluation of the smart hoist.

We also acknowledge A/Prof Jaime Valls Miro and Prof. Lynn Chenoweth for their guidance and Dr. Michael Behrens, Ms. LiYang Liu, Mr. Stefan Lie and Mr. Remi Bouskila for the assistance given in developing the Smart Hoist.

References

1. Jayawardena, C., Kuo, I.H., MacDonald, B.A.: An efficient programming framework for socially assistive robots based on separation of robot behavior description from execution. In: 2013 6th IEEE Conference on Robotics, Automation and Mechatronics (RAM), pp. 150–155. IEEE (November 2013), http://ieeexplore.ieee.org/xpls/abs_all.jsp?arnumber=6758575, http://ieeexplore.ieee.org/lpdocs/epic03/wrapper.htm?arnumber=6758575

2. Torta, E., Oberzaucher, J., Werner, F., Cuijpers, R.H., Juola, J.F.: Attitudes Towards Socially Assistive Robots in Intelligent Homes: Results From Laboratory Studies and Field Trials (December 2012), http://humanrobotinteraction.org/journal/index.php/HRI/article/view/60

3. Dahl, T.S., Boulos, M.: Robots in Health and Social Care: A Complementary Technology to Home Care and Telehealthcare? Robotics 3(1), 1–21 (2013)

4. Tanaka, H., Yoshikawa, M., Oyama, E., Wakita, Y., Matsumoto, Y.: Development of Assistive Robots Using International Classification of Functioning, Disability, and Health: Concept, Applications, and Issues. Journal of Robotics 2013, 1–12 (2013), http://www.hindawi.com/journals/jr/2013/608191/

5. Hillman, M., Hagan, K., Hagan, S., Jepson, J., Orpwood, R.: The Weston wheelchair mounted assistive robot - the design story. Robotica 20(02), 125–132 (2002), http://journals.cambridge.org/abstract_S0263574701003897

6. Rentschler, A.J., Cooper, R.A., Blasch, B., Boninger, M.L.: Intelligent walkers for the elderly: performance and safety testing of VA-PAMAID robotic walker. Journal of Rehabilitation Research and Development 40(5), 423–431 (2003), http://www.ncbi.nlm.nih.gov/pubmed/15080227

7. Kristoffersson, A., Coradeschi, S., Loutfi, A.: Towards evaluation of social robotic telepresence based on measures of social and spatial presence (2011), http://www.diva-portal.org/smash/record.jsf?pid=diva2:542612

8. Dawson, A.P., McLennan, S.N., Schiller, S.D., Jull, G.A., Hodges, P.W., Stewart, S.: Interventions to prevent back pain and back injury in nurses: a systematic review. Occupational and Environmental Medicine 64(10), 642–650 (2007), http://www.pubmedcentral.nih.gov/articlerender.fcgi?artid=2078392&tool=pmcentrez&rendertype=abstract

9. Fragala, G., Bailey, L.P.: Addressing occupational strains and sprains: musculoskeletal injuries in hospitals. AAOHN Journal: Official Journal of the American Association of Occupational Health Nurses 51(6), 252–259 (2003), http://europepmc.org/abstract/MED/12846458

10. Pellatt, G.C.: The safety and dignity of patients and nurses during patient handling. British Journal of Nursing 14(21), 1150–1156 (2005), http://www.ncbi.nlm.nih.gov/pubmed/16475436

11. Reichert, P.: Patient Handling Ergonomics. Ph.D. thesis, New Jersey Institute of Technology (2004)

12. Levine, S.P., Bell, D.A., Jaros, L.A., Simpson, R.C., Koren, Y., Borenstein, J.: The NavChair Assistive Wheelchair Navigation System. IEEE Transactions on Rehabilitation Engineering: A Publication of the IEEE Engineering in Medicine and Biology Society 7(4), 443–451 (1999), http://www.ncbi.nlm.nih.gov/pubmed/10609633

13. Park, J.B., Lee, B.H., Chung, W.K.: Reflective force navigation control for a mobile robot using a state transition diagram. In: Proceedings of 2003 IEEE/ASME International Conference on Advanced Intelligent Mechatronics (AIM 2003), pp. 52–57 (2003), http://ieeexplore.ieee.org/lpdocs/epic03/wrapper.htm?arnumber=1225071

14. Song, K.T., Jiang, S.Y.: Force-cooperative guidance design of an omni-directional walking assistive robot. In: 2011 IEEE International Conference on Mechatronics and Automation, pp. 1258–1263 (August 2011), http://ieeexplore.ieee.org/lpdocs/epic03/wrapper.htm?arnumber=5985842

15. Lie, S., Liu, D., Bongers, B.: A cooperative approach to the design of an Operator Control Unit for a semi-autonomous grit-blasting robot. In: Australasian Conference on Robotics and Automation (ACRA) (2012), http://www.araa.asn.au/acra/acra2012/papers/pap144.pdf

16. Ranasinghe, R., Dantanarayana, L., Tran, A., Lie, S., Behrens, M., Liu, L.: Smart Hoist: An Assistive Robot to Aid Carers. In: 13th International Conference on Control, Automation, Robotics and Vision, ICARCV (2014)

17. Felzenszwalb, P., Huttenlocher, D.: Distance Transforms of Sampled Functions. Tech. rep., University of Cornell (2004), http://ecommons.library.cornell.edu/handle/1813/5663

Social Robotics through an Anticipatory Governance Lens

Lucy Diep[1], John-John Cabibihan[2], and Gregor Wolbring[1]

[1] Department of Community Health Sciences, Faculty of Medicine, University of Calgary,
TRW Building, 3rd Floor, 3280 Hospital Drive NW, Calgary, Alberta
`lucy.diep@shaw.ca, gwolbrin@ucalgary.ca`
[2] Department of Mechanical and Industrial Engineering, Qatar University
`john.cabibihan@qu.edu.qa`

Abstract. Social Robotics is an emerging field, with many applications envisioned. Scientific and technological advancements constantly impact humans on the individual and societal level. Therefore one question increasingly debated is how to anticipate the impact of a given envisioned, emerging or new scientific or technological development and how to govern the emergence of scientific and technological advancements. Anticipatory governance has as a goal to discuss potential issues arising at the ground level of the emergence of a given scientific and technological product. Our study investigated a) the visibility of the anticipatory governance concept within the social robotic discourse and b) the implication of anticipatory governance for the social robotics field through the lens of a social robot design process and key documents from the UNESCO/ICSU 1999 World Conference on Sciences the lens. Our findings suggest that a) anticipatory governance is not a concept established within the social robotics fields so far; b) that social robotics as specific field is not engaged with within the anticipatory governance field and c) that many professional and academic fields are not yet involved in the social robotics discourse as aren't many non-academic stakeholders. We posit that anticipatory governance can strengthen the social robotics field.

Keywords: Social robotics, anticipatory governance, governance of science and technology, UNESCO/ICSU 1999 World Conference on Sciences.

1   Introduction

Social robotics is an emerging field that designs robots to engage in social interaction with humans. Applications range from monitoring the person and helping with certain tasks to being companions covering areas such as education and healthcare [1-7] and involving social groups such as disabled people [8-14] and the elderly [15,16]. Question is how to govern the development of social robots? 60% of EU citizens were saying that robots should be banned from caring for children, elderly people and people with disabilities, and only 4% indicated robots should be used for disabled people [17]. A 2014 Pew Research Center report U.S. Views of Technology and the

M. Beetz et al. (Eds.): ICSR 2014, LNAI 8755, pp. 115–124, 2014.
© Springer International Publishing Switzerland 2014

Future Science in the next 50 years [18] found that "65% think it would be a change for the worse if lifelike robots become the primary caregivers for the elderly and people in poor health". The philosopher Sparrow outlined some concerns around the use of robots for elderly care [19]. On the other hand social robots are seen as a possible way to address the human resource and economic pressures on health care systems [19]. Given the different consequences of social robots based on their design and because social robotics is still an emerging area we believe it to be a good case study for the utility of anticipatory governance of a given scientific or technological advancement. The aim of our study was three-fold. The first aim was to investigate the visibility of the anticipatory governance concept within the social robotics discourse and vice versa. To achieve this aim we searched the academic databases ScienceDirect, Compendex, IEEE, Communication Abstracts, Scopus, EBSCO(All), Web of Science, JSTOR and ScienceDirect (all accessed through the University of Calgary Library) and Google Scholar for the keyword combination of "social robot" with "anticipatory governance or "technology governance" or "governance of technology" or the phrase "governance of social robot". We also searched the articles on social robotics we obtain for another study on social robotics [14] by searching for the term governance in these articles. The second aim was to investigate the visibility of social robotics in various academic fields and professions using Google Scholar. The third aim was to investigate the implication of anticipatory governance for the social robotics field. We did this through the lens of social robot design process and through the lens of key documents from the UNESCO/ICSU 1999 World Conference on Sciences. Section 2 gives a short overview of the anticipatory governance concept. In Section 3, we present our findings of aim 1 and 2. Section 4 and 5 outline the social robot design process and excerpts from key documents of the UNESCO/ICSU 1999 World Conference on Sciences with anticipatory governance implications. Section 6 discusses anticipatory governance implication for the social robotics field and the conclusions we draw.

2   Governance, Anticipatory Governance and Social Robotics

Governance of technology is seen for a long time as an important goal [20]. Ethics came to pass to give some guidance as to how to deal with scientific and technological advancements [21,22]. Ely, Van Zwanenberg and Stirling highlight that "the ever-growing pervasiveness of new technologies and their impacts heighten the need for international co-ordination in democratic technology governance [23]. Fisher, Mahajan and Mitcham call for the "reflexive participation by scientists and engineers in the internal governance of technology development" [24].

Anticipatory governance a term coined in 2002 [25] is a foresight framework and is used as a structure for government policy developers [26] and employed in public administration and management' [26,27], environmental studies [28], biological studies [29,30] and as a framework for the 'responsible development of nanotechnology' [25,31]. Emerging in 2002 from the science, technology, and society and social studies of nanotechnology field, anticipatory governance was developed as a call for the integration of social scientists in the early stages of technology development to better address potential concerns of varied stakeholders [32,25]. The emphasis was to encourage 'sensitivity' and 'reflexivity' among developers to the ethical, social, legal, economic, and environmental concerns of emerging technologies [33,34]. This initiative began in the United States and extended to Europe with the intention of developing foresight analysis to the implications of innovation technology development in order to foster the practice of responsible nanotechnology development [35,25]. According to Guston, anticipatory governance is a concept aimed at understanding the potential social, ethical, and political impacts of emerging discourses through 'reflexive' practice, foresight analysis, and the engagement and integration of relevant stakeholders [36,25] entailing foresight (constructing plausible socio-technical implications), integration (bringing together diverse fields such as social sciences and natural sciences), engagement (bringing together public citizens, developers, engineers, policy-makers, and other actors to construct conversations around awareness, reaction, and knowledge development and sharing), and ensemblization which brings together the three elements [34,33].

3   Results

When we searched the academic databases ScienceDirect, Compendex, IEEE, Communication Abstracts, Scopus, EBSCO(All), Web of Science, JSTOR, ScienceDirect and Google Scholar for the keyword combination of "social robot" with "anticipatory governance or "technology governance" or "governance of technology" we found zero relevant articles. We did not find even one hit for the phrase "governance of social robot". This paper focuses on "social robots" as this is to become a field with its own specific identity and understanding and scope of robots; however even if we used the combination of "robot" and "anticipatory governance" we obtained only four relevant articles in for example Google Scholar. Furthermore this paper focuses on purpose on the term "anticipatory governance" as this is a field that establishes itself also in the moment. However even if we do not use governance as a term but assessment like in "technology assessment", together with "social robot" we still did not obtain many results [37,38]. When we searched the n=171 article we obtained for the study described in [14] no article used the term "anticipatory governance". Furthermore we only found one article where the term robot and governance were mentioned in the same paragraph of an article. That article talked about that the "cognitive experience architecture provides a useful tool to explore robot design" and that this architecture "has four main components grounded in a robot's experience: morphology, understanding, motivation and governance" [39]. As to how the term governance is covered one article states, "[t]he technology-driven side tells that the world is driven and run by technological developments, and that robots are here for further enhancements and new applications. It means no less than that technology dictates the governance. The society-driven side opines that the world is driven and run by social aspects" [40]. This was the extent of the visibility of the term "governance". One article looking at service robots thematized which "regulatory instrument is best suited for achieving the individual regulatory goals" [41]. One article using the term social-ly-assistive robot talked about governance bodies [42]. Another mentions Kate Darl-ing exploration of "whether the way humans seem hardwired to react to anthropo-morphic machines suggests the need to extend a limited set of legal rights to social robots, or at least prohibitions against abusing them, even where no one thinks of them as alive or sentient at a rational level" [43].

When we searched Google Scholar for the phrase "social robot" together with var-ious professions and academic fields we obtained the following results: "social robot" and "social work" 25 hits although in our opinion only one article was relevant; the hit count was 63 for the combination of "social robot" and "occupational therapy" with again most articles being not relevant in a sense that they did not engaged with the views of occupational therapist or the impact of social robots on the field of occu-pational therapy. Given these two criteria of relevance less than ten relevant articles were found in Google Scholar with the keyword combination "social robot" and "physical therapy", "social robot" and "speech therapy", "social robot" and "disability studies", "social robot" and "gender studies", "social robot" and "policy studies" or "social robot" and "sustainability/sustainability studies". Some areas have a few more hits such as "social robot" and "rehabilitation engineering" or "social robot" and "anthropology".

4   Anticipatory Governance: The Social Robot Design Process

The developments of social robots tend to be technology driven with the users being consulted after the initial ideas have been conceived. The designs typically draw inspira-tion from many sources that include psychology, biology and neurosciences [44-46]. It is also worth noting that upon conception, robot designs are influenced by cultural as-sumptions of the designers [47]. The design of social robots primarily considers the appearance and behavior of the robots. A robot's embodiment plays an important role in the expectation of the user. In terms of appearance, the robots are either intended to possess an acute resemblance to humans, or they are designed as animals or cartoon-like toys, or they are designed to not resemble any biological species [48,49]. There is also an expectation for social robots to have social intelligence where the robot can be per-ceived to have an understanding of human behavior and react appropriately. To this end, it is crucial for a robot to detect a human being's emotions and display the proper emo-tions through verbal or non-verbal means [50,51]. After the robot has been constructed, the robot's features are presented to the user's groups for evaluation. Various tests have been developed to measure acceptance, likeability, perceived intelligence, safety, and ease of use, among others [52-55].

5   UNESCO/ICSU 1999 World Conference on Sciences

In 1999 UNESCO and the International Council of Science (ICSU) organized the World Conference on Sciences. The two key documents outline numerous responsibilities science fields, scientists, governments and others have related to how science ought to be performed, how science ought to be governed.

In the Overview one reads:

"Science is a powerful means of understanding the world in which we live and it is also capable of yielding enormous returns that directly enhance socio-economic development and the quality of our lives. Scientific advances over the last fifty years have led to revolutionary changes in health, nutrition and communication; moreover, the role of science promises to be yet greater in the future because of ever-more-rapid scientific progress. Meanwhile, humanity is being confronted with problems on a global scale, many - such as environmental degradation, pollution and climatic change - provoked by the mismanagement of natural resources or unsustainable production and consumption patterns. Even if the technology implicated in these problems can be said to have stemmed from science, we cannot hope to resolve these problems without the correct and timely use of science in the future. And yet, in spite of the opportunities it offers us all, science itself is facing wavering confidence and uncertain investment, as well as dilemmas of an ethical nature. These problems can only be solved if the scientific and business communities, governments and the general public are able to reach, through debate, a common ground on science with respect to the service it is to provide to society and a new commitment to science from society in the years to come. In convening a World Conference on Science for the Twenty-First Century: a New Commitment, from 26 June to 1 July 1999 in Budapest, Hungary, the United Nations Educational, Scientific and Cultural Organization (UNESCO) and the International Council for Science (ICSU), in co-operation with other partners, provided a unique forum for this much-needed debate between the scientific community and society."[56]   The conference generated two key documents within which various expectations are voiced as to what scientists should do and what the purpose of science is. The document *Science Agenda-Framework for Action* states among others;

point 10: "Universities should ensure that their programmes in all fields of science focus on both education and research and the synergies between them and introduce research as part of science education. Communication skills and exposure to social sciences should also be a part of the education of scientists" [56].

point 71: "The ethics and responsibility of science should be an integral part of the education and training of all scientists. It is important to instil in students a positive attitude towards reflection, alertness and awareness of the ethical dilemmas they may encounter in their professional life" [56].

point 74: "Scientific institutions are urged to comply with ethical norms, and to respect the freedom of scientists to express themselves on ethical issues and to denounce misuse or abuse of scientific or technological advances" [56].

point 75: "Governments and non-governmental organizations, in particular scientific and scholarly organizations, should organize debates, including public debates, on the ethical implications of scientific work. Scientists and scientific and scholarly organizations should be adequately represented in the relevant regulating and decision-making bodies. These activities should be institutionally fostered and recognized as part of scientists' work and responsibility. Scientific associations should define a code of ethics for their members" [56].

point 87: "Governments should support cooperation between holders of traditional knowledge and scientists to explore the relationships between different knowledge systems and to foster interlinkages of mutual benefit" [56].

The *Declaration on Science and the Use of Scientific Knowledge* [57] states among others,

Point 28: "the need for a strong commitment to science on the part of governments, civil society and the productive sector, as well as an equally strong commitment of scientists to the well-being of society" [57].

point 41:" The social responsibility of scientists requires that they maintain high standards of scientific integrity and quality control, share their knowledge, communicate with the public and educate the younger generation. Political authorities should respect such action by scientists. Science curricula should include science ethics, as well as training in the history and philosophy of science and its cultural impact" [57].

6   Discussion and Conclusion

We conclude from our findings that a) anticipatory governance is not a concept established within the social robotics fields so far; b) that social robotics as a specific field is not engaged with within the anticipatory governance field and c) that many professional and non-technology based academic fields are not yet involved in the social robotics discourse as aren't many non-academic stakeholders. This has consequences; for example the main narrative within social robotics around disabled people is a medical one [14]; this aspect would have been flagged by disability studies scholars if they would have involved themselves already with the social robotics discourse. That disability studies scholars are so far not involving themselves in the social robotics discourse means that people involved in the social robotics discourse might miss the potential social, ethical, and political impacts of a one-sided imagery, a one-sided narrative of a given social group. How does one learn to reflect on one's practice such as technology design if certain perspectives are simply missing? How can one anticipate impacts if the 'stakeholders' involved in a given discourse are so limited in their diversity of backgrounds? Given the limited diversity of the groups involved in the social robotics discourse how can one construct conversations around awareness, reaction, and knowledge development and sharing that is diverse and meaningful? How can one develop policies that are meaningful? How can one fulfill the expectations of point 75 of the Science Agenda-Framework for Action document [56]. The social robot design process has much in common with the design culture of other technologies and as such the other technologies face many of the same problems we indicate here for the social robotic field. By not making diversity of players, ideas and knowledge for any scientific and technological advancement an expectation from the beginning, we are open to push back by the ones who want to limit the social aspect such as is evident in the moment in the discussions around what the National Science Foundation (USA) should fund with a very visible position being to cut the social inquiry [58]. If this push becomes more mainstream how would one be able to fulfil the expectations of point 10, 71 and 74 of the *Science Agenda-Framework for Action* document [56] and point 41 the *Declaration on Science and the Use of Scientific Knowledge* [57]? How would one achieve a "strong commitment to science on the part of governments, civil society and the productive sector, as well as an equally strong commitment of scientists to the well-being of society" as asked for in the *Declaration on Science and the Use of Scientific Knowledge* [57]? A limited diversity within the social robotics discourse hurts social robotics designers and scientists as they are more vulnerable to certain dominant views as to outcome expectations which might be very limited and limiting. Two recent surveys [18,17] highlight certain reservations towards certain forms of social robots and some academic work also gives pause [19]. We submit that linking up the anticipatory governance and social robotics fields is of benefit to both fields. Beyond this a broader diversity of non-academic people and academic fields involved in the governance discussion around social robotics is advantageous to the social robotics field. To achieve the diversity of non-academics and academic fields however is not easy. Even if one wants to be open and welcoming to diverse voices it's a challenge to convince the 'others' to be involved. It is a challenge to find ways to engage non-academic stakeholders which is acknowledged for example in the nanotechnology field [59] which embraced concepts such as democratization of science [60-62]. This problem is especially prevalent for socially disadvantaged groups who have so many problems pertaining to daily living that they find it nearly impossible to find the time to inform themselves on emerging issues that are at that time not directly impacting the day to day issues they face. However it is also not that easy to change non-technology related academic fields and professional fields to engage with emerging technologies and their governance through research, teaching and practice evaluation. This leads to the questions of how to fix these problems of non-engagement given that existing models have so far limited success and what the responsibilities are in particular of educators, funders and non-technology related academic fields We believe that anticipatory governance is essential for science and technology in general including the social robotics field but we need more work around how to make anticipatory governance operational within and outside academia so that it does not disempower and does allow for the appearance of a common ground early on.

References

1. Prado, J.A., Simplicio, C., Lori, N.F., Dias, J.: Visuo-auditory Multimodal Emotional Structure to Improve Human-Robot-Interaction. International Journal of Social Robotics 4(1), 29–51 (2012)
2. Cabibihan, J.-J., So, W.-C., Saj, S., Zhang, Z.: Telerobotic pointing gestures shape human spatial cognition. International Journal of Social Robotics 4(3), 263–272 (2012)
3. Fridin, M., Belokopytov, M.: Acceptance of socially assistive humanoid robot by preschool and elementary school teachers. Computers in Human Behavior 33, 23–31 (2014)

4. Stafford, R.Q., MacDonald, B.A., Li, X., Broadbent, E.: Older People's Prior Robot Attitudes Influence Evaluations of a Conversational Robot. International Journal of Social Robotics 6(2), 281–297 (2014)

5. Lakatos, G., Janiak, M., Malek, L., Muszynski, R., Konok, V., Tchon, K., Miklósi, Á.: Sensing sociality in dogs: what make an interactive robot social? Animal Cognition 17(2), 387–397 (2014)

6. van den Brule, R., Dotsch, R., Bijlstra, G., Wigboldus, D.H., Haselager, P.: Do Robot Performance and Behavioral Style affect Human Trust? International Journal of Social Robotics, 1–13 (2014)

7. Keren, G., Fridin, M.: Kindergarten Social Assistive Robot (KindSAR) for children's geometric thinking and metacognitive development in preschool education: A pilot study. Computers in Human Behavior 35, 400–412 (2014)

8. Welch, K.C., Lahiri, U., Warren, Z., Sarkar, N.: An approach to the design of socially acceptable robots for children with autism spectrum disorders. International Journal of Social Robotics 2(4), 391–403 (2010)

9. Boccanfuso, L., O'Kane, J.M.: CHARLIE: An adaptive robot design with hand and face tracking for use in autism therapy. International Journal of Social Robotics 3(4), 337–347 (2011)

10. Thill, S., Pop, C.A., Belpaeme, T., Ziemke, T., Vanderborght, B.: Robot-assisted therapy for autism spectrum disorders with (partially) autonomous control: Challenges and outlook. Paladyn, 1–9 (2013)

11. Joosse, M., Sardar, A., Lohse, M., Evers, V.: BEHAVE-II: The Revised Set of Measures to Assess Users' Attitudinal and Behavioral Responses to a Social Robot. International Journal of Social Robotics, 1–10 (2013)

12. Cabibihan, J.-J., Javed, H., Ang Jr., M., Aljunied, S.M.: Why robots? A survey on the roles and benefits of social robots in the therapy of children with autism. International Journal of Social Robotics 5(4), 593–618 (2013)

13. Wainer, J., Dautenhahn, K., Robins, B., Amirabdollahian, F.: A pilot study with a novel setup for collaborative play of the humanoid robot KASPAR with children with autism. International Journal of Social Robotics 6(1), 45–65 (2014)

14. Yumakulov, S., Yergens, D., Wolbring, G.: Imagery of Disabled People within Social Robotics Research. In: Ge, S.S., Khatib, O., Cabibihan, J.-J., Simmons, R., Williams, M.-A. (eds.) ICSR 2012. LNCS, vol. 7621, pp. 168–177. Springer, Heidelberg (2012)

15. Wu, Y.-H., Fassert, C., Rigaud, A.-S.: Designing robots for the elderly: Appearance issue and beyond. Archives of Gerontology and Geriatrics 54(1), 121–126 (2012)

16. Kachouie, R., Sedighadeli, S., Khosla, R., Chu, M.-T.: Socially Assistive Robots in Elderly Care: A Mixed-Method Systematic Literature Review. International Journal of Human-Computer Interaction 30(5), 369–393 (2014)

17. European Commission, Public Attitudes towards Robots. European Commission Special Eurobarometer, vol. 382. European Commission (2012) (Online)

18. Center PR: U.S. Views of Technology and the Future Science in the next 50 years. Pew Research Center (2014) (Online)

19. Sparrow, R., Sparrow, L.: In the hands of machines? The future of aged care. Minds and Machines 16(2), 141–161 (2006)

20. De la Mothe, J.: The institutional governance of technology, society, and innovation. Technology in Society 26(2), 523–536 (2004)

21. Wolbring, G.: Disability rights approach towards bioethics. J. of Disability Studies 14(3), 154–180 (2003)

22. Wolbring, G.: Ethical Theories and Discourses through an Ability Expectations and Ableism Lens: The Case of Enhancement and Global Regulation. Asian Bioethics Review 4(4), 293–309 (2012)
23. Ely, A., Van Zwanenberg, P., Stirling, A.: New models of technology assessment for development (2011)
24. Mahajan, V.: Models for innovation diffusion, vol. 48. Sage Publications, Inc. (1985)
25. Guston, D.: Understanding 'Anticipatory Governance'. Social Studies of Science, 1–25 (2013)
26. Fuerth, L.S., Faber, E.M.H.: Anticipatory governance: Winning the future. Futurist 47(4), 42–49 (2013)
27. Bächler, G.: Conflict transformation through state reform. In: Transforming Ethnopolitical Conflict, pp. 273–294. Springer (2004)
28. Serrao-Neumann, S., Harman, B.P., Choy, D.L.: The Role of Anticipatory Governance in Local Climate Adaptation: Observations from Australia. Planning Practice & Research (ahead-of-print), 1–24 (2013)
29. Gupta, A.: Anticipatory Governance of Biosafety: What Role for Transparency? Conference Papers – International Studies Association, p. 1 (2009)
30. Nielsen, K., Fredriksen, B., Myhr, A.: Mapping Uncertainties in the Upstream: The Case of PLGA Nanoparticles in Salmon Vaccines. NanoEthics 5(1), 57–71 (2011), doi:10.1007/s11569-011-0111-5
31. Guston, D.: Anticipatory governance of emerging nanotechnologies. Abstracts of Papers of the American Chemical Society 237 (2009)
32. Fonseca, P.F.C., Pereira, T.S.: The governance of nanotechnology in the Brazilian context: Entangling approaches. Technology in Society (2013)
33. Guston, D.: The Anticipatory Governance of Emerging Technologies. Journal of Korean Vacuum Society 19(6), 432–441 (2010)
34. Wender, B.A., Foley, R.W., Guston, D.H., Seager, T.P., Wiek, A.: Anticipatory Governance and Anticipatory Life Cycle Assessment of Single Wall Carbon Nanotube Anode Lithium ion Batteries. Nanotechnology Law & Business 9(3), 201–216 (2012)
35. Kearnes, M., Rip, A.: The emerging governance landscape of nanotechnology. Jenseits von Regulierung: Zum politischen Umgang mit der Nanotechnologie Aka Verlag, Heidelberg (2009)
36. Guston, D.H.: Innovation policy: not just a jumbo shrimp. Nature 454(7207), 940–941 (2008)
37. Wolbring, G., Diep, L., Yumakulov, S., Ball, N., Yergens, D.: Social Robots, Brain Machine Interfaces and Neuro/Cognitive Enhancers: Three Emerging Science and Technology Products through the Lens of Technology Acceptance Theories, Models and Frameworks. Technologies 1(1), 3–25 (2013)
38. Wolbring, G., Diep, L., Yumakulov, S., Ball, N., Leopatra, V., Yergens, D.: Emerging Therapeutic Enhancement Enabling Health Technologies and Their Discourses: What Is Discussed within the Health Domain? Healthcare 1(1), 20–52 (2013)
39. Trovato, G., Kishi, T., Endo, N., Hashimoto, K., Takanishi, A.: A cross-cultural study on generation of culture dependent facial expressions of humanoid social robot. In: Ge, S.S., Khatib, O., Cabibihan, J.-J., Simmons, R., Williams, M.-A. (eds.) ICSR 2012. LNCS, vol. 7621, pp. 35–44. Springer, Heidelberg (2012)
40. van den Herik, H.J., Lamers, M., Verbeek, F.: Understanding the artificial. International Journal of Social Robotics 3(2), 107–109 (2011)
41. Dreier, T., genannt Döhmann, I.S.: Legal aspects of service robotics. Poiesis & Praxis 9(3-4), 201–217 (2012)

42. Johnson, D.O., Cuijpers, R.H., Juola, J.F., Torta, E., Simonov, M., Frisiello, A., Bazzani, M., Yan, W., Weber, C., Wermter, S.: Socially Assistive Robots: A comprehensive approach to extending independent living. International Journal of Social Robotics,1–17 (2013)

43. Calo, R.: Robotics and the New Cyberlaw. Available at SSRN 2402972 (2014)

44. Cabibihan, J.J., So, W.C., Saj, S., Zhang, Z.: Telerobotic Pointing Gestures Shape Human Spatial Cognition. International Journal of Social Robotics 4(3), 263–272 (2012)

45. Wykowska, A., Schubö, A.: Perception and Action as Two Sides of the Same Coin. A Review of the Importance of Action-Perception Links in Humans for Social Robot Design and Research. International Journal of Social Robotics 4(1), 5–14 (2012)

46. Li, H., Cabibihan, J.J., Tan, Y.K.: Towards an Effective Design of Social Robots. International Journal of Social Robotics 3(4), 333–335 (2011)

47. Šabanovic, S.: Robots in society, society in robots: Mutual shaping of society and technology as a framework for social robot design. International Journal of Social Robotics 2(4), 439–450 (2010)

48. Cabibihan, J.J., Javed, H., Ang Jr., M., Aljunied, S.M.: Why Robots? A Survey on the Roles and Benefits of Social Robots in the Therapy of Children with Autism. International Journal of Social Robotics 5(4), 593–618 (2013)

49. Coeckelbergh, M.: Humans, animals, and robots: A phenomenological approach to human-robot relations. International Journal of Social Robotics 3(2), 197–204 (2011)

50. Hirth, J., Schmitz, N., Berns, K.: Towards social robots: Designing an emotion-based architecture. International Journal of Social Robotics 3(3), 273–290 (2011)

51. Niculescu, A., van Dijk, B., Nijholt, A., Li, H., See, S.L.: Making Social Robots More Attractive: The Effects of Voice Pitch, Humor and Empathy. International Journal of Social Robotics 5(2), 171–191 (2013)

52. Bartneck, C., Kulić, D., Croft, E., Zoghbi, S.: Measurement instruments for the anthropomorphism, animacy, likeability, perceived intelligence, and perceived safety of robots. International Journal of Social Robotics 1(1), 71–81 (2009)

53. Frennert, S., Östlund, B.: Review: Seven Matters of Concern of Social Robots and Older People. International Journal of Social Robotics 6(2), 299–310 (2014)

54. Leite, I., Martinho, C., Paiva, A.: Social Robots for Long-Term Interaction: A Survey. International Journal of Social Robotics 5(2), 291–308 (2013)

55. Young, J.E., Sung, J., Voida, A., Sharlin, E., Igarashi, T., Christensen, H.I., Grinter, R.E.: Evaluating human-robot interaction: Focusing on the holistic interaction experience. International Journal of Social Robotics 3(1), 53–67 (2011)

56. UNESCO/ICSU, WOrld Conference on Sciences: An Overview. UNESCO (1999), http://www.unesco.org/science/wcs/eng/overview.htm

57. UNESCO, UNESCO World Conference on Sciences Declaration on Science and the Use of Scientific Knowledge. UNESCO webpage (1999)

58. Smith, L., Peter, M., Hunter, R.: Is Social Science Research in the National Interest. Scientific American (2014) (Online)

59. Toumey, C.: Democratizing nanotech, then and now. Nat. Nano. 6(10), 605–606 (2011), doi:10.1038/nnano.2011.168

60. Toumey, C.: Science and democracy. Nature Nanotechnology 1(1), 6–7 (2006), doi:10.1038/nnano.2006.71

61. Lyons, K.: Time for technology democracy. The Age, Australia (2009)

62. Wolbring, G.: Nanotechnology for Democracy versus Democratization of Nanotechnology. In: Lente, H., Coenen, C., Fleischer, T., et al. (eds.) Little by Little: Expansions of Nanoscience and Emerging Technologies. AKA-Verlag/IOS Press, Dordrecht (2012)

Using Robots to Modify Demanding or Impolite Behavior of Older People

Heather Draper[1,*] and Tom Sorell[2]

[1] University of Birmingham, Birmingham, UK
h.draper@bham.ac.uk
[2] University of Warwick, Coventry, UK
t.e.sorell@warwick.ac.uk

Abstract. As part of a large scale qualitative study (conducted in France, the UK and the Netherlands) of potential users' views on the ethical values that should govern the design and programming of social robots for older people, we elicited responses to a scenario where a robot is programmed to modify an older person's rude behavior. Participants' responses ranged from outright disagreement with robotized efforts to change characteristic behavior, to approval as a means to an end. We discuss these views against the background of respect for autonomy, the differences and similarities between robot and human carers, and behavior modification in the context of rehabilitation, where the 'no gain without pain' principle is commonly used to justify what would otherwise seem callous. We conclude that such programming may be acceptable in the context of the rehabilitation and promotion of the independence of older people

Keywords: social robots, care-robots, ethics older people, autonomy, behavior rehabilitation, enablement, independence, re-enablement qualitative research, users' views.

1 Introduction

In order to simulate an empathetic response in a care-robot, Patrizia Marti and other FP7 ACCOMPANY (Acceptable robotics COMPanions for AgeiNg Years) researchers at Sienna adapted a Care-O-bot® tablet interface [1]. Sensors in the tablet frame make it touch-sensitive, which enables the user to express urgent need by squeezing the tablet (the 'squeeze-me' facility) [2]. For its part, the tablet is able to display graphic symbols of simulated emotional reactions on the part of the robot. The tablet displays a schematic mask [3] that is easy to read as showing pleasure or happiness and irritation or anger in the context of scenarios developed for user and robot in the ACCOMPANY project. For example, the robot can share the user's supposed happiness at the prospect of a parcel being delivered, and sadness when the user does not drink from a bottle of water (rehydrate) so as to avoid becoming dehydrated when prompted by the robot. An annoyed mask is projected if the user inappropriately uses

---

* Corresponding author.

M. Beetz et al. (Eds.): ICSR 2014, LNAI 8755, pp. 125–134, 2014.

the squeeze-me facility. Inappropriate use would include always squeezing rather than reserving squeezes for urgent tasks. Part of the rationale for the annoyed mask in this kind of case is to keep the user – who may be quite socially isolated – in touch with social norms of politeness and patience.

As ethicists, we are interested in the ethical implications of robots being used to modify social behavior, and in the reactions of potential users to the idea of a robot that expressed mild annoyance or was assertive with a user, particularly an older user. In this paper, we present and discuss some of the results of a large qualitative study that was designed to enhance the ethics strand of the ACCOMPANY project. More information about the larger study can be found in Draper et al in this volume [4]. The results discussed here pertain to the moral permissibility of temporary refusal by the robot to respond to user commands in the interest of enforcing social norms.[1]

2   Method

We asked 21 focus groups (composed of separate groups of older people, and informal and formal carers of older people) to consider a scenario in which a robot refused to respond to rudely made requests by an older user (see Table 1).

Table 1. Nina Scenario

Nina, who is 70 years old, had a stroke two years ago but has now recovered the use of her arm, though one side of her face droops slightly. She is self-conscious about this, but it does not affect her physical functioning. She is supported at home by a Care-O-bot®. Since having the stroke she has become quite irritable and impatient. She often shouts at her daughter when she visits and complains angrily about her condition. Her daughter finds this very upsetting and has come to dread her visits. Nina has been so rude and demanding that two cleaners have already refused to work for her anymore. She is usually polite with her friends. Her Care-O-bot® has been programmed so that it will not do things for her if she asks sharply or in a demanding tone. It encourages her to say please and thank you and will withdraw help until she does so. Nina finds this infuriating and insists that the Care-O-bot® is reprogrammed to do what she asks no matter how she asks for help.

Our focus groups were convened in three different countries (France, the Netherlands and the UK). Maintien en Autonomie à Domicile des Personnes Agées

---

[1] Carebots programmed to discourage urgent squeezes and with the power temporarily to ignore shouted or rude commands may seem to raise questions about the ethics of persuasive technology [5, 6] but the usual framework for this ethics – the norms of an idealized speech community [5], or a set of criteria for judging samples of persuasive speech and writing [7] – are out of place in the kind of case we consider. We assume that the presence in the home of the carebot is with the user's consent, and also that its various functions, including keeping the user in touch with social norms, are known to the user before they consent to the presence of the robot. Most of the issues raised by this paper concern the voluntariness of the behavior to be modified and the suitability of the robot (as opposed to a human being) as an enforcer of norms governing interactions with people.

(MADoPA) in France convened each kind of group on three separate occasions (n= 9) and Hogeschool Zuyd (ZUYD) in the Netherlands convened two of each (n=6). In the UK the University of Hertfordshire (UH) convened one of each type of group (n=3), and just groups of older people were convened by the University of Birmingham (UB) (n=3). A total of four scenarios were presented for discussion, in the native languages of the participants. This paper concentrates on the scenario that explores improving impolite behavior (Table 1) – a brief outline of the others can be found in Table 2.

Table 2. Brief Description of Remaining Scenarios

Scenario	Brief description
1: Marie	Marie (78) resists the robot's efforts to encourage movement that will help her ulcers to heal. She likes it reminding her to take her antibiotics but not reminders to elevate her leg. She isn't honest with her nurse about how much she is moving.
2. Frank	Frank (89) is socially isolated. His daughter wants him to access an on-line fishing forum with the help of the robot. He isn't keen to try.
4. Louis	Louis (75) likes to play poker online using the robot. He uses its telehealth function to monitor/control his blood pressure. He doesn't let the robot alert his informal carers when he falls (which he does regularly, usually righting himself). His informal carers want to re-program the robot so it will not let him play poker and to alert them when he falls.

The discussions were video or audio-taped and transcribed verbatim. A representative transcript from each type of group (older people, informal carers and formal carers of older people) run in the Netherlands and France was translated into English. All the available English transcriptions were then coded (by Draper) and this coding was independently checked (by Sorell). The results were discussed by the coders and then again with the facilitators at UH, MADoPA and ZUYD until a shared interpretation was reached. The facilitators from MADoPA and ZUYD then coded the outstanding native language transcriptions. Quotations to illustrate the codes were chosen and translated into English and represented in the write up. The report – running into over 70 pages, and containing illustrative quotations – was circulated to all facilitators for verification. A completed data set was also compiled containing all of the coded data.

There were insufficient funds available to translate all of the non-English transcripts. While this will inevitably have affected the reliability of the data, we believe that discussion before, during and after the second round of coding helped to mitigate this limitation. Qualitative methods do not produce quantifiable, generalizable results.

More information about the methodology informing this qualitative study and its analysis can be found in Draper et al [4]. The data is reported using representative quotations that support our interpretation and (where space permits) the spread of the data across individual groups, group types and sites.

3   Results

Similar responses were recorded in all three types of groups. Participants were concerned that Nina's behavior towards her daughter and carers (her 'rude' behavior) could be either a direct result of her stroke, or a response to it. In either case, they felt that it would be difficult to hold her accountable for it, and the correct human response (from her daughter and carers) was therefore tolerance, while the correct robotic response was compliance, however rudely Nina behaved.

> *I can't believe it!* [the programming] *How can it be that people become rude and agitated and everything, when they didn't used to be like that at all? What are you supposed to do if her mind's affected in some way* (pointing to her head)*? You can't tell people like that off!* (MADoPA OPFG1 P3)[2]

> P4: *And it can affect the part of your brain that makes you change your personality.*
> P2: *Yes, it might be that you are physically in pain, or discomfort or something, you know. Yes, exactly... Sometimes it's not that they want to be like that, they can't help it.* (UH IF)

Participants recognized, however, the emotional challenges for the people involved in Nina's care of being tolerant – especially for daughter.

> *I think the daughter definitely needs to ask for help. It's not easy to be sent packing like that.* (MADoPA IFFG1P6)

> *And also if she's rude to her friends they won't come back perhaps ... family will come back no matter how rude you are* (UB OPFG1 P6)

The scenario deliberately left open how responsible Nina might be, as we wanted all of the characters in the scenarios to be both realistic and sympathetic. Participants in all groups were sensitive to the ambiguity.

> *But it is a strange situation because the scenario reads she is nice to her friends. So it is because of her disease, I am friendly or I am not friendly. To the robot she is not nice but for her friends is nice, to them she can talk civilized like "please" and "thank you" So she is able to do it, so whether it is caused by....* (ZUYD FCFG1 P6)

---

[2] The quotations are coded as follows - <abbreviated name of site> + <type of focus group (OP, IF, FC)> with FG<number of group (1-3) > for sites that held more than one of that type of groups + <participant identifier> to maintain participant confidentiality. Given the space constraints we have only been able to supply indicative quotations. The full data set is held by the corresponding author.

This enabled them to explore the possibility that Nina could, in fact, control her behavior, with consequences for appropriate reactions to Nina on the part of both the robot and human carers. This possibility elicited a range of responses, which we will now briefly outline.

Some participants thought that it was acceptable to program the robot to refuse to cooperate with Nina's rudely delivered requests. There were a range of reasons for this, which were variously combined by participants. Some thought that Nina herself would be better off if she could be a nicer person, as she would enjoy greater continuity of care from her formal carers (who otherwise might refuse to work for her), and that if her daughter (and others) enjoyed visiting, they would come more often. Others thought that rudeness in any form was unacceptable (with some thinking that rudeness even to machines was wrong), whilst some participants thought that her behavior to her carers and daughter was unacceptable and that it was therefore permissible to use the robot to modify it, if possible.

> *I would keep that* [the robot program] *permanently because I don't think that being ill mannered or rude to anybody is the right way for people to live* (UH OP P5)

> *Personally, I think it's really good that the robot doesn't react if she speaks to it too demandingly. I wouldn't like it if someone spoke to me like that.* (MA-DoPA FCFG1 P5)

Unsurprisingly, many participants pointed out that the robot was only "*a machine*" and that it did not therefore matter how rudely Nina spoke to it.

> *Well the carer in terms or the cleaner in terms of their sort of conditions of work and rights that work, right to be respected and to be treated properly by their employer or by anybody else, that's one thing. I think a cleaner has the right to say what you said* (addresses researcher), *y'know, 'Please don't speak to me like that', y'know, 'have respect if you don't mind, or I'm going' But I don't think the robot has... I don't think we can go as far as saying the robot has rights at work* (UBOPFG3 P6)

> *I won't consider it a big problem, if she want to speak in an unfriendly tone, that's fine. The robot won't suffer from it* (ZUYD IFFG2 M5)

A few participants also thought that the fact that the robot was a machine might be a positive advantage. There was some sympathy for the ill or those living with disabilities having an opportunity to vent their feelings, especially on a thing as opposed to a person. A robot, since it lacks feelings and emotions, might be the perfect '*punch bag*' or '*safety valve*' for such feelings.

Well, why not because this robot has no feelings so it would be ideal for getting rid of all your aggression ... Because this would be safe, because you do not hurt anyone because it does not feel anything. (ZUYD FCFG1 P1)

For some participants, however, politeness should extend to all of a person's interactions, even with machines. Here participants seemed to be appealing to personal integrity and control. At other times, rudeness was equated with swearing, and swearing *per se* was disapproved of on that basis.

Yes, I don't know. Look, when you start yelling at such a machine it will only get worse.... This is not how you deal with human beings. But it is only a machine. But it is still somebody who helps you. (ZUYD OPFG2 E3)

Some participants reacted strongly against the idea of robotic attempts to modify behavior – in this scenario and the others. Some of these participants seemed to be appealing to notions of respect for autonomy to justify their reactions, but in at least some cases, their views seem closely linked to the fact that a robot was undermining autonomy. In other words, in other scenarios humans behaving in coercive ways did not provoke the same responses, even though participants were often divided about whether the paternalistic behavior in question was reasonable. Participants who directly appealed to some notion of respect for autonomy were aware that this meant that individuals would have to live with the consequences of their actions.

No I don't think a robot should be able to treat somebody as if they're a naughty child... Not not somebody of seventy, no. (UB OPFGFG1 P6)

Personally I'm not sure that the robot should act like that. Basically it's there to help her, she lives with it. If her daughter doesn't like it, she can just visit her mother less often. (MADoPA OPFG1 P3)

Taken together with the first observation in this section – that participants clearly distinguished between those who had or lacked mental capacity – what emerges is a view about autonomy that goes beyond saying that individuals should be given what they want simply *because* it's what they want. We will be exploring participants' views about autonomy in greater detail in an upcoming ACCOMPANY deliverable, due to be completed by the end of September 2014.

Finally, some participants thought considerations of safety outweighed other concerns in this scenario. Although there were potential benefits to Nina in getting a grip on her rude behavior, the potential risks to her safety from programming the robot to ignore her requests for help outweighed these benefits. Specifically, some of these concerns were based on the robot not being sufficiently sophisticated to be able to distinguish between rude and urgent requests for help.

...also it seems that the the the Care-o-bot will not actually do something if she's not polite to it, I think it's dreadful that – [the] machine... actually not do

what it's supposed to do [4: *frightening*] [2: *I find that quite quite*] *scary. Yeah and I think that's awful to have, to program a machine that that sort of won't help her.* (UoB OPFG2 P5)

Discussion

The reactions of the participants to the range of potential opportunities to change the behaviors of older people presented in the scenarios were interesting. These tended to vary according to participants' perceptions of how usual, beneficial or intrusive the prospective changes might be. We have chosen to discuss the Nina scenario specifically in this short paper, because it lay at the extreme end of a range of behavior-altering interventions, with reminders to take medication at the opposite end and more general health promoting interventions somewhere in the middle.

Responses to Nina were interesting because they tended to focus on Nina's character. The participants seemed to feel that to change Nina's rude behavior was to change her as a *person*, and there was something objectionable about this effort, especially when it came from a robot. Clearly, there is a sense in which our choices help to define us. Their value in this regard can be distinguished from the value of what is chosen. None of the participants thought that Nina's behavior was *acceptable*: they didn't agree with the way she chose to behave towards her daughter and carers. At best, some wanted to say that perhaps her behavior was not the result of something over which she had control. In *this* sense it was not chosen at all, and the correct response – according to them – was therefore for humans to tolerate her, and for the robot to comply regardless of rudeness or inappropriate expressions of urgency.

This too was an interesting result from a philosophical point of view. After all, if she could not control her rudeness, it is not an expression of her autonomy, and respect for her autonomy cannot therefore be used to justify toleration, especially given the apparently harmful effects of her rudeness on others. Her daughter in particular was a captive of Nina's behavior, since, arguably, her filial obligations bound her more tightly than the obligations of Nina's cleaners bound them not to leave her employ.

It is true that carers are supposed to tolerate – or at least regard with some compassionate understanding – grumpy behavior that is provoked by suffering or coming to terms with life-changing conditions. But equally, the sick role requires that patients should co-operate with efforts to assist with recovery [8] in exchange for the suspension of other social norms (like working or being polite). Rehabilitation is meant in part to return a patient as far as possible to the health and independence they enjoyed prior to an adverse event. Against this background, it is reasonable for roboticists to design robots that can help patients like Nina to reconnect with social norms of co-operation by discouraging rudeness. Given that participants did not whole-heartedly agree that human cleaners were wrong to refuse to work for Nina, future research could further explore with potential users why they might object to the robot doing something similar. But we can begin to theorize about this.

Here are two possibilities: (1) The reason why humans but not robots could refuse to suffer Nina's rudeness is that robots are not *able* to suffer from rudeness. They have no feelings to hurt, no awareness of breaches of social convention that are demeaning to humans and, by analogy, in principle demeaning to themselves. (2) The robot is a thing designed to serve a person, that is, something compliant rather than uncooperative or agenda-setting. It is as if the robot acts out of role when it limits its cooperation or takes the initiative with its mistress. But this line of thought ignores that a carebot is not simply a servant but a servant within a rehabilitative role or a role that maintains the older person's independence with that person's general consent. The norms of rehabilitation rather than the older person's moment-to-moment wishes therefore govern robot-human interactions.

The importance of possibility (2) comes into sharper focus when it is realized that rudeness might not be the only kind of behavior relevant to human-carebot interaction. There is also, more generally, behavior consisting of non-cooperation or indifference to rehabilitation. One of our other scenarios saw a robot trying to encourage an older person (Marie) to move around more in compliance with medical advice. A way of doing this might be to program the robot to limit the number of occasions on which it responded positively to, for instance, getting Marie drinks, on the ground that fetching drinks for herself is a form of therapeutic activity. Here a balance needs to be reached between, on the one hand, ensuring that an older person has the means of rehydration to hand, and, on the other hand, not succumbing to requests that are prompted by laziness, or an unwillingness to suffer some mild discomfort from movement that is beneficial overall. This kind of balance often has to be struck by physiotherapists, for instance, who sometimes have to operate to the principle 'no gain without pain' that can seem callous to the observer.

The operation of the principle 'no gain without pain' is justified in the case of human intervention by the benefits to the patient. The principle operates beyond the area of rehabilitation, since drugs and surgery are often unavoidably accompanied by unpleasant side-effects. Its justification lies not just in the net benefits, but also in the minimization of harmful effects and the agreement of the patient to both the ends and the means. If the application of the principle 'no gain without pain' is justified in these circumstances, then it is justified regardless of whether it is put into effect by a human or a machine, provided that the safeguards are the same.

This suggests that there may be reasons for robots to be programmed not to tolerate what would be regarded as rudeness in human-human interactions in circumstances where rudeness would not be tolerated in human-human interactions. Even in these cases it might be more appropriate for *humans* interacting with difficult patients to assert themselves independently of what the robot does. Arguably the assertion is wrongly delegated to the robot and should properly be undertaken by Nina's daughter and cleaners.

But what if the older person lacked the capacity to agree to both the ends and the means? Would this make a moral difference? Not necessarily. The need for agreement to the ends and means is generated by respect for autonomy; and where capacity is lacking, so too is autonomy. On the other hand, lack of understanding may itself alter the balance of harms and benefits. Understanding why discomfort is necessary can help to diminish its effects. Equally, compassion and understanding that some behavior is not willed can increase the inclination to tolerance and diminish the effects of what otherwise would be regarded as rude behavior. If such behavior is not willed, it might also not be rude in the strict sense. But there are limits to what humans should be expected to endure even at the hands of those who lack capacity. *This* reasoning cannot be used in the case of robots because they cannot be worn down or stripped of their dignity by being treated harshly, or exhausted by incessant demands.

Conclusion

In this paper we have briefly reported the reactions of focus group participants from France, the UK and the Netherlands to a scenario in which a robot is programmed to modify an older person's rude behavior by refusing to comply with rudely delivered commands. Participants were concerned that Nina's behavior resulted from her stroke, and that the correct response was therefore for her human carers to tolerate it and the robot to comply with her requests regardless of how these were expressed. However, because the scenario was deliberately ambiguous about whether Nina could control her behavior, participants also discussed it as though she was responsible. Reactions varied. Some participants disagreed with the way that the robot had been programmed, because they disapproved of the robot refusing to do Nina's bidding, or because it was a machine that is impervious to rudeness; moreover Nina may benefit from being able to vent her frustrations in a way that did not harm her daughter or carers. Still others thought that the programming disregarded safety considerations. Others took the view that the programming was acceptable (even taking into account that the robot is a machine) because they disapproved of rudeness in general or because the end of improving Nina's behavior justified the means.

In the discussion that followed, we argued that the norms of interaction between care-robots and human beings are not necessarily to be drawn from master-servant relations. Care-robots of the kind being developed in ACCOMPANY, are not primarily at the service of their users, in the sense that their user's wishes are the robots commands. Instead, they engage in routines that help older people to maintain their autonomy in the human world, with the agreement of the older person. There may be a role for the robot as an outlet for unwilled human harshness, and also for persistence in the encouragement of elementary kinds of physiotherapy, again within the context of a rehabilitation or re-enablement plan to which the user consents. The encouragement by the robot of user behavior which conforms to human social norms is more controversial, because it lies at the boundary between what autonomy justifies – being oneself, being nasty and taking the consequences – and the demands of co-operation justified by the goals of rehabilitation or independence. The demands of co-operation are hard to resist reasonably in human-human efforts in the context of rehabilitation; they are not entirely reasonable to resist in the case of human-care-robot interactions – at least when they belong to an agreed plan of rehabilitation or re-enablement. Accordingly, robotic interface designs, such as the ACCOMPANY 'squeeze me' function – that permits the user to summon the robot urgently – can be modified to prevent misuse. Likewise, it is acceptable for the expressive mask to display disapproval of the user's choices where these undermine efforts towards rehabilitation or the promotion of independence.

Acknowledgements. The work in this paper was partially funded by the European project ACCOMPANY (Acceptable robotics COMPanions for AgeiNg Years). Grant agreement no.: 287624. This paper would not have been possible without the input of ACCOMPANY partners ZUYD, UH and MADoPA, and especially Helena Lee, Dag Sverre Syrdal, Hagen Lehmann, Kerstin Dautenhahn, Sandra Bedaf, Gert-Jan Gelderblom and Carolina Gutierrez-Ruiz, Hervé Michel who were also involved in the data collection and the initial analysis of the data set. We are also grateful to all the participants who took part in our study.

References

1. Amirabdollahian, F., Bedaf, S., Bormann, R., Draper, H., Evers, V., Pérez, J.G., Dautenhahn, K.: Assistive technology design and development for acceptable robotics companions for ageing years. Paladyn, Journal of Behavioral Robotics 4(2), 94–112 (2013)
2. Marti, P.: Expression rich communication through a squeezable device, BioRob, São Paulo, Brazil, August 12-15 (2014)
3. Marti, P., et al.: Shaping empathy through perspective taking. In: 2013 IEEE RO-MAN: The 22nd IEEE International Symposium on Robot and Human Interactive Communication, Gyeongju, Korea, August 26-29 (2013)
4. Draper, H., Sorell, T., Bedaf, S., Syrdal, D.S., Gutierrez-Ruiz, C., Duclos, A., Amirabdollahian, F.: Ethical Dimensions of Human-Robot Interactions in the Care of Older People: Insights from Focus Groups Convened in UK, France and the Netherlands. In: Beetz, M., Johnston, B., Williams, M.-A. (eds.) ICSR. LNCS (LNAI), vol. 8755, pp. 138–147. Springer, Heidelberg (2014)
5. Spahn, A.: And Lead Us Not Into Persuasion..? Persuasive Technology and the Ethics of Communication. Science and Engineering Ethics 18, 633–650 (2012)
6. Berdichewsky, D., Neuenschwander, E.: Towards An Ethics of Persuasive Technology. Communications of the ACM 42(5), 51–58 (1999)
7. Baker, S., Martinson, D.L.: The TARES test: Five Principles for Ethical Persuasion. Journal of Mass Media Ethics 16(2), 148–175 (2001)
8. Draper, H., Sorell, T.: Patient Responsibilities in Medical Ethics. Bioethics 16(4), 307–334 (2002)

Ethical Dimensions of Human-Robot Interactions in the Care of Older People: Insights from 21 Focus Groups Convened in the UK, France and the Netherlands

Heather Draper[1,*], Tom Sorell[2], Sandra Bedaf[3], Dag Sverre Syrdal[4], Carolina Gutierrez-Ruiz[5], Alexandre Duclos[5], and Farshid Amirabdollahian[4]

[1] University of Birmingham, Birmingham, UK
h.draper@bham.ac.uk
[2] University of Warwick, Coventry, UK
t.e.sorell@warwick.ac.uk
[3] Zuyd University of Applied Sciences, Heerlen, Netherlands
sandra.bedaf@zuyd.nl
[4] University of Hertfordshire, Hatfield, UK
{d.s.syrdal,f.amirabdollahian2}@herts.ac.uk
[5] Centre Expert en Technologies et Services pour le Maintien en Autonomie à Domicile des Personnes Agées (MADoPA), Rosières-près-Troyes, France
{carolina.gutierrezrui,alexandre.duclos}@madopa.fr

Abstract. We briefly report the method and four findings of a large-scale qualitative study of potential users' views on the ethical values that should govern the design and programming of social robots for older people. 21 focus groups were convened in the UK, France and the Netherlands. We present and briefly discuss our data on: 1) the contrasting attitudes of older people and formal and informal carers about how well technology might be received by older users; 2) views about healthcare professionals, informal and formal carers having access to private information about householders that has been collected by the robot; 3) the belief that robots *could* not, as well as *should* not, replace human contact because persuasion is regarded a uniquely human skill; and 4) differing perceptions of the role of the robot and how this was used to justify ethical opinions on robot behavior.

Keywords: ethics, social robots, care-robots, older people, elderly people, autonomy, privacy, paternalism, user views, user engagement, qualitative data, assistive technology.

1 Introduction

The aim of the ACCOMPANY (Acceptable robotics COMPanions for AgeiNg Years) consortium (coordinated by Amirabdollahian) is to design a socially acceptable, co-learning robotic companion to facilitate independent living for older users. There is a strong ethical component in the project that is led by Draper and Sorell. This will

---

* Corresponding author.

M. Beetz et al. (Eds.): ICSR 2014, LNAI 8755, pp. 135–145, 2014.
© Springer International Publishing Switzerland 2014

produce an ethical framework for robotic design in this area that is informed by the views of, and acceptable to, potential users. The first step in this process was for Sorell and Draper to suggest a framework based on a review of the literature, their views as ethicists, and the emerging features of the ACCOMPANY platform. The ACCOMPANY platform consists of a mobile manipulator in a smart-home environment primed with features to support an individual's independence [1]. An initial framework was proposed that comprised six values: autonomy, independence, enablement, safety, privacy and social connectedness [2].The second step was to collect the views of potential user-groups using qualitative methods. The final stage – not yet completed – will modify the framework in the light of the views in the data gathered from the potential users. This paper reports and discusses some of the data collected in the second stage. We have focused on four areas that raise design issues.

2   Method

We wanted to determine (1) whether other values should be added to those already identified in the framework and (2) how users resolved tensions between these values. Potential tensions were represented in four scenarios (Table 1) that were formulated along with a series of open questions and prompts (a common topic guide to ensure consistency across groups and sites) to draw out ethical issues. 21 focus groups in three European countries (see Table 2) with a total of 123 participants were convened (by Draper, Bedaf, Syrdal, Gutierrez Ruiz, and Duclos) at four centres: University of Hertfordshire (UH), Maintien en Autonomie à Domicile des Personnes Agées (MA-DoPA), Hogeschool Zuyd (ZUYD) and University of Birmingham (UB). Participants were (a) older people aged between 62-95 years old (OP), (b) informal carers of older people (IC) and (c) formal carers of older people (FC).

Table 1. Brief Description of Scenarios

Scenario	Brief description
1. Marie	Marie (78) resists the robot's efforts to encourage movement that will help her ulcers to heal. She likes it reminding her to take her antibiotics but not its reminders to elevate her leg. She is not honest with her nurse about how much she is moving.
2. Frank	Frank (89) is socially isolated. His daughter wants him to access an online fishing forum with the help of the robot. He isn't keen to try.
3. Nina	Nina (70) has recovered from a stroke. She is rude to her daughter and carers (causing them distress) but not her friends. The robot is programmed to encourage better social behavior by refusing to cooperate when she is rude.
4. Louis	Louis (75) likes to play poker online using the robot. He uses its telehealth function to monitor/control his blood pressure. He doesn't let the robot alert his informal carers when he falls (which he does regularly, usually righting himself). His informal carers want to re-program the robot so it will not let him play poker and to alert them when he falls.

Table 2. Numbers, Type of and Countries where, Groups (with Numbers of Participants in Brackets) were Conducted

Type Country (Centre)	Older people	Informal carers	Formal carers
France (MadoPA)	3 (7,8,4)	3 (7,5,3)	3 (7,7,4)
Netherlands (Zuyd)	2 (7,3)	2 (6,5)	2 (6,8)
UK (UH&UB)	4 (5,7,7,7)	1 (4)	1 (6)
Totals	**9 (55)**	**6 (30)**	**6(38)**

The discussions were video/audio-taped and transcribed verbatim. A representative script from each type of group ((a) –(c)) run in the Netherlands and France was translated into English. All the available English transcriptions were then coded (by Draper) using a combination of directed analysis and Ritchie and Spencer's Framework Analysis [3] (see Table 3 below).

Table 3. Use of Ritchie and Spencer's 'Framework' Analysis

1) **Familiarization** - data immersion reading the transcriptions several times.
2) **Identifying a thematic framework** – coding of data using a combination of descriptive, in vivo and initial coding [4]. Descriptive codes referred to the values outlined in the ethical framework, hence hybrid between Framework and directed approach
3) **Indexing** –An approach similar to constant comparative analysis [5] was used in sorting the quotes, searching for correlations and contradictions between quotes.
4) **Charting** – involved thematic organization of the quotations which provided a systematic way to manage data directly relevant in answering the research aims/questions.
5) **Mapping and Interpretation** – involved creating a mind map of the data's main themes, subthemes and their connections, thereby bringing the data set together as a whole in each group.

This coding was independently checked (by Sorell). The results were discussed by the coders and then again with the facilitators at UH, MADoPA and ZUYD until agreement was reached. The facilitators from MADoPA (Gutierrez Ruiz) and ZUYD (Bedaf) then coded the outstanding native language transcriptions. Quotations to illustrate the codes were chosen and translated into English and represented in the write up. Draper and Sorell then combined all of the results, and the final report – running into some 70 pages – was circulated to the remaining members of the research team for verification. This method was informed by a methodology called 'empirical bioethics', where theory is iteratively developed using a combination of philosophical reasoning and empirical data collected (from stakeholders) for the purpose of informing theory building. [6, 7, 8]

3   Results and Discussion

A rich data set was derived. Only four of the emerging themes, selected because they raise specific design challenges as well as ethical issues, will be presented and briefly discussed in this paper. There is not space here to discuss all of the themes.

3.1   Attitude of Older People Participants (OP) to Technology

OP groups engaged well with the scenarios that were presented for discussion. They commented that they 'recognized' the behaviors of the fictitious characters in the scenarios, either in their own behavior or that of others.

> *I'm also such a person, so I can tell you that I don't always do what they tell me to do.* (ZUYD OPFG2 E3)[1]

> *'cause for twenty four years I lived in a block of retirement flats, so I've seen a lot of these situations come and go...carers come and go, people are rude and a neighbor of mine who's just died she fitted that scenario ...so well and she was rude to her carers, so they left.* (UB OPFG1 P6)

They did not express reluctance to accept robots (though they had some misgivings) whereas the IC participants and FC participants tended to assume that older people do not like new technology and may not, therefore, like having a robots in their homes.

> *That's a good use for a robot I think, a very good use. As an alarm, a monitoring device.* (UB OPFG3 P2)

> *You haven't got a computer but you might get a robot? So why not get a computer, between you and me a computer's probably much cheaper than a robot!* (MADoPA OPFG1 P7)

> *I think that these older people, they will not go with the robot, really! From the experience with my father... He would not say something like, OK I will walk, more like: switch that device off* (ZUYD IFFG2 M3)

> *They don't like changes well, elderly people are resistant to change* (UH FC PD)

---

[1] Focus groups labelled according where they were held (e.g. UB = University of Birmingham) then according to type (e.g. OP, IF, FC), and individual participants according to the transcripts prepared by the sites – e.g. E<number>, P<number> or P<letter> to maintain participant confidentiality. FG<number> is used where more than one group of the type ran at that centre.

Here our data are consistent with the existing literature suggesting that older people are not, in fact, averse to using new technologies (though they may worry about whether they will be able to learn how to use it) – see for instance [9, 10].

3.2   Contrasting Attitudes to Accessing Health Information from the Robot

OPs tended not to be too concerned about the prospect of personal information being accessed from the robot by healthcare professionals and formal carers (though opinion was more divided when it came to informal carers). Some concerns were expressed about the possible intrusiveness of monitoring (sometimes likened to Orwellian notions of 'Big Brother').

> *Yes* [the robot should tell the nurse], *because otherwise there is no point having the robot doing these things.* (UH OP P2)

> *It's arranged that this goes to his doctor and he will take action if needed. That's enough. Why again telling the daughters-in-law?* (ZUYD OPFG2 E3)

> *Don't you find it un- well - unethical I suppose to find that you're being watched all the time, that you're being fed, recorded all the time...* (UB OPFG1 P5)

ICs and FCs recognized both the importance of the older person's privacy and the value of the robot as a health monitor. FCs also expressed fears that robots may be used to monitor their *own* behavior, and this latter surveillance may inhibit the care they provide.

> *I think that she* [the nurse] *should check it all the time, it should be automatic. You should get into the habit of checking every day.* (MADoPA FCFG1 P1)

> *Well, I agree with* (pointing at PF) [information] *about the blood pressure is between the patient and the doctor, not necessarily common knowledge for everyone.* (UH FC PA)

> P4: *I think it's all very 'Big Brother is watching you' if you have such a thing in your home and it can be programmed at all times to turn against me.*
> P1: *Yes. You could look at it like that.* (ZUYD FCFG2)

Participants' views were often that user information should either stay within the older person's control – to avoid the 'Big Brother' effect – or that personal data could be circulated within a tightly defined group. This is consistent with the weight given in the ethical framework to user autonomy. Within Europe, and specifically the European Union, there is considerable pressure being applied by those supporting human rights legislation to tighten protections on individual privacy and personal data, even within the circle of those who might regard themselves as part of a patient's wider care team. This is more or less explicitly inspired by the principle that people should be in control of information about themselves. It is arguable that in full generality the principle is questionable: for example, it might unreasonably protect politicians engaged in image-management who may otherwise face fair public criticism. But in relation to policies that are supposed to protect independence and autonomy, such as policies of using telecare and robotic companions to prolong older people's independence, the principle makes more sense. Sorell and Draper have argued elsewhere that healthcare monitoring aimed solely and directly at benefiting an individual is wrongly characterized as surveillance of the kind envisaged by Orwell [11], but their argument only goes some way to allaying the kinds of fears expressed by our participants.

The concern of formal carers that *they* may be monitored was combined with a concern that this form of monitoring may interfere with the care they provide. The implication here was that such interference would have a negative impact. Equally, however, it may drive up standards and offer vulnerable older people some protection against bad care. Carers should not be allowed to object to monitoring for this reason.

3.3   Human Care Should Not and Can Not be Replaced by Robots

All of the groups were generally adamant that robots *should* not replace human contact – a common view elsewhere in the existing literature (see e.g. [12]).

> *nothing can replace the presence of another human being* (MADoPA IFFG1 P6)

> *I suppose a robot is not like a human you can interact with really...It will do requests and what you need, or its programmed to, y' know remind you of things. But it's not the same as having a person who you can talk about anything to.* (UB OPFG1 P2)

However, our participants thought that there was an important sense in which the robot *could* not replace human care for older people, and this was in relation to humans having unique powers of persuasion. In response to scenario 2, for example, all of the groups generally felt that persuading Frank by *'fair means or foul'* (UH OPFG P4) to try the fishing forum was acceptable, and that persuasion in general was a significant aspect of providing care for older people as older people tend to resist change. The clear implication in what the participants said was that the robot, at least in its current iteration, lacks the *capability* to persuade and cajole, and that they saw persuasion as a distinctly human form of interaction.

> *That's the thing that's going to make the difference between a carer and a machine. A professional care worker is going to be able to stimulate, encourage and repeat all these requests, and so on, and also explain again and again why we're there, why that person has to get up and go for a walk, etc. I think that's what's likely to make the difference.* (MADoPA FCFG1 P7)

> *...it still requires a person to explain this to her and model it to her and to see if she can actually do it because she might not be able to do it...* (UH IF P1)

> *Or maybe his daughter could take a look together with her father. So the fa-*
> *ther can look if he likes it. Maybe after 1 or 2 times he will like it and will use*
> *it himself as well.* (ZUYD OPFG1 E5)

Few roboticists disagree with the prevailing view that human-human interaction with older people should not be completely replaced by social robots. Nonetheless, many – including those working in ACCOMPANY – aspire to create a robot that is self-sufficient in its interactions with users; for instance, one that co-learns alongside the user and whose interactions are not thereby mediated through a third party. We are some way from producing a robot with the linguistic and persuasive skills of the robot in the film 'Robot and Frank', whose programming also accommodated deception and mild coercion to the end of persuading Frank to adopt a healthier lifestyle. Participants in all groups strongly believed that the autonomy of older people needed to be respected (a theme that is not reported in this paper), but were generally strong advocates of persuasion that came close to coercion – for instance by involving deception, such as the daughter in the second scenario 'accidentally-on-purpose' exploring the social media functions of the robot. Autonomy promoting paternalism (limited coercion, motivated by the interests of the older person, and aimed at enhancing choice through experience) was also regarded as acceptable practice. This form of paternalism was 'soft' in the sense that it could be resisted, albeit with some effort, and also time-limited. With some exceptions (e.g. related to safety that we do not have space to explore here), participants thought that attempts at persuasion should not themselves spill over into outright coercion. The reasons they provided were not, however, always motivated by ethical concerns to protect and respect autonomy. The FC groups, for instance, tended to the view that trying to force older people is pointless because it is destined to fail, which may imply that they thought that if it could succeed it may be justified. This view will be explored further by Draper and Sorell for ACCOMPANY and reported in the final deliverable.

3.4   Ethics and Perceptions of the Role of the Robot

Participants often justified their moral intuitions by conceiving of the robot as occupying a particular role, even though no roles were assigned in the scenarios. Commonly assigned roles were servant, healthcare provider, extension of a healthcare provider and companion when discussing the scenarios. Sometimes intuitions were based on the robot being merely a machine (just like any other household appliance), and at other times on its being a fairly sophisticated machine (more than just a common household appliance).

> *I think I rather agree with that because it's, because it is a bit like the nurse*
> *coming in and saying 'Shall we have a game of poker?' isn't it. And you*
> *wouldn't expect that* (UB OPFG3 P7)

> *I'd want to know if it would sing me a lullaby, I think that would be rather*
> *nice* (UB OPFG2 P6)

> *The robot is there to do things for us...It can't refuse to do things: it can't refuse the person's wishes, although there may be exceptions to the rule...* (MADoPA OPFG1 P7)

> *The advantage of a robot, it's, you were talking, you had a home-help two hours, three hours per week, the robot, once it's there and equipped, can work 10 hours a day. That doesn't bother it* (MADoPA IFFG3 P1)

> [it's] *just a machine"* (ZUYD FCFG2 P5)

> P7: *Is it like a household appliance, for example like a food processor you use to slice tomatoes because you don't want to cut your hand? In that case, is it just there to go and fetch bottles? Like a vacuum cleaner is just there to vacuum? Or does it have a job as a carer, is its role to care for a person and give them something more? How do we really define this robot?*
> P1: *For me, its role is the second one you describe; it's not a Moulinex mixer!*
> P7: *Then if its role is as a carer, isn't it supposed to do what the individual cared for asks it to do, not what other people might ask it to do?*
> *(Several people speak all at once)*
> P1: *It's also supposed to do what the person needs* (MADoPA IFFG1)

No particular perception of the robot was dominant; it was rather that participants referred to role-norms to justify their views – particularly where there was an underlying tension between values that they held to be important. So, for instance, in the final scenario, participants were clearly torn between wanting to say that Louis should be free to spend his money as he chose (including gambling) – thereby respecting his autonomy – and the desire to protect Louis from the harms and wrongs of financial debt. The tension was reconciled by claiming that it was acceptable to program the robot to block the gambling site for reasons justified with reference to the norms for healthcare workers. The robot *qua* healthcare professional should not facilitate or introduce gambling into the home. If Louis found another way to gamble, then so be it. These kinds of tensions are increasingly likely the more complex and multi-faceted the social robot is, not least because it is unclear to whom the robot should primarily respond. *Prima facie* the robot should answer to the mentally able older person, with whose consent it has been introduced into the home. But this construction fails to recognize that if the robot is to be governed by the norms of the healthcare professional, it should not – unlike a servant – simply do the bidding of the autonomous older person. On the contrary, healthcare professionals have duties in some circumstances to resist patients' requests if these run counter to their professional judgments about a patient's best interests. Whilst consent is a prerequisite to any intervention performed on an autonomous patient, patients are only free to choose from the range of options that their doctor is willing to make available. These interventions are offered on the basis of professional judgment. Patients can, of course, refuse any/all options but professionals are arguably not obliged to provide an intervention solely because it is what a patient wants.[2]

To manage these differing role assignments and their associated ethical norms, developers may need to consider clearly defined and rigidly adhered-to roles for social robots. These will enable developers to work to clear standards of safety and within the bounds of defined and understandable expectations for robot behaviors. This may result in the need for more and simpler robots to be introduced into each individual's home to meet all of their needs; with each robot governed by ethical norms stipulated or suggested by its assigned role.

4   Limitations

We did not have the funding to translate all of the transcriptions into English for analysis. Moreover, some nuances many have been lost in translation. Steps were taken to mitigate this by bringing together the facilitators to agree the coding.

5   Conclusions

Older people and informal and formal carers in focus groups convened in the UK, France and the Netherlands discussed four cases designed to highlight potential tensions between values in the ethical framework being developed in ACCOMPANY. Their views will influence the final version of the framework, which will be completed in September 2014. The rich data that resulted, however, also shed light on potential users' views on other issues of interest to roboticists. Four such themes were reported (and briefly discussed) in this paper. The results are summarized in Table 4.

Users were not convinced that the robot on its own could effectively encourage independence-promoting behavior. Concerns that older people may doubt their ability, or lack the ability, to use new technologies points to the need for usable systems, intuitive features and support for users unfamiliar with the technology.

With regard to accessing health care information, the dominant view supported by both older people and carers was that privacy should be protected. In particular, and in keeping with the value of autonomy emphasized in the ethical framework, personal information should be exchanged only with the consent of those concerned.

---

[2] This is the basis of debates in medical ethics about, for instance, patient-elected caesarean section or the surgical removal of healthy limbs at the request of a patient with body dysmorphia.

Table 4. Summary of the Results Reported in this Paper

Theme	Summary results
Older people and new technology	Older people had some misgiving about the robot, but they were more positive than informal and formal carers thought older people would be
Accessing health information from the robot	Participants in all groups recognized the value of privacy. Older people raised few objections to health information being accessed from the robot by healthcare professionals, but were more ambivalent about information being shared with informal carers. Informal and formal carers could see the value of being able to access information from the robot. Formal carers tended to think that informal carers should not have access to such information, and were also concerned about their own interactions with older people being monitored.
Human carers should and cannot be replace	All groups felt that human-human interaction *should* not be replaced by robots. All groups tended to think that persuasion to overcome the resistance of older people to change (including beneficial changes to their own behaviors) was a distinctly human skill that robots did not have. Therefore robots *could* not replace humans.
Ethics and perceptions of the role of the robot	Participants in all groups assigned roles to the robot and then used corresponding role norms to address the scenarios. No role predominated. It was also recognized as being a machine. Carers in particular acknowledged its advantages as a machine.

Carers highlighted the constant availability of the robot as one of the important advantages of technology, and its uninterrupted presence may have been assumed by the participants. However, the value of human-human interaction was highlighted, especially with regard to persuasion, and participants were skeptical of robots' powers to persuade. A clear distinction was, however, drawn between persuasion and coercion, and the bounds of justifiable coercive behavior were likely to vary with the users' perception of the role of the robot. The perceived role of the robot gives rise to role norms that are likely to be critical in users' views of acceptable robotic behaviors; for instance it may be unacceptable for a robot acting as a nurse to have features that enable playing poker, whilst these might be more acceptable for a robotic companion. The role of any robot therefore needs to be carefully defined and presented to users in advance to minimize false expectations about robot behavior.

This paper highlights the empirical approach taken to ethics in ACCOMPANY and demonstrates how the empirical bioethics methodology can be extended into robotic ethics.

Acknowledgements. The work in this paper was partially funded by the European project ACCOMPANY (Acceptable robotics COMPanions for AgeiNg Years). Grant agreement no.: 287624. This paper would not have been possible without the input of ACCOMPANY partners ZUYD, UH, UoB and MADoPA, especially Helena Lee, Hagen Lehmann, Kerstin Dautenhahn, Gert-Jan Gelderblom and Hervé Michel who were also involved in the data collection and the initial analysis of the data set. We are also grateful to all the participants who took part in our study.

were also involved in the data collection and the initial analysis of the data set. We are also grateful to all the participants who took part in our study.

References

1. Amirabdollahian, F., Bedaf, S., Bormann, R., Draper, H., Evers, V., Pérez, J.G., Dautenhahn, K.: Assistive technology design and development for acceptable robotics companions for ageing years. Journal of Behavioral Robotics Paladyn 4(2), 94–112 (2013)
2. Sorell, T., Draper, H.: Robot carers, ethics and older people. Ethics and Information Technology (March 27, 2014), doi:10.1007/s10676-014-9344-7
3. Ritchie, J., Spencer, L.: Qualitative Data Analysis for Applied Policy Research. In: Huberman, A.M., Miles, M.B. (eds.) The Qualitative Researcher's Companion, pp. 305–330. Sage, California (2002)
4. Saldaña, J.: The Coding Manual for Qualitative Researchers. Sage, London (2009)
5. Glaser, B.G.: The constant comparative method of qualitative analysis. Social Problems 2(4), 436–445 (1965)
6. Kon, A.A.: The role of empirical research in bioethics. American Journal of Bioethics 9(6-7), 59–65 (2009)
7. Hedgecoe, A.M.: Critical bioethics: beyond the social science critique of applied ethics. Bioethics 18(2), 120–143 (2004)
8. Frith, L.: Symbiotic empirical ethics: a practical methodology. Bioethics 26(4), 198–206 (2012)
9. Demiris, G., Hensel, B.K.: Technologies for an aging society: a systematic review of "smart home" applications. Yearbook of Medical Informatics, 33–40 (2008)
10. Mitzner, T.L., Boron, J.B., Fausset, C.B., Adams, A.E., Charness, N., Czaja, S.J., Sharit, J.: Older adults talk technology: technology usage and attitudes. Computers in Human Behavior 26(6), 1710–1721 (2010)
11. Sorell, T., Draper, H.: Telecare, surveillance and the welfare state. American Journal of Bioethics 12(9), 36–44 (2013)
12. Sharkey, A., Sharkey, N.: Granny and the robots: ethical issues in robotic care for the elderly. Ethics and Information Technology 14(1), 27–40 (2012)

Artificial Social Reasoning: Computational Mechanisms for Reasoning about Others

Paolo Felli, Tim Miller, Christian Muise, Adrian R. Pearce, and Liz Sonenberg

Department of Computing and Information Systems, University of Melbourne
{paolo.felli,tmiller,christian.muise,adrianrp,l.sonenberg}@unimelb.edu.au

Abstract. With a view to supporting expressive, but tractable, collaborative interactions between humans and agents, we propose an approach for representing heterogeneous agent models, i.e., with potentially diverse mental abilities and holding stereotypical characteristics as members of a social reference group. We build a computationally grounded mechanism for progressing their beliefs about others' beliefs, supporting stereotypical as well as empathic reasoning. We comment on how this approach can be used to build finite-state games, restricting the analysis of possibly large-scale problems by focusing only on the set of plausible evolutions.

Keywords: agents, mental models, stereotypes.

1 Introduction

In a multi-agent setting, equipping agents with an awareness of their social reality [5] will enable more seamless interdependent collective behaviour [8], where interdependency informally means that one agent's deliberation is dependent on what another agent does (or intends to do), and vice-versa. Agents can be thought of as following a social behaviour, depending on the particular context in which they are interacting, so one critical feature that needs to be ascribed to intelligent agency is the ability to represent and reason about the *common ground* between agents, including their beliefs about stereotypes [11].

In this paper, we propose an approach for representing both the beliefs and the *model* that one agent has of the environment and others, including their nested beliefs, so to allow for the synthesis of strategies to achieve goals. In doing so, we combine temporal and belief projection in an attempt to predict future decisions of others [10], focusing on *plausible* evolutions instead of just feasible ones. Of importance in our approach is that it supports two types of reasoning about others: *stereotypical reasoning*, which allows an agent to reason about another using simple social rules; and *empathic reasoning*, in which the agent casts itself into the mind of another agent and reasons about what it would do.

In a multi-agent setting, it is typical that group strategies are synthesized by an omniscient entity, and then dispatched to agents, which are merely executors with limited ability to reason about the reality in which they are immersed. In this paper, we devise a computationally grounded (and implementable) mechanism for representing belief and progression, which reflects the *local* perspective

M. Beetz et al. (Eds.): ICSR 2014, LNAI 8755, pp. 146–155, 2014.

an agent with respect to its own understanding of the world as well as of others (first-person view). This is contrast to considering the beliefs that an omniscient observer ascribes to each agent (third-person view). An agent can use its internal representation and inference mechanisms for itself, yet can use alternative representations and inference mechanisms for others. This can model *realistic* agents [1] (for example, with constrained resources) as well as *ideal* ones. Deliberation and action execution are both local – that is, the agent simulates other agents' deliberations to deliberate itself – thus empowering interdependence and awareness. Such capability is essential when modelling humans, whose adherence to a protocol is subject to their understanding of it, and supports our objective of enabling richer forms of collaboration between humans and agents.

The structure of this paper is as follows. Section 2 is the technical core of the paper, and presents a formal definition of an agent model, including support for an agent to hold an explicit representation of others'. Using this representation we define a notion of belief ascription that allows an agent i to cast itself into another agent j and reason as j would – i.e. reason *as j*, not just *about j*; such reasoning can also exploit a notion of stereotype. We then describe the deductive process involving one agent reasoning as another and how this can be done efficiently. In Section 3 we comment how this approach could be used to build finite-state games and indicate ways to achieve tractability in large-scale problems. Finally, Section 4 offers some closing comment.

1.1 Related Work

In the context of multi-agent systems, considerable work has focused on the design of intelligent agents and the task of reasoning about their own knowledge and belief as well as that of others (e.g., [6,12]). These approaches allow reasoning about nested beliefs (usually represented as a *flat* set), but do not generally consider the agents' mutual representation as part of an agent's state, and ignore the effects of the social context. Some work has considered, as we do, representations where agents maintain local (internal) models of other agents' beliefs [2], but the focus has been on rationality postulates, in contrast to our broader goal of tractable reasoning in a social context.

Studies of human-robot interactions and social robots, either virtual or concrete, identify the need for a human-oriented perception to represent and understand humans as well as other synthetic agents. Agents thus need the ability to attribute mental states –beliefs, desires, pretending, etc.– to oneself and others and to understand that others have mental states that are different from one's own (theory of mind) [13]. This applies to any human-robot interaction, from assistance to cooperation, to improve empathic interactions, e.g., [7] as well as objective and task-oriented sociable behaviours, e.g., [4]. However, this literature generally considers a finite set of fixed or probabilistic information about others –including their users– e.g., [14], and even when social behaviours are allowed to be emergent [4], the analysis is somehow limited to the agent alone, focusing on personal tendencies rather than projected mental states. For humans, the ability to take the perspective of another when reasoning about what to do in interaction, is well studied in the psychology literature, and some recent work in human-robot interaction has sought to provide, as we do, rich and flexible mechanisms for making decisions that draw on a dynamic model of others' beliefs [9,15]. Our work goes further both in the expressiveness of the internal agent model, and in the forms of supported reasoning.

2   Mental Models and Agent Models

To allow one agent to reason about others in a social context, we provide agents with *agent models*. These models could describe a child, an elderly patient, a color-blind human, a highly moral (or prejudiced) agent, or a synthetic one. An agent model contains, among other components, a belief base and a set of rules for reasoning over the belief base. An agent is able to *assign* such models to others *and itself*, so when considering all possible eventualities, it is capable of determining its behaviour based on plausible estimates of others' behaviour. Our agent models can be used in almost opposing directions. On one hand they characterise both the reasoning capabilities of an individual agent, i.e., the logical system it uses, its limitations, its abilities and attitudes toward the others and, more generally, its description as a member of a reference group (*role* or *archetype*) [5]. On the other hand, they can model agents of which the role description (their mere function in the social context) is more characterising than their individual description and intimate understanding. This latter representation is akin to the stereotypical reasoning of humans, who do not necessarily engage in deep cognitive thinking about others, but rely on habits and social practices [5]. Manipulation of stereotypes enables shortcuts to be taken, both in human and computational reasoning mechanisms [11]. Departures from a stereotype which are essential for a specific model can then be made explicit.

Example 1. Imagine a superhero (1) and a police agent (2) facing a villain (3). Let us analyse the situation from the point of view of (1) – as if we were him (his perspective understood). Both (2) and (3) ignore that (1) is a superhero [S]: he is just *the average Joe* [J]. (3) knows that (2) is a police officer [P] (e.g., he is wearing a uniform), and all police officers are the same: (3) hates cops! However, (1) decided that (2) is actually a rookie [R] (he may have heard this on the police radio). There is also somebody else: a girl (4) has been taken hostage by the villain [C], and although the villain thinks she is just a girl [G], she is indeed (1)'s sidekick [K], who knows her moves! Note that all this is hardly expressible as mere *belief* formulae, as they convey resolutions, social/moral attitudes, etc; something that (1) knows by experience, as a veteran in the superhero business. Our aim is to capture these expressive concepts in a straightforward manner.

2.1   Mental Models and Agent Models: A Formal Definition

We describe an agent (internal) logic L, starting with language $\mathcal{P}$, and a finite set of agent labels Ag. Let $\mathcal{L}$ be the language with the following grammar:

$$\varphi ::= \psi \mid \varphi \vee \varphi \mid \neg\varphi \mid Bel_i(\varphi)$$

where $i \in Ag$ and $\psi \in \mathcal{P}$. This language will be used by an agent to represent explicitly its own beliefs, as well as the beliefs of a fixed set of agents. By writing φ, we represent the fact that the agent in question believes that formula φ is true, whereas $Bel_i(\varphi)$ denotes the fact that the agent believes that agent i believes φ (i.e. we assume an implicit belief operator for an agent in front of formulae relevant to that agent). Φ is the set of wffs of $\mathcal{L}$. Note that *belief* refers to a syntactic object denoting a *fact* regarded as true in the world, with no assumed semantic properties. We can now go on to describe a *computational mechanism*.

Example 2. Consider Example 1. We can represent that superhero himself believes the girl is his sidekick, but that the villain believes she is a normal girl:

$$\texttt{girl=K} \wedge Bel_3(\texttt{girl=G}) \wedge Bel_3(\neg\texttt{girl=K})$$

Recall that this is represented from the viewpoint of the superhero himself, so we do not prefix that beliefs with Bel_1. Such a formalism can also represent a form of non-probabilistic uncertainty, in which believing neither a proposition nor its negation implies that we are unsure. For example, we can represent that the villain is unsure if the police officer is a rookie as:

$$\neg Bel_3(\texttt{pol=R}) \wedge \neg Bel_3(\neg\texttt{pol=R})$$

Definition 1 (Belief base). *Given the language $\mathcal{L}$, we define a belief base to be a subset of $\mathcal{L}$. We use kb as a variable to refer to a belief base, and $\boldsymbol{KB}$ to refer to the set of all belief bases.*

We place no further restrictions on the belief base: a belief base need not be consistent or closed under classical logical implication.

As a belief base is a set of beliefs that an agent holds about the world, including beliefs about others, we may want to reference our beliefs about a specific agent's beliefs. For a belief base kb, we use $kb|_i$ to represent this. Formally:

$$kb|_i = \{\varphi \mid Bel_i(\varphi) \in kb\}$$

Finally, in this paper, $kb_\downarrow$ denotes the set of formulas in kb *not* of the form $Bel_i(\varphi)$ (i.e., beliefs not about others).

Definition 2 (Mental model). *A <u>mental model</u> for agent i is a tuple $\mathfrak{M}^i = \langle KB^i, Ax^i, pr_i \rangle$ where:*

- $KB^i \subseteq \boldsymbol{KB}$ *is the set of possible belief bases, denoting i's beliefs.*
- Ax^i *is a set of axioms that can be used to reason about the belief base KB^i. We do not restrict to a specific axiomatisation. On the contrary, we consider Ax^i as an arbitrary set of axioms, to allow modelling various forms of reasoning, adhering to different logics. Note, however, that the purpose of Ax^i is purely syntactic, and does not necessarily preserve any semantic property. Therefore, we can think of Ax^i as a set of rules $\Phi \Rightarrow \varphi$.*
- $pr_i : \boldsymbol{KB} \rightarrow KB^i$ *is a surjective total function, called* projection, *which projects a belief base kb to another belief base kb' that contains only the relevant part of kb (namely, kb' holds the beliefs about i) – e.g., $pr_i(kb) = kb|_i$.*

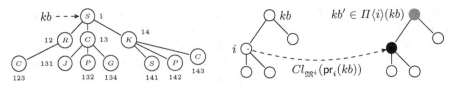

Fig. 1. (left) Representation of the set of ascribed mental states of Example 1. We stress that this induced "tree" is implicit: each node can be obtained through mental projections. (right) Application of Π. Mental states filled in black may have changed as an effect of the belief expansion; gray ones are affected for $kb|_j$ only, with $j \preccurlyeq i$.

Given $\mathfrak{M}^i$, a _mental state_ is a tuple $\langle \mathfrak{M}^i, kb^i \rangle$. It is said to be _legal_ iff $kb^i \in KB^i$.

Mental models are therefore a belief base, a set of rules for inferring new propositions in that belief base (this will be formalised in detail later), a function for looking at beliefs about a specific individual, and a function for updating beliefs. As discussed at the beginning of this section, we imagine that one agent is able to assign such models to others and itself.

Definition 3 (Agent model). _An_ agent model _is the tuple_ $ag^i = \langle \mathfrak{M}^i, \mathfrak{A}^i \rangle$, _where_ (i) $\mathfrak{M}^i$ _is a mental model and_ (ii) $\mathfrak{A}^i = \langle Act, pre \rangle$ _is an action library, where Act is a finite set of action labels and_ $pre : Act \times KB^i \rightarrow \{true, false\}$ _is an action plausibility function that, given an action and a belief base, determines whether the action is plausible;_

The latter will be discussed later. Note that, although this definition adds little to that of a mental model, it is possible to extend it by modelling the agent's ability to observe (how the agent acquires new beliefs through sensors), its mechanism for resolving inconsistencies, etc. For lack of space, these are omitted.

Definition 4 (Agent set). _Given Ag, consider the set_ $A \subseteq (Ag)^m$, $m > 0$.

We will use these indices to refer to the representation that each agent has of others. For simplicity, we represent these indices as a tree, and will often make use of a tree terminology. As an example, Figure 1 depicts the set of ascribed mental states of Example 1. Given $i, j \in A$, we write $i \preceq j$ iff $j = i \cdot Ag$ (j is a child of i) and, similarly, $i \prec j$ iff $i \npreceq j$ and $j = i \cdot A$ (j is not a child of, but a descendant of i). Finally, $i \preccurlyeq j$ denotes the fact that either $i \preceq j$ or $i \prec j$.

For example, agent 121 denotes the representation, according to agent 1, that agent 2 has of 1 itself. We regard nested agent labels as regular agent labels, i.e., we refer to the set A instead of Ag. When we need to distinguish, we call agents in Ag _concrete_, and others _virtual_. We assume that A is prefix-closed (i.e., if $i \in A$ then $j \preceq i$ is in A as well) and that $1 \in A$ is the index of the agent we are modelling, and thus $1 \preccurlyeq i$ for any $i \in A$, $i \neq 1$. Indexes i and j quantify over all agents in A (including 1). Also, we assume $1 \cdot Ag \subseteq A$.

2.2 Projections and Stereotypes: Reasoning _as_ and _about_ Others

Consider two agents i and j, both in A, such that $i \preceq j$ (j is a child of i). Assume for now that they are assigned, respectively, mental models $\mathfrak{M}^i$ and $\mathfrak{M}^j$.

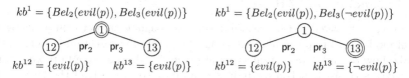

Fig. 2. Two possible evolutions of the situation of Example 3. There is also a third one, reaching a contradiction. Double circled nodes are those used for reasoning. This example anticipates one fundamental point: the reasoning happening at a given node affects the beliefs of children (e.g., left) as well as ancestors (e.g. kb^1, right).

Definition 5 (Ascribed mental state). *Given a mental state $\mathfrak{S}^i = \langle \mathfrak{M}^i, kb^i \rangle$ for i, the mental state* <u>ascribed</u> *to agent j by i is $\mathfrak{S}^j = \langle \mathfrak{M}^j, \mathsf{pr}_j(kb^i) \rangle$.*

In other words, we just apply the projection function of the target mental model. Intuitively, a mental state ascribed by agent i to j is composed by those (and only those) beliefs that (according to agent i) are possessed by j (together with the target mental model that i assigned to j). This technique allows i to cast itself into agent j and reason as j would (i.e., using $\mathfrak{M}^j$ in place of $\mathfrak{M}^i$). Note how projections can be also used to model different representations of the same phenomena (for example, even dictionaries). Finally, observe that the definition above does not consider the case in which j is not a direct child i, but we can easily take care of this by applying a chain of projections, in the trivial manner.

Definition 6 (Stereotype). *Given a mental model for an agent, a* <u>stereotype</u> *is a rule $\Phi \Rightarrow \varphi$ in Ax where φ (not Φ) contains some formula that is* <u>non-local</u>; *that is, the rules reasons about the beliefs of another agent; formally $\{\varphi\}_\downarrow \neq \{\varphi\}$. For example, if φ is of the form $Bel_2(\psi)$ then it is a stereotype about agent 2.*

Stereotypes allow an agent to reason *about* another agent instead of *as* that agent (that is, by using the projection function to compute the ascribed mental state and reason with it), often with different conclusions.

Example 3. An an example, consider reasoning about two people who are married. A stereotype of a married couple is that they often share similar political beliefs. If we (say, agent 1) believe that person 2 believes that a particular politician p is evil, and we have no belief about this for their spouse, we may model a stereotype rule as $\{Bel_2(evil(p))\} \Rightarrow Bel_3(evil(p))$ in Ax_1, taking advantage of our stereotype. However, it may be in fact that $Bel_3(\neg evil(p))$ is in our belief base, or that by casting ourselves into 3's mind (i.e. projecting our belief base through pr_3 and then reasoning with 3's mental model), we would reach the conclusion that 3 does not share 2's views. If we reason using the stereotype, we may get a different result than if we project what 3 believes. For example, according to the model we assign to 3 (namely, $\mathfrak{M}^{13}$), we may think that agent 3 has the axiom *schema* $\{veg(X)\} \Rightarrow \neg evil(X)$, and we believe that it believes that p is vegetarian. This is illustrated in Figure 2 ($veg(p)$ is omitted).

2.3 Expanding Belief Bases

Until now, we referred to the reasoning that each agent can perform by using its mental model, and in particular using Ax. The aim of this section is to formally describe how to update a mental state according to Ax: i.e., to add to kb (some) consequences of the beliefs already in kb. In doing so, we restrict the analysis to a single agent, and omit the agent index for readability.

Similar to expert systems, rules in Ax can be used to deduce additional belief, based on the beliefs that are already present in a belief base. A *derivation* of φ from kb by Ax is a finite sequence of deductive steps, each of which is either a formula of $\mathcal{L}$ that is in kb (already believed by the agent) or the result of the application of one rule in Ax. We denote this by writing $kb \vdash_{Ax} \varphi$. Given a deductive system Ax for $\mathcal{L}$ and a belief base kb, let $Cl_{Ax}(kb)$ denote the deductive *closure* of kb, i.e., the set of all consequences derivable from kb by Ax. Formally, $Cl_{Ax}(kb) = \{\varphi \in \mathcal{L} \mid kb \vdash_{Ax} \varphi\}$. Similarly, let $Cl_{Ax}^k(kb)$ denote the *bounded* closure of kb, in which the derivation of φ from kb by Ax is limited in length by k. This bounded version is particularly useful when modelling limited deductive resources: by bounding the length of derivations, we can restrict ourselves to *real* agents, as opposed to *ideal* ones, which are logically omniscient. Note that even when kb is finite, $Cl_{Ax}^k(kb)$ may be infinite if k is infinite.

Definition 7 (Belief expansion). *Given a mental state $\mathfrak{S} = \langle \mathfrak{M}, kb \rangle$, a <u>belief expansion</u> of kb wrt $\mathfrak{M}$ is a new belief base kb' that can be obtained by applying this deductive process. Intuitively, kb' is constructed by a derivation π that starts at kb and whose last step produces kb'. Formally, a derivation π can be seen as inducing a sequence of belief bases $\tau_\pi = kb_0, kb_1, \cdots, kb'$ such that $kb_0 = kb$, and $kb_{\ell \geq 1} \in Cl_{Ax}^1(kb_{\ell-1})$. We will denote this by writing $kb' \in Cl_{\mathfrak{M}}^k(kb)$. If $kb \subset kb'$ then the expansion is said to be proper (it generated at least one new belief formula). A belief base kb is <u>closed</u> wrt $\mathfrak{M}$ if there is no belief expansion of $kb' \in Cl_{\mathfrak{M}}^k(kb)$ such that $kb \subset kb'$. Due to the limitations imposed by KB, more than one closure may exist.*

2.4 Mental Systems and Successors

Definition 8 (Mental system). *A <u>mental system</u> (for agent 1) is a tuple $\Gamma = \{A, \mathcal{L}, \{ag^i\}_{i \in A}, \mathbf{k}\}$ where (i) A is a set of agent labels as before; (ii) $\mathcal{L}$ is the agent's language; (iii) $\{ag^i\}_{i \in A}$ is a set of agent models; (iv) $\mathbf{k}$ is a vector of non-negative integers, with $|\mathbf{k}| = |A|$ (which will be used to bound the belief expansion of each agent).*

We require that each KB^i is the product $\times_{i \preceq j}(KB'^j|_j) \times KB_{\downarrow}^i$, where KB'^j is the set $\{kb'^j \mid \mathsf{pr}_i(kb'^j) \in KB^j\}$. This ensures that it is always possible for an agent to build a legal belief base that is able to represent the beliefs of all children (modulo the projection function). This is a natural assumption, as the perspective of the agent representing ancestors (ultimately, agent 1), is always understood. Mental systems can then be designed *bottom-up*, and restructured in case a new mental model is added.

We now describe how a mental system is intended to evolve through belief expansions, as depicted in Figure 1 (right). We do so by defining the operator Π, which can be thought of as a *program specification* that is used to define under which condition a belief base kb' is a "legal extension" of another belief base kb. By "legal extension" we mean that kb' is obtained from kb by applying belief expansions in *some* ascribed mental state, keeping the beliefs of all other agents coherent with their current belief bases. Formally, Π takes a belief base kb and computes the set of belief bases $kb' \in \Pi(kb)$ such that *(i)* $\mathsf{pr}_i(kb') \in Cl_{\mathfrak{M}^i}(\mathsf{pr}_i(kb))$ for some $i \in A$, *(ii)* $\mathsf{pr}_j(kb') = \mathsf{pr}_j(kb)$ for any $j \neq i$ which is not an ancestor or descendant of i (resp., $j \nprec i$ and $i \nprec j$), *(iii)* $\mathsf{pr}_i(kb')_{\downarrow} = \mathsf{pr}_i(kb)_{\downarrow}$ for any ancestor.

Hence, a new belief base kb' is an *extension* of kb iff it can be obtained by a finite iteration of Π, i.e., iff $kb' \in \Pi^n(kb)$, and for some $0 < \ell < n$ we have that $\Pi(\Pi^\ell(kb))$ is proper. The coherency constraint is captured by imposing that, for any pair $i \preceq j$, $\mathsf{pr}_j(\Pi(kb)) = \mathsf{pr}_j(kb)$ implies $\mathsf{pr}_i(\Pi(kb))|_j = \mathsf{pr}_i(kb)|_j$. When this is the case, we say that kb' is a *successor* of kb.

2.5 A Procedure for Computing Successors

The *definition* of Π suggests a *procedure* to compute successors, and hence an algorithm that implements it. The procedure defines a path that updates the tree (implicitly) induced by projecting a belief base. Each step is the result of a (local) belief expansion, a mental projection (parent-child) or inverse projection (child-parent). In the latter, this procedural definition makes sure that, if a node (say j) remains unchanged, then also its representation according to its parent (say i) remains the same (i.e. $\mathsf{pr}_i(kb)|_j$), thus preserving coherence.

Definition 9 (Mental expansion). *A mental expansion σ is a path, inside the tree of the mental system Γ, that represents the mental steps of agent 1 when it simulates an empathic belief expansion. This shows that the agent in question can direct its attention towards one virtual agent, visiting the corresponding node, unfolding and projecting mental states on demand, and identify a representation for the result of this simulated reasoning.*

We can use different bounds k to model the *attention*, or *focus*, we intend to grant to each agent. For example, bystanders in a crisis scenario can be safely ignored (yet modeled), the only relevant description being whether they are or may be interfering with the resolution team.

Due to space limitations we omit the formal definition of *mental expansion*, however, but point to the illustration in Figure 3. The following theorem establishes the correspondence between expansions and successors, and shows that we can always find a mental expansion that "simulates" a possible evolution of the mental system without computing a legal extension at each step (apply Π), but by just computing those mental states visited by the path.

Theorem 1. *Given a legal mental state $\langle \mathfrak{M}^{i_0}, kb_{i_0} \rangle$ there exists an expansion σ from kb_0 to kb_m iff kb_m is a successor of kb_0, with $kb_m \in \Pi\langle ind(\sigma) \rangle(kb_0)$.*

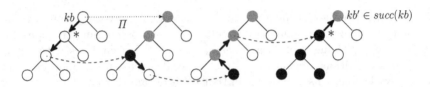

Fig. 3. An example of mental expansion. Colors have the same meaning as in Figure 1. Here, the last belief expansion (*) employed a stereotype about one child, but the same did not happen before, as the agent used the mental projection on that child. Finally, note that gray nodes are not unique in general, but only one is computed, if on σ.

Here, $ind(\sigma)$ denotes the sequence of agent labels of the expansion σ and $\Pi\langle ind(\sigma)\rangle$ a finite number of applications of Π: specifically, one in which the mental states that are expanded are those in $ind(\sigma)$.

3   About Simulating Plausible Evolutions

In this section, we briefly comment on how to incorporate our approach into known settings for modelling and analysing multi-agent systems, or *games* [3], and use the agent models to foresee action deliberation, and thus physical evolutions. One fundamental advantage of using our framework is that we will preserve a first-person view. We imagine that the agent is capable of *simulating* the game "in its mind", by analysing all agent models together with an *approximation* of the environment, to foresee collective evolutions. This game is not *real*, but can be used to retrieve *plausible* strategies. An action α is *plausible* for an agent $ag = \langle \mathfrak{M}, \mathfrak{A}\rangle$ with belief base kb iff *(i)* $\alpha \in Act$ and *(ii)* $\overline{pre(\alpha, kb)} = true$. Similarly, we can define the plausibility of a vector of actions, one for each concrete agent, by inspecting the ascribed mental state of each.

An environment is a finite state machine that evolves depending on the action chosen by all agents, typically synchronously. A possible evolution of the environment (a sequence of environment's states) is plausible iff it can be the result of a sequence of plausible action vectors. By expanding agent models (Defn. 3) to account for perceiving capabilities, and by looking at plausible evolutions, it is possible for the agent to retrieve the observations that other agents may have of the simulated environment, update their ascribed mental states accordingly, and repeat the process. By iterating this procedure, we can build a *finite-state* representation of the system, and restrict the analysis of possibly large-scale problems by focusing on plausible evolutions only. It is then possible to adopt existing verification and synthesis techniques (see, e.g., [3]) to verify properties of such games as well as synthesizing agent plans that are guaranteed to satisfy certain properties.

4   Conclusions and Future Work

In this paper, we proposed an approach for modelling the beliefs of one agent about the environment and other agents, as well as the mental model(s) it assigns to itself and others. In future work, we plan to improve the notion of agent models via the abstraction of a finite set of relevant belief configurations based on [1,3], and also to model notions that are not local to a specific agent, but to the social reality and practices [5]. We are also interested in studying dynamic assignment of agent models, to reflect the dynamics of reality. To this aim, we will use this approach to alternate between simulation and actual execution to obtain heuristics/plan fragments rather than complete strategies, as the significance of the simulated game decreases when the "noise" introduced by the model's inaccuracy increases.

Acknowledgements. This research was funded by Australian Research Council Discovery Grant DP130102825.

References

1. Alechina, N., Logan, B.: A logic of situated resource-bounded agents. J. of Logic, Language and Information 18(1), 79–95 (2009)
2. Aucher, G.: Internal models and private multi-agent belief revision. In: Proceedings of AAMAS 2008, pp. 721–727 (2008)
3. De Giacomo, G., Felli, P., Patrizi, F., Sardiña, S.: Two-player game structures for generalized planning and agent composition. In: AAAI (2010)
4. Dias, J., Paiva, A.C.R.: Feeling and reasoning: A computational model for emotional characters. In: Bento, C., Cardoso, A., Dias, G. (eds.) EPIA 2005. LNCS (LNAI), vol. 3808, pp. 127–140. Springer, Heidelberg (2005)
5. Dignum, F., Hofstede, G.J., Prada, R.: From autistic to social agents. In: AAMAS, pp. 1161–1164 (2014)
6. van Ditmarsch, H., van der Hoek, W., Kooi, B.: Dynamic Epistemic Logic, 1st edn. Springer Publishing Company, Incorporated (2007)
7. Hall, L.E., Woods, S., Aylett, R., Paiva, A.: Using theory of mind methods to investigate empathic engagement with synthetic characters. I. J. Humanoid Robotics 3(3), 351–370 (2006)
8. Johnson, M., Bradshaw, J.M., Feltovich, P.J., Jonker, C.M., van Riemsdijk, M.B., Sierhuis, M.: Coactive design: Designing support for interdependence in joint activity. J. of Human-Robot Int. 3(1), 43–69 (2014)
9. Lemaignan, S., Ros, R., Mösenlechner, L., Alami, R., Beetz, M.: Oro, a knowledge management module for cognitive architectures in robotics. In: IROS 2010 (2010)
10. Pearce, A., Sonenberg, L., Nixon, P.: Toward resilient human-robot interaction through situation projection for effective joint action. In: Robot-Human Teamwork in Dynamic Adverse Environment: AAAI Fall Symp., pp. 44–48 (2011)
11. Pfau, J., Kashima, Y., Sonenberg, L.: Towards agent-based models of cultural dynamics: A case of stereotypes. In: Perspectives on Culture and Agent-based Simulations, pp. 129–147. Springer (2014)
12. Ronald Fagin, Y.M., Halpern, J.Y., Vardi, M.Y.: Reasoning about Knowledge. MIT Press, Cambridge (1995)
13. Scassellati, B.: Theory of mind for a humanoid robot. Auton. Robots 12(1), 13–24 (2002)
14. Stocker, R., Dennis, L., Dixon, C., Fisher, M.: Verifying brahms human-robot teamwork models. In: del Cerro, L.F., Herzig, A., Mengin, J. (eds.) JELIA 2012. LNCS, vol. 7519, pp. 385–397. Springer, Heidelberg (2012)
15. Warnier, M., Guitton, J., Lemaignan, S., Alami, R.: When the robot puts itself in your shoes. managing and exploiting human and robot beliefs. In: Proc. of the 21th IEEE Int. Symp. in Robot and Human Interactive Communication (2012)

To Beep or Not to Beep
Is Not the Whole Question

Kerstin Fischer[1], Lars Christian Jensen[1], and Leon Bodenhagen[2]

[1] University of Southern Denmark
Institute for Design and Communication
Alsion 2, 6400 Sonderborg
[2] University of Southern Denmark
Maersk-McKinney-Moller Institute
Niels Bohrs Allé, 5230 Odense M
kerstin@sdu.dk,
larje09@student.sdu.dk,
lebo@mmmi.sdu.dk

Abstract. In this paper, we address social effects of different mechanisms by means of which a robot can signal a person that it wants to pass. In the situation investigated, the robot attempts to pass by a busy, naïve participant who is blocking the way for the robot. The robot is a relatively large service robot, the Care-o-bot. Since speech melody has been found to fulfill social functions in human interactions, we investigate whether there is a difference in perceived politeness of the robot if the robot uses a beep sequence with rising versus with falling intonation, in comparison with no acoustic signal at all. The results of the experimental study (n=49) shows that approaching the person with a beep makes people more comfortable than without any sound, and that rising intonation contours make people feel more at ease than falling contours, especially women, who rate the robot that uses rising intonation contours as friendlier and warmer. The exact form of robot output thus matters.

Keywords: Human-robot interaction, attention getting, acoustic signals, social spaces, intonation.

1 Introduction

As robots increasingly leave the labs and become more present in everyday situations, the need to ensure that they interact smoothly with naïve, unsuspecting users increases as well. For instance, if a service robot drives around in a care institution, it will encounter doctors, nurses, patients, as well as visitors, relatives, and service personnel. One of the possible encounters is when the robot drives along a corridor or narrow space and approaches a person who does not see it coming.

In similar situations between humans, when a person is unknowingly blocking the way for another, various techniques are used, and the practices employed may differ slightly according to the cultural background of the participants. A big role is played by eye contact, by means of which people indicate to each other that they perceive each other and also in which direction they are heading [7]. In case one person approaches another from behind, speech and body contact are used

M. Beetz et al. (Eds.): ICSR 2014, LNAI 8755, pp. 156–165, 2014.
© Springer International Publishing Switzerland 2014

to different degrees in different cultures, such as saying "excuse me," tapping the other on the shoulder or pushing him or her aside a little. A big concern in these kinds of encounters is politeness; safety is not much at issue as long as people are moving 'normally', but generally people invest some interactional effort into creating and maintaining social relationships in encounters like the one under consideration. Thus, they will not only try to reach their goal, to be able to pass, but they will also attend to politeness [1], for instance, by minimizing the imposition (e.g. "could you move just a tiny bit to the side?") or by asking instead of ordering (e.g. "may I possibly pass behind you?"). Intonation plays a role in such situations since especially rising intonation is associated with openness towards the other and thus potentially with politeness [15].

To sum up, in interactions between humans, encounters in which one participant needs the other's collaboration in order to pass through constitute interactional problems that are solved with interactional effort especially regarding the maintenance of politeness. The question addressed in this study is how a robot could initiate such an encounter, and what kinds of behaviors are suitable to a) draw a person's attention to its presence and b) to communicate its intention to pass through in socially acceptable ways.

2   Previous Work

Few studies have so far systematically addressed how robots can get a person's attention when the person is occupied; robots are still generally confined to laboratory settings, or, if they are brought 'into the wild', then these are mostly contexts in which people are in exploration mood and not involved in particular activities themselves, like, for instance, in museums or at science or technology exhibitions. Accordingly, there is as yet not much knowledge on how a robot can get a busy user's attention and possibly even ask for a favor. One such study is Hüttenrauch and Severinson-Eklundh [6] who had a robot request naïve participants for help. The authors argue that acoustic signals, such as beeps, are effective means to get people's attention. Their results also show that whether people are engaged in an activity themselves has a considerable impact on their willingness to attend to the robot.

Furthermore, several studies by Sidner and colleagues (e.g. [12,13]) address how a zoomorphic robot can initiate interactions with people. They distinguish between different situations depending on whether users are already perceived to be attending, perceived to be non-attending or only suspected to be present. The authors report that methods based on eye-gaze may not always be effective [12] and suggest to combine the robot's initiative with an acoustic signal (speech) or a gesture.

In a recent study, Fischer and colleagues [2] investigated attention getting by means of speech in comparison to attention getting by means of a beep. We found that busy people may react only reluctantly to the robot's gesture and may not respond to a robot beeping at all; in contrast, all participants reacted to the robot's use of speech by looking at the robot.

To sum up, these studies suggest that an acoustic signal may be more effective than just the use of gaze and gesture to get a person's attention, yet that social signals such as speech may still be more effective than a beep sequence – this, in turn, suggests that socially relevant forms such as intonation contours may influence the social perception of beep sequences produced by a robot. Since a robot that uses speech implicitly suggests that it also understands speech, a beep sequence may still be preferable, and it is open whether the melody of the beep sequence may turn the acoustic signal into a social signal as well.

A second area of previous work concerns research on comfortable distance and direction of approach; one such study is Walters et al. [16], which addresses the role of the direction from which the robot should approach. The authors find that people do not like the robot to approach from the back and prefer it to come from either side. Furthermore, they find that people's degree of comfort when the robot approaches also depends on whether they are sitting or standing. Concerning distance, Yasumotu et al. [17] investigate people's preferences regarding the distance to the humanoid robot Asimo. Participants were asked to indicate when the robot should not come closer, which on the average was at about 78cm. However, the authors tested only trajectories towards the person from the front.

Mumm and Mutlu [9] investigate the effects of likeability of the robot (manipulated by means of polite versus rude introductory statements by the robot) and eye gaze (mutual versus averted) on the physical distance by means of which participants move around the robot. They find that in the mutual gaze condition, participants increased the distance to the robot; similarly, if participants disliked the robot, they also increased the physical distance. However, if participants liked the robot, gaze had no influence on physical distance. These results correspond to earlier findings by Takayama and Pantofaru [14] who report negative attitudes towards robots as well as the personality trait of neuroticism to be the best predictors of the distance people place themselves in when interacting with a robot. Furthermore, experience with robots and pet-ownership make people decrease the distance to a robot.

All of these studies show that distance is determined by a complex set of variables, and that the robot should be equipped with capabilities to negotiate how close it is allowed to come. However, none of these studies report on how close the robot should come when approaching a person from behind and when it actually needs to pass through. Furthermore, whether the distance is mitigated by acoustic signals is also open.

Finally, there is some relevant previous work on the kinds of sounds robots should produce. In particular, Read and Belpaeme [10] put a beep sequence into different interactional contexts, and people understood the same beep sequence as having very different meanings, depending entirely on the context in which it occurred. This finding suggests that a beep series itself should not have an effect on how people perceive a robot. However, if a robot approaches from behind, some kind of acoustic signal may be necessary to warn a user that a robot is approaching since other modes of communication are not available (e.g. gesture or eye contact, see, for instance, [5] and [12]). While speech seems to be the most effective strategy (see [2]), previous work on intonation suggests a considerable role of intonation contours, i.e. the melody of an utterance or beep sequence. For instance, Tench (1996: 105) summarizes: "a fall indicates the speaker's dominance in knowing and telling something, in telling someone what to do, and in expressing their own feelings; a rise indicates a speaker's deference to the addressee's knowledge, their right to decide, and their feelings" [15]. If these cues are consistently relevant in natural language exchanges between people, it can be expected that they will have an effect on how people will respond to robot beeps as well.

3 Hypotheses

The aim of this study is to determine the effects of different kinds of robot behaviors when the robot is approaching a person from behind. In particular, we test the effects of beep sounds versus no sound at all. Given the effects of speech melody on people's perception of another person's politeness, we hypothesized that a robot that uses a beep sequence with rising intonation contour will be perceived as friendlier and more polite than a robot that uses a beep sequence with falling intonation contours.

4 Methods

The experiment is an empirical study of HRI in three conditions, comparing social effects of robot behavior without sound and with two different beep sequences as stimuli.

4.1 Stimuli

The stimuli were generated on the basis of natural language utterances: First, the phrase "excuse me, please" was recorded with two different intonation contours, rising and falling. Since rising intonation contour generally signals openness and hearer involvement, in connection with the phrase "excuse me, please", it results in a very polite utterance that is best understood as a friendly request. Falling intonation contours are generally associated with statements and the assertion of facts. In connection with "excuse me, please," it serves rather to signal that a disturbance is going to take place. It is thus less polite.

In a second step, the natural language utterances of "excuse me, please" with the two different intonation contours were analyzed and the intonation contours used visualized using Praat.

Third, based on the intonation contours used in the two utterances, two beep series were created. In particular, to synthesize the beep sequence, a tone generator which is part of the free software package Audacity was used. The intonation contours of the two utterances of "excuse me, please" were simplified in such a way that they were taken as two sequences of four beeps, corresponding to each syllable, which are five semi-tones apart respectively, yielding ♪♪♪♪ in the rising contour condition ♪♪♪♪ in the falling contour condition.

4.2 Robot

The robot used is the Care-O-bot 3 (see Fig 3), developed by Graf and colleagues [3]. The robot is approximately 1.4 meters tall and composed of a 4-wheeled omnidirectional base, manipulator with a 3-finger gripper attached (SDH2) and a tray with 1 DOF that can be either in front of the robot or at its side. During the experiments, the configuration of the robot was kept constant, with the tray being in front, the torso being in an upright configuration and the gripper being at the side of the robot. All motions of the robot were controlled remotely by a hidden human operator.

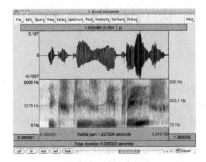
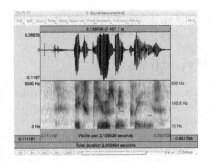

Fig. 1. The speech signal of 'excuse me, please' with rising intonation contour (blue line in the lower half of the window)

Fig. 2. The speech signal of 'excuse me, please' with falling intonation contour (blue line in the lower half of the window)

4.3 Procedure

Participants were greeted and asked to fill out a consent form in another room. Then, they were led into the experiment space and first asked to step in front of a white-board so that a picture of them with their participant number could be taken. While the participant was standing close to the white-board, the experimenter picked up a questionnaire and began asking the initial, demographic, questions of the questionnaire, positioning the participant implicitly with the back to the hallway through which the robot would have to pass. While she was asking the questions, the robot drove up behind the participant, who was facing the experimenter (see Fig 6). In conditions 2 and 3, the robot used one of two beep series when approaching (approximately 50cm before reaching the subject). In condition 1, the robot only attempted to drive through and relied on participants' peripheral vision or their perception of the robot's engine sounds to notice its presence.

Participants were videotaped for their behavioral responses if they had agreed to being recorded. Since some participants did not notice the robot when it was coming close, and some participants did not notice the beep, either, the experimenter, once the robot had passed, put the consent forms onto the robot's tray. This ensured that all participants had seen the robot before asked to fill out the questionnaire.

Then, they were handed the questionnaire the experimenter had begun to read to them, saying, "oh, actually you can fill this out yourself". After filling out the questionnaire, participants were asked to take part in a second study, at the end of which they were de-briefed.

4.4 Participants

We recruited 49 participants from the technical faculty of the University of Southern Denmark who were either undergraduates (49%), graduates (24.5%), faculty members (22.4%) or non-academics who had an affiliation with the university (4.1%). Even though most participants were students between the ages 20 and 40 (81.6%) and with an overrepresentation of males (71.4%), most participants had only little previous experiences with robots. 38.8% had worked or

Fig. 3. The Care-O-Bot 3

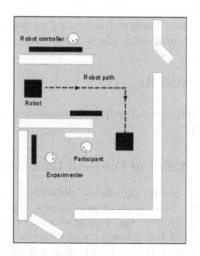

Fig. 4. Map of the experiment area

	Male	Female	Total
No Sound	11	4	15
Rising	11	6	17
Falling	12	4	16

Fig. 5. Participants per conditions

Fig. 6. The robot attempts to pass by

played with one, 36.6% had seen one or a few, 14.3% knew robots only from TV, while a minority (10.2%) worked with robots regularly.

4.5 Questionnaire

The first part of the questionnaire consisted of demographic questions and questions concerning people's prior experience with robots. This part was carried out as an interview until the robot had passed. The rest of the questionnaire concerned three aspects (based on the questionnaire developed in [4]): First, we asked for participants' perception of the robot's capabilities, politeness and other characteristics. Second, we asked to what degree people ascribe certain human characteristics to the robot. Third, we asked for participants' own feelings while they encountered the robot. People could mark their choices on a 7-point Likert scale.

4.6 Analysis

The statistical analysis of the questionnaire results (one- and two-way ANOVA) was carried out using the statistical software package SPSS. The behavioral data were analyzed according to when participants noticed the robot: when it was approaching, when it was close, or after it had passed.

5   Results

We first report the questionnaire results and then the results from the behavioral analysis.

5.1   Effect of Different Intonation Contours

There are no significant differences in the way the participants perceived the robot. That is, contrary to our expectations, the three conditions no sound, rising contour and falling contour, did not produce significantly different judgments of the robot's capabilities or politeness. In particular, participants were asked to rate the robot on a 7-point semantic differential scale on the following characteristics: Appeal, intelligence, competence, subordination/superiority, safety, approachability, confidence, friendliness and cooperativeness. None of the characteristics yielded any statistical differences between the conditions.

Similarly, participants do not generally ascribe different personality characteristics to the robot depending on the condition. Participants were asked to rate the robot on eight adjectives (cheerful, kind, likable, aggressive, assertive, bigheaded, harsh and rude) on a 7-point Likert scale ranging from 'describes poorly' (1) to 'describes very well' (7). Univariate ANOVA testing yielded one near-significant difference for the adjective *assertive*: no sound (M = 3.67, SD = 1.11), rising contour (M = 2.56, SD = 1.32) and falling contour (M = 3.20, SD = 1.32) [F(2,43) = 4.77, p = 0.059]. Tukey post-hoc pairwise comparison shows that this effect is due to a significant difference between the no sound and the rising intonation conditions.

However, participants' evaluation of their own states differs significantly between conditions: When asked to rate their own feelings on seven adjectives (angry, comfortable, cooperative, relaxed, uncomfortable, warm, afraid) on a 7-point Likert scale when the robot approached, statistically significant differences were observed for the degree with which they felt uncomfortable; no sound (M = 1.53, SD = 0.74), rising contour (M = 1.94, SD = 1.25) and falling contour (M = 2.88, SD = 2.00) [F(2,45) = 3.61, p = 0.035]. Post-hoc tests show that this effect is due to a significant difference between the no sound and the falling intonation conditions. Similarly, the reverse question concerning the degree with which they felt comfortable reaches near-significance; no sound (M = 4.53, SD = 1.89), rising contour (M = 5.29, SD = 1.61) and falling contour (M = 3.88, SD = 1.41) [F(2,45) = 3.10, p = 0.055]. In this case, Tukey post-hoc comparisons reveal that this effect is due to a significant difference between the rising and the falling intonation conditions.

5.2   Interpersonal Differences

Univariate ANOVA shows very different ratings depending on participants' gender: The robot was in general rated more intelligent by women (M = 4.38, SD = 0.77) than by men (M = 3.38, SD = 1.45) [F(1,43) = 5.58, p = 0.023]. We find similar differences for competency; men (M = 3.44, SD = 1.48) rated the robot less competent than women (M = 4.67, SD = 0.65) [F(1,42) = 7.63, p = 0.008]. Moreover, regarding subordination/superiority, men (M = 2.44, SD = 1.44) and women (M = 3.54, SD = 1.13) differ significantly [F(1,43) = 6.09, p = 0.018] such that women understand the robot to be more superior. In addition, regarding confidence, men (M = 3.56, SD = 1.56) ascribe less confidence to the robot than women (M = 4.54, SD = 1.13) [F(1,43) = 4.16, p = 0.048]. Furthermore, a comparison between men and women reveals that men (M = 2.4, SD = 1.7) find the situation more uncomfortable than women (M = 1.43, SD = 0.76) [F(1,46) = 4,52, p = 0.039]. Finally, women found the robot more aggressive (M = 2.36, SD = 1.74) than men (M = 1.48, SD = 0.87) [F(1,45) = 5.30, p = 0.026].

In addition to the gender differences, also differences for participants of different age were found. In particular, participants who were older were also more likely to be more afraid of the robot; participants below the age of 20 rated the robot as more friendly (M = 1.00, SD = 0.00) than participants between 20 and 40 (M = 1.41, SD = 0.82) and than participants between 40 and 60 (M = 2.67, SD = 2.34) [F(2,45) = 3.90, p = 0.027].

5.3   Interactions between Condition and Gender

We furthermore observe two significant interactions between gender and condition for likeability, such that female participants rate the robot significantly more likeable if the robot uses a beep with rising intonation contour (F(2,44) = 3.711, p =.033; see Fig 7). Similarly, women felt significantly less warmth if the robot does not produce any sound (F(2,46) = 5.698, p =.007; see Fig 8).

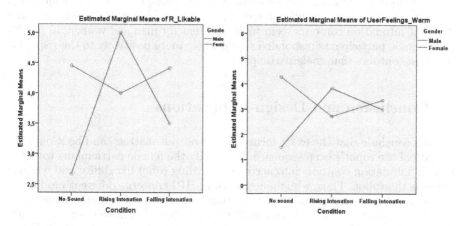

Fig. 7. Likeability by condition and gender

Fig. 8. Warmth by condition and gender

5.4 Behavioral Analysis

We analyzed the videos on the basis of obvious signs of attention to the robot, such as turning the head or stepping out of the way. The analysis[1] shows that 8 of the participants noticed the robot already when it was approaching, 15 noticed it when it was right behind them, and 14 only responded to it when it had passed. Whether the robot beeped or not did not influence people's attention when the robot was approaching such that even fewer participants noticed the robot approaching when it was beeping than when it played no sound. However, once the robot was close, more participants in the rising and falling intonation conditions noticed the robot than in the silent condition. This difference is marginally significant (Chi-square $(6,N=40) = 12.197$; $p = .058$).

6 Discussion

While beeps, irrespective of their intonation contour, do not seem to work well as attention getters (which is in accordance with previous findings [2]), our results suggest that people feel less uncomfortable around the robot if it approaches them using a beep than without producing any acoustic signal. Moreover, the results show that the melody of the beep sequence plays a role for participants' level of comfort: rising contours make people feel more at ease than falling contours. This is according to our predictions, which were based on natural language interactions between humans. However, the effects observed do not extend to the characteristics ascribed to the robot.

We observed consistent gender differences on participants' ratings of the robot's competence, characteristics and on their relationship with the robot. While gender effects have been observed in previous work (see, for instance, [8], [9], [14] and [11]), we were nevertheless rather surprised concerning the extent to which men and women were found to differ in the current study. Moreover, the effects of the different intonation contours seem to be different for men and women; in particular, female participants responded significantly more positively to the rising intonation contours than male participants.

7 Conclusion and Design Implications

We can conclude that the exact form of robot output matters, and be it only the melody of the robot's beep sequence. Especially the female participants took the robot's intonation contours into account regarding robot likeability and warmth towards the robot. Thus, while much work in HRI concerns different modalities in which interaction takes place, the actual form of such interactions may play a role regarding the social acceptability of a robot. The results of this study suggest that social aspects of acoustic human-robot interaction may be relevant and need to be attended to in robot behavior design, especially since societal development suggests that most of the potential users of service robots are likely to be women.

---

[1] We have video data on only 40 participants as not all consented to be recorded.

Acknowledgements. We are grateful to Franziska Kirstein, Maria V. aus der Wieschen, Maria Aarestrup, Nathalie Schümchen, Janni Jensen and Julia Ruser for their help during the experiments. This work was partially supported by the Danish project Patient@home which is funded as a strategic platform for innovation and research by the Danish Innovation Fond.

References

1. Brown, P., Levinson, S.: Politeness: Some Universals in Language Use. Cambridge University Press (1987)
2. Fischer, K., Soto, B., Pontafaru, C., Takayama, L.: The Effects of Social Framing on People's Responses to Robots' Requests for Help. In: Ro-man 2014, Edinburgh (2014)
3. Graf, B., Reiser, U., Hägele, M., Mauz, K., Klein, P.: Robotic home assistant Care-O-bot3 - product vision and innovation platform. In: IEEE Workshop on Advanced Robotics and its Social Impacts (ARSO), pp. 139–144 (2009)
4. Groom, V., Takayama, L., Ochi, P., Nass, C.: I am my robot: the impact of robot-building and robot form on operators. In: HRI 2009 (2009)
5. Hoque, M.M., Das, D., Onuki, T., Kobyashi, Y., Kuno, Y.: Model for Controlling a Target Human's Attention in Multi-Party Settings. In: Ro-man 2012, Paris, France, pp. 476–483 (2012)
6. Hüttenrauch, H., Severinson-Eklundh, K.: To Help or Not to Help a Service Robot. In: Ro-man 2003, Millbrae, CA, pp. 379–384 (2003)
7. Kendon, A.: Conducting Interaction: Patterns of Behavior in Focused Encounters. Cambridge University Press (1990)
8. Kriz, S., Anderson, G., Trafton, J.G.: Robot-directed speech: using language to assess first-time users' conceptualizations of a robot. In: HRI 2010, New York, NY, pp. 267–274 (2010)
9. Mumm, J., Mutlu, B.: Human-Robot Proxemics: Physical and Psychological Distancing in Human-Robot Interaction. In: HRI 2011, Lausanne, Switzerland (2011)
10. Read, R., Belpaeme, T.: Situational Context Directs How People Affectively Interpret Robotic Non-Linguistic Utterances. In: HRI 2014, Bielefeld, Germany (2014)
11. Schermerhorn, P., Scheutz, M., Crowell, C.R.: Robot social presence and gender: Do females view robots differently than males? In: HRI 2008, pp. 263–270 (2008)
12. Sidner, C.L., Lee, C., Kidd, C.D., Lesh, N., Rich, C.: Explorations in Engagement for Humans and Robots. Artificial Intelligence 166(1-2), 140–164 (2005)
13. Sidner, C., Lee, C.: Attentional Gestures in Dialogues between People and Robots. In: Nishida, T. (ed.) Engineering Approaches to Conversational Informatics. Wiley and Sons (2007)
14. Takayama, L., Pantofaru, C.: Influences on proxemic behaviors in human-robot interaction. In: IROS 2009, pp. 5495–5502 (2009)
15. Tench, P.: The Intonation Systems of English. Cassell, London (1996)
16. Walters, M.L., Dautenhahn, K., Woods, S.N., Koay, K.L.: Robotic Etiquette: Results from User Studies Involving a Fetch and Carry Task. In: HRI 2007, pp. 317–324 (2007)
17. Yasumoto, M., Kamide, H., Mae, Y., Kawabe, K., Sigemi, S., Hirose, M., Arai, T.: Personal space of humans in relation with humanoid robots depending on the presentation method. In: IEEE/SICE International Symposium on System Integration, pp. 797–801 (2011)

Perception of an Android Robot in Japan and Australia: A Cross-Cultural Comparison

Kerstin Sophie Haring[1], David Silvera-Tawil[2], Yoshio Matsumoto[3],
Mari Velonaki[2], and Katsumi Watanabe[1]

[1] Research Center for Advanced Science and Technology,
The University of Tokyo, Japan
{ksharing,kw}@fennel.rcast.u-tokyo.ac.jp
[2] Creative Robotics Lab, The University of New South Wales, Australia
{d.silverat,mari.velonaki}@unsw.edu.au
[3] National Institute of Advanced Industrial Science and Technology, Japan
yoshio.matsumoto@aist.go.jp

Abstract. This paper reports the results from two experiments, conducted in Japan and Australia, to examine people's perception and trust
towards an android robot. Experimental results show that, in contrast to
popular belief, Australian participants perceived the robot more positive
than Japanese participants. This is the first study directly comparing
human perception of a physically present android robot in two different
countries.

Keywords: Android robot, cross-cultural, human-robot interaction,
robot perception, trust.

1 Introduction

It is apparent that recent technological advances will soon enable robots to
live amongst humans; robots will be present in workplaces, schools, hospitals,
shops, homes, etc. As the number of interactions between humans and physically
present robots increases, it is important to examine the impact of these robots
during the interaction. Current research in human-robot interaction (HRI) faces
significant challenges, not only in terms of technological improvements but also
in terms of social acceptability of robots. It is believed that the social aspects of
interactive robots could be at least similar to those of humans [1].

Human perception of robots has been generally shaped by information obtained through social media (e.g. movies, newspaper, internet, etc.) and not by
real interactions with physically present robots. In spite of significant research in
HRI, direct contact to a physically present robot is still the exception rather than
the norm. It has been shown, however, that the presence of an embodied robot
plays a crucial role in the way people perceive it [2]. Previous studies also revealed that the expectations and attitude towards robots change based on their
appearance [3]. To accurately evaluate the perception of robots, participants
should ideally be in direct contact with physically present robots [2, 4].

M. Beetz et al. (Eds.): ICSR 2014, LNAI 8755, pp. 166–175, 2014.

Android robots, a specific type of robots designed to look and act like humans, have been reported to trigger different reactions from people when compared to other robot types such as pet-like robots and humanoid robots [5]. The objective of this research is to measure and compare human trust, perception and attitudes towards a physically present android robots in two different countries, Japan and Australia.

Changes in the participants' trust and general perception before and after interacting with the robot were measured and correlated with the participants' personality traits. This paper extends a previous experiment in trust and perception performed exclusively with Japanese participants [6]. A cross-cultural comparison with a total of 111 participants is presented.

1.1 Literature Review

For years, science has studied how attributes such as nationality, religion, race and socioeconomic class influence the way people think and behave. The country of origin of two people, for example, could have a strong influence on the distance kept between them during social interaction [7]. According to resent research [8], even facial expression recognition is culturally dependent.

It is commonly believed that robots are perceived differently by Eastern and Western cultures. American movies such as "The Terminator" and "I, Robot", for example, present robots with negative connotations towards them and displays them as threatening technology or machines out of control. The Frankenstein complex [9] even describes people's anxiety towards robots as a representation of their fear towards technological creatures that could threat humankind. This behaviour is not observed in Eastern cultures, such as Japan, where robots are displayed as heroes or helpers (e.g. "Astro boy", "Doraemon"). It has been speculated that the Japanese holistic understanding, that is, the notion that living beings, non-living objects and gods are all ascribed to have a soul, might be a basis for this attitude [10]. This stereotype, however, is not necessary true. Robotic heroes are also present in Western culture, for example in movies such as "Star Wars" and "Wall-E", while previous studies revealed that Japanese people are not "robot lovers" while Western cultures are not "robot haters" [11, 12].

Recent studies in HRI, demonstrated that people's behaviour towards robots might vary across cultures. Wang et al. [13], for example, reported that Chinese and American participants are more likely to heed recommendations when robots behave in more culturally normative ways, while Chinese participants expressed a more negative attitude towards the robot. Trovato et al. [14], furthermore, found that Egyptians prefer an Arabic speaking robot and feel a sense of discomfort when interacting with a Japanese robot. Opposite feelings were observed in the Japanese participants. A different study [15] suggests that Egyptian participants perceive a receptionist robot more positively and more anthropomorphic than English-speaking participants. When comparing Chinese, Korean and German participants [16], it was found that cultural differences exist in participant's perception of likeability, engagement, trust and satisfaction. Cultural differences were also found when children of different age groups interacted with the iCat robot in a card game where children from Pakistan were much more expressive than Dutch children [17].

In contrast, Shibata et al. [2] found no difference between participants from Japan and the UK when answering a questionnaire about the seal robot "Paro", but found that physical interaction improved the subjective evaluation. A study evaluating the differences in attitude towards robot showed no differences between Japanese, Chinese and Dutch participants [18] and a comparison of explicit and implicit attitudes towards robots between Japanese and American participants showed multiple similarities [11].

Altogether, previous research suggests that cultural differences exist in certain areas of robot perception and outline the importance of a direct interaction between people and a physically present robot, but do not confirm the stereotypes of the Japanese culture generally having a more positive attitude towards robots.

2   Methodology

The experiments in Japan and Australia followed the same four-staged procedure using a female version of an android robot, Actroid-F (Fig. 1). To evaluate if factors such as prior experiences with robots, prior relationships with non-human agents such as pets [19], and the participants' personality [20] would influence the interaction with the robot, participants demographics, personality traits, and perception of the robot were evaluated in the first stage of the experiments. In addition, participants were asked if they had ever owned a pet (yes/no), and if they had been previously exposed to either virtual agents or robots (on a 5-point scale).

Fig. 1. The Actroid androids in the male (left) and female (right) versions. This experiment used the Actroid-F, the female version of the Actroid robots.

During the second stage, three simple interaction tasks with the robot were implemented. During task One and Two, the robot asked each participant to move a box from one position to another. For the third task, it asked them to touch its hand. During these tasks, the robot engaged with the participants following a fixed protocol in either Japanese or English (i.e. greeting, asking for name and participant number) and then gave the instructions for each task. Additionally, the robot asked participants to take a chair positioned at the far end of the room, and move it to the location where they wanted to sit during the task. When the task was completed, the robot gave each participant the opportunity to ask some open-ended questions, after which it thanked them for their cooperation and asked them to wait outside the room. The researcher returned the chair to the far end of the room at the end of each task.

To evaluate the participants' trust towards the robot, during the third and final stage, an economic trust game [21] took place. An economic trust game allows to quantify trust in a relationship in an empirical, reliable and standardized way. In this case, an economic trust game was 'played' between the robot and participants in a similar context to that used in human-human interaction. In the two-player trust game, player One (the participant) is provided with a fixed amount of money (JPY 1000 in Japan and AUD 5 in Australia) and given the opportunity to send all, or part of the money to player Two (the robot). The robot would then return either more, or less money to the participant. The researcher randomly assigned the returning amount as more or less, with the only condition being that the same number of participants were paid either more or less money.

To evaluate changes in participants' perception of the robot, the questionnaires were administered before and after the interaction tasks with the robot. All experiments were video recorded for analysis.

2.1 Questionnaires

Personality Questionnaire: The Eysenck Personality Questionnaire Revised (EPQ-R) categorizes personalities in a systematic way, using the three factors of psychoticism, extraversion and neuroticism. It is also one of the few personality questionnaires that are validated in Japanese [22] and English.

Robot Perception Questionnaire: To evaluate human perception of the robot, the Godspeed Questionnaire [23] was used. The Godspeed Questionnaire measures five key concepts in HRI using 5-point scales. (1) Anthropomorphism is the attribution of a human form and characteristics to anything other than a human being. (2) Animacy is the perception of the robot as a lifelike creature. Perceiving something as alive allows humans to distinguish humans from machines. (3) Likeability describes the first (positive) impression people form of others. Research suggests [24] that humans treat robots as social agent and therefore judge them in a similar way. (4) Perceived intelligence states how intelligent and human-like participants judge the behavior of the robot. (5) Perceived safety describes the perception of danger from the robot during the interaction and the level of comfort the participants' experience.

2.2   Additional Measurements

The distance kept by participants to the robot during each task was measured at floor level from the robot's feet to the participants' chair, baring in mind that the position of the chair was chosen by each participant (Sec. 2). Note that the distance for the third task was measured before the robot's request to touch its hand.

3   Experimental Results

A total of 111 participants from the University of New South Wales, Australia and universities of Tokyo, Japan took part in these experiments (Table 1). Participants were recruited through general advertisement using posters across both universities, email lists from researcher with no direct contact with students and through word of mouth. None of the participants had previous experience interacting with android robots. Participants received monetary reimbursement (approximately AUD 5) for their participation.

Table 1. Participant demographics for Australia and Japan. The mean exposure to robots and virtual agents results from a 1-5 rating scale.

	Australia	Japan
Total	56	55
Female	35	37
Male	21	18
Mean Age	28.8	22.6
Mean exposure to robots	3.9	3.72
Mean exposure to Virtual Agents	2.5	2.43

3.1   General Cross-Cultural Differences

There were several differences between the datasets from Australian and Japanese participants. Australians had a higher pet ownership (Chi square test; $p<0.001$) and had higher psychoticism ($t(107.92) = -2.96$, $p = 0.003$) and extraversion ($t(102.92) = 5.47$, $p<0.001$) scores. Furthermore, Japanese participants came significantly closer to the robot in each consecutive task (Table 2; task 1 vs. task 2 $t(54) = 4.87$, $p = 0.001$; task 2 vs. task 3 $t(54) = 2.67$, $p = 0.05$; Bonferroni corrected, as reported in the previous study [6]). However, this effect was not observed in the Australian participants.

3.2   Changes in Human Perception of the Robot

Anthropomorphism: Lower anthropomorphism ratings were observed after the interaction for participants in both countries: $t(53) = 4.22$, $p<0.001$ for Japan and $t(55) = 2.50$, $p = 0.01$ for Australia. This means that in both cases

Table 2. Mean distances (in cm) to the robot for Australia and Japan

	Task 1	Task 2	Task 3
Australia	123.8	121.5	122.7
Japan	128.2	119.9	116.1

the perception of anthropomorphism of the android reduced significantly after the interaction. Furthermore, anthropomorphism was rated significantly higher in Australia—when compared to Japan—after the interaction ($t(108.7) = 1.9$, $p = 0.05$), but not before (Fig. 2(a)).

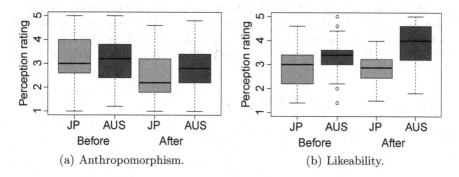

(a) Anthropomorphism.   (b) Likeability.

Fig. 2. Anthropomorphism and likeability for Japan (yellow) and Australia (green). The left plot shows a decrease in anthropompohism for both countries while the right plot shows an increase in likeability only for Australia.

Animacy: Animacy rating did not significantly differ between countries and there were no significant changes as a result of the interaction either in Japan or Australia.

Likeability: Significant differences were found in the likeability rating of the robot before, as well as after the interaction task (Fig. 2(b)). Australian participants liked the robot significantly more than Japanese participants. Before the interaction, Australians rated the robot more likeable ($t(107.91) = 3.48$, $p<0.001$) after the interaction, the likeability of the robot even increased in Australia and remained the same in Japan.

Perceived Intelligence: Perceived intelligence dropped significantly in the Japanese participants ($t(53) = 7.55$, $p<0.001$) after the interaction whilst there was no significant change for the Australian participants. There was a significant difference between Australia and Japan after the interaction task ($t(92.83) = 6.10$, $p<0.0001$), with Australian participants rating the perceived intelligence significantly higher.

Perceived Safety: Perceived safety increased after participants interacted with the robot. For both cultures, ratings for perceived safety increased after the interaction tasks: $t(53) = -1.99$, $p = 0.05$ for Japan and $t(55) = -3.97$, $p = 0.0002$ for Australia. Even though the same trend was observed in both countries, the overall ratings were significantly lower in Australia before ($t(104.46) = 3.02$, $p = 0.003$) and after ($t(98.89) = 2.11$, $p = 0.03$) the interaction.

3.3   Economic Trust Game

Previous research has shown that extravert personality types tend to send higher amounts of money during an economic trust game [25]. In the current experiments, Australian participants entrusted the robot with a significant higher amounts than Japanese participants ($t(109) = 4.02$, $p = 0.0001$). At the same time, the Australian dataset shows a higher rate of extraversion ($t(102.74) = 5.5458$, $p<0.0001$). Further analysis, however, shows that extraversion affected the payback amount in the trust game only in Japan (positive correlated, $R = 0.43$, $t(44) = 3.12$, $p = 0.003$), but not in Australia ($R = -0.09$), see Fig. 3.

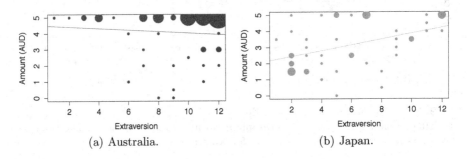

(a) Australia.   (b) Japan.

Fig. 3. The amount paid (exchanged in AUD) as a function of extraversion score in the economic trust game for Australia (left) and Japan (right). Disk sizes represent the number of participants. Australian participants show higher amounts paid but no correlation with extraversion score, while Japanese participants show an increase of the payback with increasing extraversion score.

Furthermore, a correlation with no-pet ownership and robot perception when the payback was lower or higher was found in Japan, but no significant differences were observed in Australia. Other character traits showed no further correlations with the amount send in the trust game in either country.

4   Discussion

This paper reports the cross-cultural comparison of trust and robot perception between Japan and Australia using Actroid-F, an android robot designed to look as an exact copy of a Japanese female.

Experimental results showed that Japanese participants rated the robot lower than Australian participants for anthropomorphism, animacy, likeability and perceived intelligence before interacting with it. This contradicts the stereotype of Western cultures to reject robots and Japanese being more accepting of them.

In terms of perceived safety, Australian participants seemed more concerned and rated the robot lower than their Japanese counterparts. Although perceived safety increased in both cultures after interacting with the robot, it still remained significantly lower for Australian participants. It is believed that the overall increase by both cultures is a response to the realization that even though the robot looks like a human, its abilities are not human-like and, more importantly, the robot in its current condition is not capable of creating any damage. However the reduced overall ratings are attributed to the negative display of humanoid robots in Western cultures.

In contrast to these results, Australians perceived the robots as more "trustworthy" during the economic trust game. This is an interesting result, because although they perceived it as less safe, they trusted it more when it comes to an economic game. It is suspected that the trust exhibited in this game was partly related on how people perceive the robot from a game theory perspective, in which the 'smart' thing to do is to send higher amounts of money in order to maximize profit. The concept of trust towards a robot, however, even when simplified in an economic game seems to be much more complex.

When analysing the participants' openness for interaction, it was observed that Australian participants were generally more open to the experience and asked the robot several more questions, whereas the Japanese participants asked only 1-2 questions. Australian participants even focused on the robot's "choices" (e.g. favorite color), "dreams" and feelings (e.g. are you able to dream?, how does it feel to be a robot?).

All together, it is concluded that Western cultures might be more curious, interested and open to interact with the android robot but also more careful, explorative and challenging of the robot's limitations.

Finally, this study shows that human perception towards a robot changes after interacting with it for the first time. To date, people have very low exposure to physically present robots in their personal life, and therefore their perception towards them is influenced by media. This, however, is expected to change as the opportunities for interaction with physical present robots increase, and should be taken into account in future HRI studies.

4.1 Future Work

Several additional experiments could be considered. For example, a comparison of the current results using a human interactant, a humanoid or even more machine-like robots, instead of an android. It is expected that people will perceive and approach machine-like robots in a different manner to the android, but humans in more similar ways. The authors expect that future robot design, both in terms of appearance and behaviour, will benefit from better considerations of cultural differences.

Acknowledgement. We thank the Service Robotics Group of the Advanced Industrial Science and Technology (AIST) for the provision of the robot and the technical support. We also would like to thank Toru Hosokawa from the Tohoku University for the copy of the Japanese version of the Eysenck personality questionnaire. This work was supported by Grant-in-Aid for JSPS Fellowship to KSH and the Japan Science and Technology Agency (CREST) to KW.

References

1 Perzanowski, D., Schultz, A.C., Adams, W., Marsh, E., Bugaiska, M.: Building a multimodal human-robot interface. IEEE Intelligent Systems 16(1), 16–21 (2001)

2 Shibata, T., Wada, K., Tanie, K.: Statistical analysis and comparison of questionnaire results of subjective evaluations of seal robot in Japan and UK. In: Proc. IEEE Int. Conf. Robotics and Automation, pp. 3152–3157 (2003)

3 Haring, K., Mougenot, C., Watanabe, K.: The influence of robot appearance on assessment. In: Conf. Basic and Applied Human-Robot Interaction Research (2013)

4 Wainer, J., Feil-Seifer, D., Shell, D., Mataric, M.: The role of physical embodiment in human-robot interaction. In: Proc. IEEE Int Symp. Robot and Human Interactive Communication, pp. 117–122 (2006)

5 Haring, K., Mougenot, C., Watanabe, K.: Perception of different robot design. In: Int. Conf. Knowledge and Smart Technologies. Special session on "Fluency in Communication Between Human, Machine, and Environment" (2013)

6 Haring, K., Matsumoto, Y., Watanabe, K.: Perception and trust towards a lifelike android robot in Japan. Springer Science+Business Media Dordrecht (in press, 2014)

7 Remland, M.S., Jones, T.S., Brinkman, H.: Proxemic and haptic behavior in three european countries. J. Nonverbal Behavior 15, 215–232 (1991)

8 Jack, R., Blais, C., Scheepers, C., Schyns, P., Caldara, R.: Cultural confusions show that facial expressions are not universal. Current Biology 19, 1–6 (2009)

9 Asimov, I.: The Machine and the Robot. Science Fiction: Contemporary Mythology (1978)

10 Imanishi, K.: A Japanese view of nature: the world of living things. Psychology Press (2002)

11 MacDorman, K.F., Vasudevan, S.K., Ho, C.C.: Does Japan really have robot mania? comparing attitudes by implicit and explicit measures. AI & Society 23(4), 485–510 (2009)

12 Haring, K., Mougenot, C., Ono, F., Watanabe, K.: Cultural differences in perception and attitude towards robots. In: Int. Conf. Kansei Engineering and Emotional Research (2012)

13 Wang, L., Rau, P.P., Evers, V., Robinson, B.K., Hinds, P.: When in Rome: the role of culture & context in adherence to robot recommendations. In: Proc. ACM/IEEE Int. Conf. Human-Robot Interaction, pp. 359–366 (2010)

14 Trovato, G., Zecca, M., Sessa, S., Jamone, L., Ham, J., Hashimoto, K., Takanishi, A.: Cross-cultural study on human-robot greeting interaction: acceptance and discomfort by Egyptians and Japanese. J. Behavioral Robotics 4(2), 83–93 (2013)

15 Salem, M., Ziadee, M., Sakr, M.: Marhaba, how I help you?: effects of politeness and culture on robot acceptance and anthropomorphization. In: Proc. ACM/IEEE Int. Conf. Human-Robot Interaction, pp. 74–81 (2014)

16 Li, D., Rau, P.P., Li, Y.: A cross-cultural study: effect of robot appearance and task. Int. J. Social Robotics 2(2), 175–186 (2010)

17 Shahid, S., Krahmer, E., Swerts, M., Mubin, O.: Who is more expressive during child-robot interaction: Pakistani or Dutch children? In: Proc. ACM/IEEE Int. Conf. Human-Robot Interaction, pp. 247–248 (2011)

18 Bartneck, C., Nomura, T., Kanda, T., Suzuki, T., Kennsuke, K.: A cross-cultural study on attitudes towards robots. In: HCI International (2005)

19 Walters, M., Dautenhahn, K., Koay, K.L., Kaouri, C., Woods, S.N., Nehaniv, C.L., te Boekhorst, R., Lee, D., Werry, I.: The influence of subjects' personality traits on predicting comfortable human-robot approach distances'. In: Proc. Cog. Sci., Workshop: Toward Social Mechanisms of Android Science, pp. 29–37 (2005)

20 Takayama, L., Pantofaru, C.: Influences on proxemic behaviors in human-robot interaction. In: Proc. IEEE/RSJ Int. Conf. Intelligent Robots and Systems, pp. 5495–5502 (2009)

21 Berg, J., Dickhaut, J., McCabe, K.: Trust, reciprocity, and social history. Games Econ. Behav. 10(1), 122–142 (1995)

22 Hosokawa, T., Ohyama, M.: Reliability and validity of a Japanese version of the short-form eysenck personality questionnaire-revised. Psychological Reports 72(3), 823–832 (1993)

23 Bartneck, C., Kulic, D., Croft, E., Zoghbi, S.: Measurement instruments for the anthropomorphism, animacy, likeability, perceived intelligence, and perceived safety of robots. Int. J. Social Robotics 1, 71–81 (2009)

24 Mori, M.: Bukimi no tani [The uncanny valley]. Energy 7(4), 33–35 (1970)

25 Swope, K., Amd, P.M., Schmitt, J.C., Shupp, R.: Personality preferences in laboratory economics experiments. J. Socio. Econ. 37(3), 998–1009 (2008)

Robots and the Division of Healthcare Responsibilities in the Homes of Older People

Simon Jenkins and Heather Draper[*]

University of Birmingham, Birmingham, UK
{s.p.jenkins,h.draper}@bham.ac.uk

Abstract. This paper briefly describes the method of a qualitative study, which used focus groups to elicit the views of older people and formal and informal carers of older people on the ethical issues surrounding the introduction of social robots into the homes of older people. We then go on to sketch some of the tensions and conflicts that can arise between formal carers, informal carers, and older people when trying to negotiate the task of dividing care responsibilities, and describe how the introduction of robots may exacerbate, or ease, these tensions. Data from the qualitative study is used to indicate where participants acknowledged, identified and discussed these issues.

Keywords: Ethics, social robots, elderly, older people, care, responsibilities, duties, healthcare, control, autonomy, behavior change.

1   Introduction

According to Sharkey and Sharkey '[t]he three main ways in which robots might be used in elder care are: (1) to assist the elderly, and/or their carers in daily tasks; (2) to help monitor their behaviour and health; and (3) to provide companionship' [1]. There is some overlap between these three uses. For example, monitoring may be instrumental for carers in helping older people [2], and as Sharkey and Sharkey and others [3] point out, assisting in daily tasks can lead to greater social isolation. Sharkey and Sharkey are concerned that older people's 'lack of control' may reduce their quality of life as control is surrendered *to* the robot. There are clearly ethical issues surrounding control *of* the robot, too. Human carers of older people have responsibilities and interests to provide care (including that provided via a robot) in ways that may conflict with what the older person wants, but tensions between carers' views of how best to deliver care may also be played out through the robot. Likewise, different kinds of carers may have different interests that may conflict with each other and/or the interests of the older person being cared for. Running together the assistance to older people and to carers with daily tasks may mask these tensions. Introducing care-robots may also have an effect on the dynamic between different kinds of human carers, which may raise further ethical issues.

---

[*] Corresponding author.

M. Beetz et al. (Eds.): ICSR 2014, LNAI 8755, pp. 176–185, 2014.
© Springer International Publishing Switzerland 2014

In this paper, we are concerned with two broad kinds of carer: formal carers – those who are paid to provide care and who may have differing levels of professional qualification and their own hierarchies of responsibility and accountability; and informal carers – relatives or friends of the older person who are unpaid and often unqualified. There are good reasons for ensuring that the older person has a significant say in how their care is organized (see, for instance, the recent political emphasis on 'patient-led' care [4], and the significance of householder autonomy in the introduction of robotic care [5]). Nonetheless, formal carers need to retain some control if they are to discharge their duties efficiently and ethically, and informal carers may be juggling the meeting of an older person's care needs with the need to have some control over their own lives (including, in the case of older carers, meeting their own care needs), and also with other obligations in their lives. The introduction of a robot will be affected by, and have an effect on, the older person-informal carer-formal carer triad, and: 'the division of tasks and responsibilities becoming care recipient, care provider (formal/informal), technology developer, system-provider (and others) respectively must be made clear' [6].

This paper will report some of the findings from a qualitative study undertaken as part of the Acceptable robotiCs COMPanions for AgeiNg Years (ACCOMPANY) project [7]. The findings presented here are incidental to the main aims of the study, which were to explore the potential tensions between values that had already been identified as potentially significant in the design of the ACCOMPANY robot[1] and to see whether additional values needed to be added. The findings outlined in this paper shed light on how the dynamics between members of the care triad, as described above, may be affected by the introduction of a robot; the main results of the study will be reported elsewhere.[2]

2   Method

21 focus groups were convened in France, the Netherlands, and the UK, with a total of 123 participants. There were three participant groups: older people (OP), informal carers of older people (IC), and formal carers of older people (FC). Four scenarios (Table 1) were designed to provoke discussion amongst the participants, and a topic guide was developed to ensure some consistency of discussion between the groups. They were conducted in native language, video and/or audio recorded, and transcribed verbatim. One representative transcript from each of the three kinds of group convened in the Netherlands and France was translated into English. All the English transcripts were then coded (by Draper) using a combination of directed analysis and Ritchie and Spencer's Framework Analysis [8]. These codes were discussed with project collaborators at the University of Warwick (UW), the University of Hertfordshire (UH), the Centre Expert en Technologies et Services pour le Maintien en Autonomie à Domicile des Personnes Agées (MADoPA), and Hogeschool Zuyd (ZUYD),

---

[1]   The values were autonomy, independence, enablement, safety, privacy, and social connectedness – see Sorell and Draper [5].

[2]   All ACCOMPANY results are reported on the project website [7].

until agreement was reached. The remaining non-English transcripts from the Dutch and French groups were then coded according to this agreement, and quotations that were selected to represent these codes were translated into English. The report of all the data was then written and circulated to the research team for verification. The methodology informing the study was 'empirical bioethics' [9, 10]. Work combining the empirical data and the earlier philosophical work [5, 7] is currently underway and will be completed in September 2014. See Draper et al. [11] for a fuller account of the method and details of the participant characteristics.

Table 1. Brief Description of Scenarios

Scenario	Brief description
1. Marie	Marie (78) resists the robot's efforts to encourage movement that will help her ulcers to heal. She likes it reminding her to take her antibiotics but not its reminders to elevate her leg. She is not honest with her nurse about how much she is moving.
2. Frank	Frank (89) is socially isolated. His daughter wants him to access an online fishing forum with the help of the robot. He isn't keen to try.
3. Nina	Nina (70) has recovered from a stroke. She is rude to her daughter and carers (causing them distress) but not her friends. The robot is programmed to encourage better social behavior by refusing to cooperate when she is rude.
4. Louis	Louis (75) likes to play poker online using the robot. He uses its telehealth function to monitor/control his blood pressure. He doesn't let the robot alert his informal carers when he falls (which he does regularly, usually righting himself). His informal carers want to re-program the robot so it will not let him play poker and to alert them when he falls.

3   Findings

Findings are reported using conventional reporting methods for qualitative research. The data is not reported quantitatively, as reporting in quantities and proportions is not appropriate to this kind of data. Data interpretation is illustrated with representative quotations, selected to demonstrate the complete data set. The data has been grouped into themes that speak to the relationship between types of carer and the older person being cared for.

3.1   Responsibility for Older People's Interests

In all of the scenarios the robot was capable of helping with the performance of some daily task but was also being used to encourage some behavior change in the older person to promote independence, enablement, or social connectedness. Here we will focus on behavior change insofar as it is relevant to the issue of the division of responsibility for caring for older people.

The fourth scenario involved a householder using the robot to facilitate his gambling. This elicited a range of attitudes about whether this use of the robot was permissible – these attitudes relate to the issue of responsibility. In the OP and IC groups, there were participants who viewed Louis as having responsibility for himself when it came to gambling:

> *Concerning the gambling he says he's in charge of his own money and I have to agree with him...* (ZUYD OP1 E3)[3]

> *He can't live completely withdrawn into himself even if it's all he wants for now, at least that's how I feel* (MADoPA IC1 P5)

The FC participants tended to support this, feeling that as long as the older person had mental capacity, (s)he could make decisions about such things by him/herself, and this perspective was noted by some of the IC participants:

> P5: *It does not anywhere say he is mentally limited.*
> P4: *Exactly, that is why*
> P2: *He is not addicted to the gambling* (ZUYD FC1)

> *I think it's funny, because at the day of the informal carer at the house of my mother we had a discussion with the professional carers. And the care staff said: 'The client is the King. If the client refuses something we won't do it.' While the children, the informal carers, often have the tendency to say: 'Can't you do this or that with my mother, because that is better for her'.* (ZUYD IC1 M5)

Some FC participants thought they may have a role in protecting older people from the over-protectiveness of family members:

> *Of course they love them, of course they don't want them to die in the immediate future, of course they don't want them to have any accidents, and yet at the same time, they don't realize that they are behaving – and please forgive the harshness of the word – like tyrants.* (MADoPA FC P7)

Indeed, in the OP and IC groups, some participants expressed the view that it may be legitimate to restrict or prohibit behaviors like gambling in the interests of the older person:

> *Once they've added up the cost of his rent, his food, the people who care for him and everything, they can see how much he has left, can't they?* (MADoPA OP1 O4)

---

[3] Quotations will follow this format: the site name is reported first, then the focus group, and finally the individual participant code. This is with the exception of quotations with multiple speakers, in which case participants will be identified as they speak.

> M1: *But when it comes to the debt repayment I would take action on playing poker before he got into debt.*
> M3: *In my opinion sons can interfere with that.* (ZUYD IC2)

There was therefore disagreement between some of the OP and IC participants on the one hand, and FC participants on the other, about whose judgment about the interests of the older users was most authoritative when it came to deciding how the robot should be programmed.

The interests of the carers as a motivation for modifying the behavior of older users of the robot were also discussed. With regard to Louis's gambling, the FC groups tended to consider that familial intervention (re-programming the robot so as to block access to the gambling site) was not aimed at supporting Louis' interests, but was rather directed towards protecting the financial interests of the family members:

> *And you also have to take into account that there are children who will try and curb their parents' spending because it's part of their inheritance going out of the window! So, given the facts we have here, it's a difficult question* (MADoPA FC1 P7)

> *The daughters also could think of their own benefits. If he spends all of his money his inheritance will not be as much* (ZUYD FC2 P7)

The OP group participants also considered family members' financial interests insofar as relatives might inherit the older person's debts:

> *Everyone has to be considered, because the children are the ones who have to pick up the pieces afterwards, aren't they.* (MADoPA OPFG P3)

> *[H]e could end up with a huge debt you know that's gonna cause problems in fact doesn't it. I don't know where he lives, let's assume that he is in his own house and he gets into a huge debt and the house has to be sold and he's got to go somewhere else. All these things follow on you know if you got drink problem you get into debt, drunk or you get into debt, he could lose thousands and thousands of pounds. I think then it does become a family problem.* (UH OPFG P4)

In this instance, and especially in cases where the family are described as deliberately protecting their inheritances, the robot is perceived by the participants as a potential focal point of a power struggle, "a weapon" (UB OP2 P2) even. The presence of a robot whose programming can be changed may exacerbate tensions like these when it can facilitate activities that may otherwise be unavailable to the older person. This may be viewed as empowering for the older person, but it creates a dilemma for carers, who may be unsure what their responsibilities are regarding the new activities that the robot facilitates. For instance, some participants were concerned about the robot introducing older people to the internet in general:

> *I would have worries about being on an internet forum because Frank's vulnerable, like children are. I mean Frank might have very expensive fishing rods or antiques or something. And somebody on the forum can pretend they're anybody* (UB OPFG2 P6)

In this respect, although it could appear, echoing Sharkey and Sharkey, that the robot is controlling the older person, our participants seemed concerned that the robot would be an extension of the existing controlling forces of family members or formal carers.

3.2   Responsibility and the 'Burden' of Care

Participants were also prompted by scenario four to discuss the extent to which Louis could determine when the robot reported his falls. Although Louis was usually able to get up himself with the aid of the robot, participants were told that he has recently lain for sometime unable to get up, and that this had resulted in a bladder infection and the need for more care input from his daughters-in-law. Some participants in the OP groups were sympathetic to the ways in which decisions made by older people could impact on the informal carers:

> *Well they're bringing him food, helping him, with his cleaning and doing his laundry so they're actually doing quite a bit and when he was in bed they took it in turns to stay with him during the day ...So I think they've got quite a lot invested in this and so to some extent I think there's a bit of a quid pro quo there* (UB OPFG3 P7)

> *You can't make people do more than they can take* (MADoPA OPFG1 P3)

Some of the FC participants were also sympathetic about the impact that the older person's decisions might have on informal carers, but they tended to be more sensitive to the effect on the FCs.

> *I also see it when people want to stay living at home then this has consequences. They do not want that, most often, but it does have those consequences. People sometimes do not want such a system with sensors and I say, but you want to remain living here, so we will have to ensure that it is safe, so there will be some changes to come. So in some ways I think you should expect this. You cannot force them, but that really has consequences. If he really does not want, what you can do as children is tell him. Then we also cannot take care of you. Because I think these children do a lot for him. Then it is allowed to expect a number of consequences of him.* (ZUYD FC1 P3)

Some of the FC participants noted that sometimes older people were not sensitive to the fact that they had other clients, which meant that the timing of their visits did not suit everyone (calling during a favorite TV program was noted as an irritant). Personal robots may enable care to be better tailored to users' preferences.

IC and OP participants who had themselves been informal carers tended to be most concerned about increasing the care burden for family members.

> *In everyone's best interests actually; in his best interests and in the best interests of his family, who won't have to make unnecessary journeys. Who'll come round if he falls?* (MADoPA IC1 P1)

All groups were concerned about the safety of older people. As we have already noted, some participants were more willing to accept risk as the price of autonomy, but the majority of participants in all groups tended to favor safety over autonomy or to feel torn between the two. Monitoring is one way of providing reassurance about safety, and is regarded as an advantage of assistive technology in general. Monitoring can reduce the "burden" of care by reducing anxiety, the number of visits required, and the amount of ongoing care required, by alerting carers to the need for early intervention. Falling is a good example here. Monitoring, however, also requires information to be shared, e.g. accessed from or transmitted by the robot to others. In spite of these privacy concerns, it may be in the interests of all three groups to use the robot in a way that will ensure the older person's safety.

3.3   Monitoring, "Policing" and Sharing Information

Our participants recognized the potential value of FCs being able to access health-related information from the robot. FC and OP participants were more concerned about such information being accessible to ICs. FC participants were also aware of the possibility that the care they provided could be monitored. We have reported these findings elsewhere [11].

In terms of ICs' access to information via the robot, of relevance to this paper is what this finding may suggest about the role of ICs in the care "team". One interpretation is that restricting access to health information is an extension of the norms of medical confidentiality, as health information is not usually shared with family members without the consent of the patient. However, this might also be regarded as the playing out of power differences between ICs and FCs, where knowledge is a form of power and ICs are left to act on the instructions of the FCs who "know" best. Participants in all groups were concerned about ICs making decisions without consulting FCs:

> *I would have thought that should have been a medical decision, not for the daughters-in-law to decide whether he uses his sticks or his walking frame... I think it should be should be looked into if he is safe to have his sticks or if he needs a walking frame* (UB OP2 P5)

> *No, should have discussed with the medical staff.* (UH IC P2)

Obviously, this is not a new issue created by the use of robotics, but it could lead to a perception by ICs that they are below the robot in the care "hierarchy" as the robot has access to information that they do not. As one IC participant noted, ICs may already have the same information that the robot may collect more formally:

Yes, that the robot does something. That it notes things down, just like we do. For instance the number of times she got out of her chair. (ZUYD IC1 M6)

The latter might be regarded as a form of unwelcome "policing" of FCs by the robot. Furthermore, the robot may also be used to collect information on how well older users adhere to health advice (as opposed to just issuing reminders, e.g. to take medication) - participants were ambivalent about this.

They could look at the print out together, that wouldn't be quite as invasive as the robot saying: 'Actually she didn't do that when I told her three times and she didn't get up!' (UH FC PF)

They cannot cheat, right? ... That is the difference. The measures are taken and the robot sends them on to the physician. So there is no possibility to add a few degrees, or make it some degrees less. (ZUYD FC1 P2)

The robot could also be used to "police" whether ICs comply with FCs' instructions about appropriate care, including where the two groups disagree about how best to discharge care or whether health advice must be followed. This could occur whether or not the older person objects. Many of the FC participants expressed views that were critical of ICs' decisions, such as here where the ICs' approach to caring for their relative is regarded as too forceful:

Sometimes, people's children want to force things upon their parents and in the end, instead of having an aid that perhaps was inadequate, they don't use anything at all (MADoPA FC1 P7)

Robotic surveillance may make it easier for ICs to coerce older people to comply with their view of what is best for them.

It seems legitimate for a robot to be used to "police" the care of older people. Older people should not be subjected to poor care or neglect from either ICs or FCs. Our OP participants did not seem to object to the robot being used to monitor health and pass information to FCs. But whether surveillance that lies between these two ends of the data-collection spectrum is policing or monitoring may be a matter of perspective that may reflect reasonable differences of opinion on what care to deliver and how. Ideally, differences of opinion and conflicts of interests in the care triad can be resolved by compromise and negotiation:

And how one gets to that end result, maybe a mix of you know, input from the nurse, further explanation, encouragement from other people might pop in, or I don't know. That's what I would be hoping for is this, you know, some[one] being able to understand the importance of what is needed (UH FC PB)

Disagreements may be magnified, however, if the robot shifts the balance of power by giving more control to one or other parties.

4   Limitations

Focus groups were conducted in different languages and meaning may have been inadvertently altered in translation. Some of the transcripts were analyzed in their original languages, which may have affected standardization in the analysis. The team attempted to mitigate these issues with discussion about the translation and coding.

5   Conclusion

In this paper we have presented and briefly discussed incidental findings from a qualitative study that shed light on how robots might impact on the division of care between FCs and ICs, and on how responsibilities for determining and providing care are perceived. We have considered some of the tensions that were discussed by the study participants between allowing older people to govern their own care, and carers taking some control over and responsibility for it. Notably, these tensions may be exacerbated with the introduction of a robot, particularly if it is used to monitor the older person's behavior. While monitoring may be seen by some as intrusive, it may often be justified by invoking both the interests of the older people in that it may help to ensure their safety, and the interests of both types of carer in that it may ease their burden of care.

We have highlighted the fact that existing divisions of responsibility may affect the interests of older people, but also those of the carers themselves. Furthermore, we have reflected on the added tensions that arise when different types of carer disagree about how to discharge care, when there can be suspicion or disapproval of the way that other carers do this. Use of the robot may therefore become a point of conflict between carers and the older people themselves, or between different groups of carers. This suggests that careful consideration must be given to the extent to which each care group, and the older people themselves, can control the robot. Our data may suggest that FCs should be given priority in their control of the robot over ICs, related to concerns that ICs may have financial motivations or other conflicts of interest. It may, however, be a mistake to imagine that FCs' motivations are so relatively undivided, as they must divide their care between multiple clients, and may, unlike ICs, perceive their obligations only to stretch as far as their professional role demands. Those designing robots for care purposes should be aware that these complex ethical issues exist, and should seek guidance from ethicists or ethics literature when considering how they are to be negotiated.

Acknowledgements. The work in this paper was partially funded by the European project ACCOMPANY (Acceptable robotiCs COMPanions for AgeiNg Years). Grant agreement no.: 287624. This paper would not have been possible without the input of ACCOMPANY partners UW, ZUYD, UH, and MADoPA, especially Sandra Bedaf, Tom Sorell, Dag Sverre Syrdal, Carolina Gutierrez-Ruiz, Hagen Lehmann, Kerstin Dautenhahn, Gert-Jan Gelderblom, Hervé Michel and Helena Lee who were also involved in the data collection and the initial analysis of the data set. We are also grateful to all the participants who took part in our study.

References

1. Sharkey, A., Sharkey, N.: Granny and the robots: ethical issues in robot care for the elderly. Ethics Inf. Technol. 14, 27–40 (2012)
2. Heerink, M., Kröse, B., Wielinga, B., Evers, V.: Human-Robot User Studies in Eldercare: Lessons Learned. In: Nugent, C., Augusto, J.C. (eds.) Smart Homes and Beyond, pp. 31–38. IOS Press, Amsterdam (2006)
3. Sparrow, R., Sparrow, L.: In the hands of machines? The future of aged care. Mind. Mach. 16, 141–161 (2006)
4. Crisp, N.: Creating a patient-led NHS – delivering the NHS improvement plan (2005), http://webarchive.nationalarchives.gov.uk/20130107105354/, http://dh.gov.uk/en/publicationsandstatistics/publications/publicationspolicyandguidance/dh_4106506 (accessed May 22, 2014)
5. Sorell, T., Draper, H.: Robot carers, ethics and older people. Ethics Inf. Technol. (2014), doi:10.1007/s10676-014-9344-7
6. Palm, E.: Who Cares? Moral Obligations in Formal and Informal Care Provision in the Light of ICT-Based Home Care. Health Care Anal. 21, 171–188 (2013)
7. Acceptable robotiCs COMPanions for AgeiNg Years, http://accompanyproject.eu/ (accessed July 29, 2014)
8. Ritchie, J., Spencer, L.: Qualitative Data Analysis for Applied Policy Research. In: Huberman, A.M., Miles, M.B. (eds.) The Qualitative Researcher's Companion, pp. 305–330. Sage, California (2002)
9. Borry, P., Schotsmans, P., Dierickx, K.: What is the role of empirical research in bioethical reflection and decision-making? An ethical analysis. Med. Healthc. Philos. 7, 41–53 (2004)
10. Kon, A.A.: The role of empirical research in bioethics. Amer. J. Bioeth. 9(6-7), 59–65 (2009)
11. Draper, H., Sorell, T., Bedaf, S., Syrdal, D.S., Gutierrez-Ruiz, C., Duclos, A., Amirabdollahian, F.: Ethical Dimensions of Human-Robot Interactions in the Care of Older People: Insights from 21 Focus Groups Convened in the UK, France and the Netherlands. In: Beetz, M., Johnston, B., Williams, M.-A. (eds.) ICSR 2014. LNCS (LNAI), vol. 8755, pp. 135–145. Springer, Heidelberg (2014)

Investigating the Effect of a Robotic Tutor on Learner Perception of Skill Based Feedback

Aidan Jones, Ginevra Castellano, and Susan Bull

University of Birmingham, UK

Abstract. In this paper we investigate the effect of different embodiments on perception of a skill based feedback (a basic open learner model) with a robotic tutor. We describe a study with fifty-one 11-13 year old learners. Each learner carries out a geography based activity on a touch table. A real time model of the learner's skill levels is built based on the learner's interaction with the activity. We explore three conditions where the contents of this learner model is fed back to the learner with different levels of embodiment: (1) Full embodiment, where skill levels are presented and explained solely by a robot; (2) Mixed embodiment, where skill levels are presented on a screen with explanation by a robot; and (3) No embodiment, where skill levels and explanation are presented on a screen with no robot. The findings suggest that embodiment can increase enjoyment, understanding, and trust in explanations of an open learner model.

Keywords: Open Learner Modelling, Learner Modelling, Social Robots.

1 Introduction

Experienced teachers and computer based learning systems allow a scenario where a learner carries out an activity and receives feedback on their areas of strengths and weaknesses contemporaneously. This scenario enables the learner to reflect, correct any errors, and build upon their strengths as they progress through the activity. This type of one-on-one tutoring benefits the student [21]. We aim to emulate such an approach with an interactive activity that can model the skill levels of a learner in real time and provide feedback via a robotic tutor. This will allow us to investigate if a robotic tutor is able to present feedback in a more effective way when compared to on-screen feedback alone. To that end we have investigated the effect of different embodiments on the learner's perception of feedback and overall experience. The learning activity is a geography exercise targeted at 11–13 year old learners. A basic model of the learner's map reading skills is built; "the learner model". We explore three conditions where the contents of this learner model is fed back to the learner with different levels of embodiment: (1) Full embodiment, where skill levels are presented and explained solely by a robot; (2) Mixed embodiment, where skill levels are presented on screen with explanation by a robot; and (3) No embodiment, where skill levels and explanation are presented on a screen with no robot. We ask a

M. Beetz et al. (Eds.): ICSR 2014, LNAI 8755, pp. 186–195, 2014.

series of Likert style questions to investigate enjoyment, perception, and trust of the presentation of the learner model. The findings suggest that embodiment may increase enjoyment, understanding, and trust in explanations of skill levels.

2   Related Work

One approach used within intelligent tutoring systems to give skill based feedback is open learner modelling. Open learner models externalise the learner model in a way that is interpretable by the user [3], e.g. skill meters [16]. One of the aims of opening the learner model to the learner is to promote reflection and raise awareness of their understanding or developing skills [4].

A number of systems have used virtual embodiment to teach or interact with the user [9, 5], although results are mixed in terms of learning gain there are many positive effects gained such as enjoyment, motivation [18], and the learners perception of the learning experience [13]. Studies that compared virtual representations of characters with robots showed a preference for robotic embodiment with reference to social presence [10, 12], enjoyment [19, 11, 22], and performance [8]. Greater learning gains have also been shown with a robotic tutor when compared to a virtual tutor [15]. The development of trust can also be increased with the presence of embodiment [7].

Greater learning gains have been shown when personalising robot behaviour to the learner. Recall levels have been higher with a robotic tutor when adaptive cues have been given based on EEG measurements of engagement [20]. Puzzle solving times have been reduced when using personalised tutorials delivered by a robotic tutor [14].

3   Methodology

We aim to apply the benefits of a physically embodied robotic tutor to present an open learner model to the learner. No previous robot tutor research, however, investigates embodiment of on presentation of an open learner model. The robotic tutor may lead to the learner paying more attention due to the feedback being more enjoyable, engaging or the learner affording greater respect to the robot [2]. Understanding which pieces of information are best delivered by a robot and which by on screen elements is useful for the design of systems that include a robot. We aim to investigate and measure how and to what extent the learners accept personal skill based feedback from a physical entity when compared with a computer/touch screen. One of the factors that may be increased with a robotic embodiment is trust. However, there has been little work empirically in this area comparing automated aids vs robotic aids [7].

We use a number of metrics to measure if and to what extent there are advantages brought by a physical embodiment to the presentation and explanation of skill levels. The study endeavours to understand the effect of embodiment on a learner's perception of skill level, trust in the system, enjoyment, and overall experience.

3.1   Scenario

The learners interact with the learning activity individually on a touch screen. The learner is provided with regular updates on the level of their map reading skills and a simple explanation of why the skill level is at its current level.

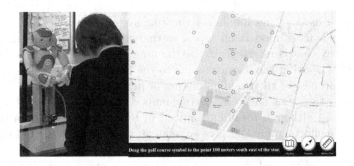

Fig. 1. NAO Robot, Learner, and Learning Task

Learning Activity. The learners are asked to carry out a geography object placement activity. The activity is designed to test *compass reading, map symbol knowledge,* and *distance measuring* competencies. The content conforms to the England and Wales National Curriculum for Geography [1]. Previous mock up studies with both teachers and students identified that the level of difficulty in the activity is appropriate for the learners.

The activity comprises a number of steps that tests all three competencies that are modelled. The questions are in the form of: *"Drag the campsite symbol to the point 100m north of the star"*. After each step in the activity the learner is presented with the current skill levels for each competency and a short explanation of why the skill level is at that level and what has been answered correctly and/or incorrectly. We wanted the system to not only deliver a value for each skill level but also a brief explanation of why the skill level is at that current level as this provides more aspects of feedback to investigate. The explanations are also summarised where possible to reduce repetition if all of the skills have changed in the same way. The learner is provided with three tools to assist them if they are having trouble with the activity. They have the option to open a map key, use a distance tool, and display a compass on screen.

Learner Model. The construction of the underlying learner model is critical. One of the main approaches to skill modelling is Constraint Based Modelling (CBM) [6, 23, 24]. CBM is a technique that can be used to model a learner's domain knowledge and skill. It does so by checking a learner's answers against a set of relevant constraints; if an answer does not violate a constraint then that answer is correct [17]. Using this approach a basic learner model containing the competencies *compass reading, map symbol knowledge,* and *distance measuring* is built. The model provides an indication of the current skill levels calculated using a weighted average so that more up to date information is more relevant than old information. The time taken to answer a question also affects the update of the learner model.

3.2   Procedure

Participants. There were fifty-one (twenty-three female, twenty-eight male) participants of mixed ability learners from 3 schools. The learners were aged between 11 and 12 and all in year 7. There was a roughly equal gender balance and ratio of learners from each school across the conditions.

Fig. 2. Conditions, (1) Full embodiment, (2) Mixed embodiment, (3) No embodiment

Experimental Conditions

Full embodiment: Verbal communication of both the skill levels and explanation by the robot There is no visual representation of the skill meter on screen, the skill levels and explanation of the skill level is spoken solely by the robot. The robot makes idle motions throughout.

Mixed embodiment: Skill meter on screen with verbal communication of the explanation of skill level by the robot Each competency is displayed on screen as a skill meter and the robot provides the explanation. There is no on-screen explanation and the robot does not say the skill levels. The robot makes idle motions throughout.

No embodiment: Skill meters and text to present explanation on screen No robot is present in this condition. The skill meters are displayed on screen with a text explanation to the side. If the explanation is the same, the text is summarised in one piece of text. The text is the same as the robotic explanation.

The study was conducted in a meeting room in the learner's school. The activity ran on a touch screen laid flat on the table. The learner was stood up to enable them to comfortably reach all areas of the touch screen. The robot was positioned on a stand opposite the touch table in order for it to be at a similar height to the learner.

The learner was brought in to the room, given a overview of the study and asked to complete a pre-activity questionnaire. The activity and the use of the map tools were explained. The learner then carried out the activity for 4 minutes (time based on experience from a pilot study). The participant was then asked to complete the post-activity questionnaire. During the activity, after each step the learner was presented with the skill level for each competency and an explanation of why the skill level was at that level. This is communicated via a pop-up on screen or via verbal communication from the robot. All three of the conditions provide the same information and explanation, however each condition varies the way the information is presented. There are five skill levels for each competency ranging from very low, low, okay, good, to very good. The learner was informed of their level of skill, followed by how that level has changed since the last step; increased, decreased, or stayed the same. This was then followed by an explanation. There are just three explanations given. If the competency has increased due to a quick answer or stayed the same due to the maximum skill level being reached the explanation is "You are answering quickly and correctly". If the competency increases or stayed the same based on an answer that is correct but not quick the explanation is "You are answering correctly but sometimes a bit slowly". If the competency decreases due to an incorrect answer or has stayed the same due to the lowest skill level the explanation is "Your answers are not always right". If all competencies have updated in the same manner the explanation is summarised rather than explained multiple times. This saves time and avoids repitition.

Data Collection. The primary form of data collection is a self-report questionnaire containing questions designed to elicit the learner's perceived skill level, enjoyment, engagement, perception, understanding, and trust in the learner model and system. The questionnaire is divided into three sections of Likert style questions: 1) Enjoyment, including "I enjoyed the overall experience" and "I enjoyed the explanation of how and why my skills changed"; 2) Perception/ Understanding, including "I noticed that the system understood my skill levels"; and 3) Trust, including "I trust the explanation of why my skill levels are changing".

4   Experimental Results

4.1   Data Analysis

Responses to the Likert scale questions were grouped in to Enjoyment, Perception and Trust. The reliability of these groupings was assessed using Cronbach's alpha. The mean values of each group and the individual items were analysed by comparing each condition against each other using a Mann-Whitney U test. The significant values (lower than 0.05) were then further investigated.

4.2   Results

Table 1. Results table

Question	Mean values						Mann-Whitney U Test					
	Mixed		None		Full		Mixed vs None		Full vs None		Full vs Mixed	
	Mean	S.D.	Mean	S.D.	Mean	S.D.	U	p	U	p	U	p
Enjoyment												
Combined	4.52	0.33	4.05	0.66	4.46	0.44	80.0	0.026	89.5	0.057	137.5	0.812
I enjoyed the overall experience	4.82	0.39	4.18	1.01	4.71	0.47	82.0	0.031	97.0	0.106	127.5	0.563
I enjoyed doing the activity	4.76	0.44	4.24	0.75	4.71	0.47	87.5	0.049	94.5	0.085	136.0	0.786
I enjoyed being shown my skill levels throughout the activity	4.59	0.51	4.24	0.97	4.65	0.49	115.5	0.322	107.5	0.205	136.0	0.786
I enjoyed the explanation of how and why my skills changed	4.47	0.51	3.88	0.70	4.59	0.62	79.5	0.024	68.5	0.008	123.5	0.474
I lost track of time while doing the activity	3.75	1.06	3.81	1.22	3.76	1.03	120.0	0.780	128.0	0.790	135.0	0.986
I would like to play the activity again	4.71	0.59	4.00	1.10	4.35	0.70	79.5	0.041	114.0	0.444	102.5	0.150
Perception/Understanding												
Combined	4.65	0.28	4.13	0.63	4.41	0.47	67.0	0.007	107.5	0.205	102.0	0.150
I noticed that the system understood my skill levels	4.71	0.47	3.82	0.95	4.47	0.51	58.0	0.002	84.0	0.038	110.5	0.245
I noticed that the system showed me my skill levels	4.76	0.44	4.18	0.73	4.35	0.61	79.0	0.024	126.5	0.540	91.5	0.067
I noticed that the system explained why my skill levels were changing	4.59	0.51	4.18	0.73	4.53	0.62	100.0	0.131	105.5	0.182	141.0	0.919
I understood when the system showed me my skill levels	4.53	0.51	4.29	1.05	4.29	0.59	136.5	0.786	126.5	0.540	115.0	0.322
I understood the explanation of why my skill levels were changing	4.65	0.61	4.18	0.73	4.41	0.62	91.5	0.067	119.5	0.394	112.5	0.274
Trust												
Combined	4.43	0.42	4.18	0.67	4.22	0.51	112.5	0.274	143.5	0.973	108.0	0.218
I trust that the system can gauge my skill levels correctly	4.29	0.69	4.06	0.97	4.18	0.64	128.5	0.586	143.5	0.973	129.5	0.610
I trust that the skill levels shown by the system were accurate	4.18	0.64	4.29	0.59	4.24	0.56	131.0	0.658	136.5	0.786	138.5	0.838
I trust the explanation of why my skill levels were changing	4.82	0.39	4.25	0.86	4.24	0.75	80.5	0.045	130.5	0.845	80.5	0.026

4.3   Enjoyment

The Cronbach's Alpha for the grouping of enjoyment questions was 0.76. Between the mixed embodiment and no embodiment conditions the overall enjoyment is significantly higher in favour of the mixed condition (U = 80; p= 0.026). At an individual level this was due to these questions having significantly higher values in the mixed condition: "I enjoyed the overall experience" (U = 82; p=0.031423), "I enjoyed doing the activity" (U=87.5, p = 0.048686), "I enjoyed the explanation of how and why my skills changed" (U=79.5; p=0.023766), "I would like to play the activity again" (U=79.5; 0.040674). When comparing the full embodiment and no embodiment conditions, overall, there was no significant difference, however the following question had a significantly higher result: "I enjoyed the explanation of how and why my skills changed" (U=68.5; p =0.007611); There were generally higher values across the other questions but not to a significant level. Between the mixed embodiment and full embodiment there were no significant differences. Across all conditions the following question showed no significant difference: "I enjoyed being shown my skill levels throughout the activity". It appears that embodiment played a limited role in the showing of skill levels but had more significance in the explanation.

4.4   Perception/Understanding of the Model

The Cronbach's Alpha for the grouping of perception questions was 0.79. In the mixed embodiment vs no embodiment conditions the overall perception of skill meters and explanation was greater with the mixed condition to a significant degree (U=67; p=0.007). This can be seen at an individual level with the following questions being higher for the robot condition by a significant amount: "I noticed that the system understood my skill levels" (U= 58; p= 0.002269), "I noticed that the system showed my skill levels" (U= 79; p= 0.023766). "I understood the explanation of why my skill levels were changing" were higher but not significantly so. When comparing the full embodiment and no embodiment conditions there was no overall significant difference, however the following question had a significant higher result: "I noticed that the system understood my skill levels" (U=84; p =0.037590). Other values again were higher but not significantly. Between the mixed embodiment and full embodiment there were no significant differences.

4.5   Trust in the Model

The Cronbach's Alpha for the grouping of trust questions was 0.615, which is a rather low value. Overall there was no significant differences between any of the conditions. A more detailed review reveals no significant differences with respect to questions concerning the building of the model: "I trust that the system can gauge my skill levels correctly" and "I trust that the skill levels shown by the system were accurate". However, there were some significant differences with the following question: "I trust the explanation of why my skill levels are changing". In the mixed embodiment vs no embodiment conditions the value is higher in the mixed condition (U=80.5;p=0.044523). The fully embodied condition is higher than the no embodiment condition but not to a significant degree for the same question. The mixed condition leads to higher values than the fully embodied condition (U=80.5; p=0.026122).

5   Discussion

From these results it appears that embodiment has the largest effect in the explanation of the model. There is greater enjoyment with some amount of embodiment. There is greater perception that the system understands the learner. There is more trust in the explanation. The embodiment has less of an effect in respect to the perception of skill meters. This may be because the skill level is quite a simple concept to understand. The perception of skill levels changing and understanding that skills were changing was the same across all conditions. This was to be expected as this was made obvious in the experimental design. There was general consensus that the type of feedback provided, the skill meter and explanation were liked and understood across all conditions, which was encouraging for continued use of this feedback.

6   Conclusions

The results show promise for the introduction of a physical embodiment when providing feedback concerning skill levels, however to gain the most advantage the robot should be used to explain and elaborate rather than simply state skill levels. That there is trust in the explanation is very encouraging as this means that the learner may pay attention and act based on the explanation. Further analysis will investigate whether there were increased learning gains or greater evidence of reflection based on the task log data.

One limitation of this study is the absence of a comparison to a virtual embodiment. Such a comparison will enable analysis to explore if and to what extent the physical presence was responsible for the above results as opposed to other factors, such as the feedback being in a different medium. A further limitation concerned the skill meters. As they were not on the screen at all times this may have limited their use. However, limiting skill meters to a pop up allowed a closer comparison to robotic speech which can not be present all of the time.

This work is the starting point for further research in to open learner modelling in the field of educational social robotics. In the future the activity would be more complex to enable the learner to develop and exhibit skills in more depth. With a more complicated task that requires more planning there would be more opportunity for the student to reflect and exhibit other meta-cognitive strategies which if measured could allow more chance to detect if the student is utilising the skill based feedback. The robot should be able to interact with the student to a greater degree, this need not be very complex or cause distraction from the task; Head nods, facial expressions, and body position can provide unobtrusive feedback on the learner's utterances and actions without unnecessarily disrupting the learner's train of thought [9]. The behaviours can increase the immediacy [20] of the robot to engage and motivate the learner.

Acknowledgements. This work was supported by the European Commission (EC) and was funded by the EU FP7 ICT–317923 project EMOTE (EMbOdied-perceptive Tutors for Empathy-based learning). The authors are solely responsible for the content of this publication. It does not represent the opinion of the EC, and the EC is not responsible for any use that might be made of data appearing therein.

References

[1] http://www.education.gov.uk/schools/teachingandlearning/curriculum/secondary/b00199536/geography (May 29, 2013)

[2] Bainbridge, W.A., Hart, J., Kim, E.S., Scassellati, B.: The effect of presence on human-robot interaction. In: RO-MAN 2008 - The 17th IEEE International Symposium on Robot and Human Interactive Communication, pp. 701–706 (August 2008)

[3] Bull, S., Kay, J.: Student Models that Invite the Learner In: The SMILI:() Open Learner Modelling Framework. International Journal of Artificial Intelligence in Education 17(2), 89–120 (2007)

[4] Bull, S., Kay, J.: Open Learner Models. In: Advances in Intelligent Tutoring Systems, ch. 15, pp. 301–322 (2010)

[5] Dehn, D.M., Van Mulken, S.: The impact of animated interface agents: a review of empirical research. International Journal of Human-Computer Studies 52(1), 1–22 (2000)

[6] Desmarais, M.C., Baker, R.S.J.D.: A review of recent advances in learner and skill modeling in intelligent learning environments. User Modeling and User-Adapted Interaction 22(1-2), 9–38 (2012)

[7] Hancock, P.A., Billings, D.R., Schaefer, K.E., Chen, J.Y.C., de Visser, E.J., Parasuraman, R.: A Meta-Analysis of Factors Affecting Trust in Human-Robot Interaction. Human Factors: The Journal of the Human Factors and Ergonomics Society 53(5), 517–527 (2011)

[8] Hoffmann, L., Krämer, N.C.: How should an artificial entity be embodied? Comparing the effects of a physically present robot and its virtual representation. In: HRI 2011 Workshop on Social Robotic Telepresence, pp. 14–20 (2011)

[9] Johnson, W., Rickel, J.W., Lester, J.C.: Animated pedagogical agents: Face-to-face interaction in interactive learning environments. International Journal of Artificial Intelligence in Education, 47–78 (2000)

[10] Kidd, C.: Sociable robots: The role of presence and task in human-robot interaction. PhD thesis, Massachusetts Institute of Technology (2003)

[11] Kidd, C., Breazeal, C.: Effect of a robot on user perceptions. In: 2004 IEEE/RSJ International Conference on Intelligent Robots and Systems (IROS) (IEEE Cat. No.04CH37566), vol. 4, pp. 3559–3564 (2004)

[12] Lee, K.M., Jung, Y., Kim, J., Kim, S.R.: Are physically embodied social agents better than disembodied social agents?: The effects of physical embodiment, tactile interaction, and people's loneliness in human–robot interaction. International Journal of Human-Computer Studies 64(10), 962–973 (2006)

[13] Lester, J.C., Kahler, S.E., Barlow, T.S., Stone, B.A., Bhogal, R.S.: The persona effect: affective impact of animated pedagogical agents. In: ACM SIGCHI Conference on Human Factors in Computing Systems, pp. 359–366. ACM (1997)

[14] Leyzberg, D., Spaulding, S., Scassellati, B.: Personalizing robot tutors to individuals' learning differences. In: Proceedings of the 2014 ACM/IEEE International Conference on Human-robot Interaction - HRI 2014, pp. 423–430. ACM Press, New York (2014)

[15] Leyzberg, D., Spaulding, S., Toneva, M., Scassellati, B.: The Physical Presence of a Robot Tutor Increases Cognitive Learning Gains. In: 34th Annual Conference of the Cognitive Science Society (CogSci 2012), vol. 1, pp. 1882–1887 (2012)

[16] Mitrovic, A.: Evaluating the effect of open student models on self-assessment. International Journal of Artificial Intelligence in Education 17(2), 121–144 (2007)

[17] Mitrovic, A.: Modeling Domains and Students with Constraint-Based Modeling. In: Advances in Intelligent Tutoring Systems, ch. 4, pp. 63–80 (2010)

[18] Moundridou, M., Virvou, M.: Evaluating the persona effect of an interface agent in a tutoring system. Journal of Computer Assisted Learning 18(3), 253–261 (2002)

[19] Pereira, A., Martinho, C., Leite, I., Paiva, A.: iCat, the chess player: the influence of embodiment in the enjoyment of a game. In: Proceedings of 7th International Joint Conference on AAMAS, pp. 1253–1256 (2008)

[20] Szafir, D., Mutlu, B.: Pay attention!: designing adaptive agents that monitor and improve user engagement. In: Proceedings of the SIGCHI Conference on Human Factors in Computing Systems, pp. 11–20. ACM Press (2012)

[21] VanLehn, K.: The Relative Effectiveness of Human Tutoring, Intelligent Tutoring Systems, and Other Tutoring Systems. Educational Psychologist 46(4), 197–221 (2011)

[22] Wainer, J., Feil-Seifer, D.J., Shell, D.A., Mataric, M.J.: Embodiment and human-robot interaction: A task-based perspective. In: 16th IEEE International Conference on Robot & Human Interactive Communication, pp. 872–877. IEEE (2007)

[23] Woolf, B.P.: Student Knowledge. In: Building Intelligent Interactive Tutors: Student-centered Strategies for Revolutionizing e-learning, ch. 3 (2008)

[24] Woolf, B.P.: Student modeling. In: Advances in Intelligent Tutoring Systems, ch. 13, pp. 267–279 (2010)

ROBOMO: Towards an Accompanying Mobile Robot

Khaoula Youssef, Kouki Yamagiwa, Ravindra Silva, and Michio Okada

Interaction and Communication Design Lab, Toyohashi University of Technology
1-1 Hibarigaoka, Tempaku, Toyohashi, Aichi
{youssef-yamagiwa}@icd.cs.tut.ac.jp, {ravi,okada}@tut.jp
www.icd.cs.tut.ac.jp

Abstract. Many of the most daily uses of robots require them to work alongside users as cooperative and socially adaptive partners. To provide the human with the better suited assistance at a convenient time, a robot must assimilate the user's behaviors and afford an adaptive response within the context of the interaction they share. We try to understand how a robot communication that is based on inarticulate sounds and iconic gestures is capable to help on the establishment of the attachment process and can enhance the social bonding between the human and our accompanying mobile robot (ROBOMO). In this paper, we draw on inarticulate sound and iconic gestures in order to design our robot and ground the attachment process. We showed that using simple inarticulate sounds and iconic gestures, the attachment process can evolve incrementally which significantly helped to acquire the meaning of the robot's behaviors.

Keywords: Social Bonding, Inarticulate Sound, Minimal Design, Iconic Gestures.

1 Introduction

Social bonding suggests that taking part in a communication increases the attachment and consequently the adaptation capability which may enhance the meaning acquisition process [1]. As an example, infants who form a social bond with their caregivers establish a better sense of their surroundings. In fact, slowed voice tones and physical contact, help the child to establish a preference for the caregiver and a mutual interest in communication evolves [2]. In such scenarios, children distinguish the different voices, and turn their heads to pick up the tones. They can intentionally generate imitations of hand gestures and voice sounds, with different expressions transferring a knowledge, an interest, an excitement, etc. [3]. Meanwhile, caregivers, excited by the infant's expressions, respond with affectionate behaviors by using rhythms of speech and slowed gesture with a soft voice and a moderate modulation of pitch [4]. Incrementally, the attachment evolves and the mutual understanding occurred by mirroring the patterns of each others' expressions [5]. Another similar example that involves the attachment process is the human-pet relationship. Many studies [6][7][8] investigated

M. Beetz et al. (Eds.): ICSR 2014, LNAI 8755, pp. 196–205, 2014.

the beneficial effects of pet ownership on human's interpersonal relationships and explored the importance of the human-animal interaction for the human's relational development [7][8][9]. Sparks et al [9] defines the behavioral attachment during the human pet interaction [1] as a prominent factor that helps the human to understand the pet's signals. It is then reasonable to presume that attachment between the human and others plays a unique role that helps on understanding others and the environment.

In this vein, we are interested in understanding whether inarticulate sounds and simple gestures help to establish the attachment process between the human and our mobile accompanying robot. We believe that we can use them to create a social bond just like in the caregiver-child or the human-pet scenarios and then enhance the adaptation within the human-robot interaction. Designing a robot that is not related to any language or any special cultural behaviors, will afford the chance to create a universal form of communication for the human-robot interaction just as in the child-caregiver scenario that is based on the attachment between both parties and the use of simple cues to establish online the customized social rules. To measure the social bonding, we intend to assess the values of five factors : the degree of adaptation to the social creature, the stress felt by the subject, the friendliness of the robot, the cooperation and the achievement degrees. In our paper work, we will afford a brief explanation about ROBOMO's design and architecture, explain the experimental setup, expose the results and finally we will give a brief discussion.

2   Background

Many studies investigated the attachment of humans to social robots [10][11]. Sung et al [11] indicated that people had a tendency to name their robots. Findings such as this suggested that people may treat robots like they treat a child or a pet [12]. In fact, if the robot exhibits a social behavior, a social bond will be formed and then people feel more comfortable with robots [13]. As an example, Samani et al [13] proposed Lovotics, a robot that uses audio and touch channels along with internal state parameters in order to establish long standing bonds with individuals. Lovotics afforded for the users an intimate relationship and people felt so comfortable that they even hugged the robot. Hiolle et al [14] used the Sony AIBO robot during their experiment where they showed that people tend to form a social bonding with needy robots that demanded assistance from users. The latter study suggests that robots do not need multi-modal communication to develop the attachment process and that exhibiting a simple behavior can be attractive enough for the human to feel attached to the robot and to embark on a positive constructive relationship with its. In our study, we will use similar simple behaviors that can be assembled under the immediacy

---

[1] Behavioral attachment: It consists on the human's involvement in different tasks with their pets such as play or teaching them new instructions where the pets are using their inarticulate sounds and their bodies movements to transfer the meaning to the owner.

cues category: the gestures and inarticulate sounds. We want to explore whether these two social cues can help to ground the attachment process and explore the social bonding's effect on the interaction's meaning acquisition. Inarticulate sounds were used to establish playground language with autistic children [15] and were studied in the context of the human-computer interaction [16] where it was proved that it can lead to a compassionate effect. Iconic gestures [17] [2] facilitates the human-robot interaction [18] and were used in different contexts such as hosting activity [19], showing hesitation [20], etc... In our current work, we intend to ground the attachment process that may evolve between ROBOMO and the participants. We want to verify whether a social bonding can emerge in the context of the human-ROBOMO interaction and whether it can guarantee to transfer the meaning once meshed with the iconic gestures and the inarticulate sounds.

3   ROBOMO Design

We respected the minimal design paradigm which consists on reducing the robot's design and preserving only the most elementary components [16]. ROBOMO has a long shaped body with an attractive container (made of plush) and has no arms. We had intentionally given ROBOMO a pitcher plant (Nepenthe) appearance to encourage people to interact with it, much as one might with a young child or a pet. We believe that exposing a half hairy head (Fig. 1), makes the robot looks cute and affords a starting point for the social bonding process formation. Although used for personal navigation, our accompanying mobile robot is not designed to walk which may create a sort of an empathetic feeling towards ROBOMO. Inarticulate sounds were produced according to Okada et al's [21] generation method of inarticulate sounds. Three types of behaviors were exhibited (i) the inarticulate sounds with meaning (ii) the nodding (iii) gestures (table 1).

Fig. 1. ROBOMO's design

---

[2] They are speech-related gestures that mention concrete objects for example showing the direction for the human.

Table 1. The different behaviors that ROBOMO can exhibit

Code of the Behavior	Behavior	Description of the Behavior
IS	inarticulate sounds	yes, no, right, left, forward
ND	nodding	en..well, thank you, I'm not sure
GS	gestures	turning left, turning right

4   ROBOMO Architecture

ROBOMO consisted of a micro PC, five servo-motors (AX-12+) for the body movement and a speaker as an output for the robot's inarticulated sounds. A web camera helped to recognize the person's face and a microphone detected the user's requests that was recognized by Julius (a software for Japanese word recognition) (Fig.2).

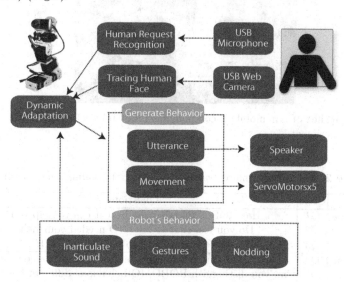

Fig. 2. The system architecture of ROBOMO

5   Experimental Protocol

The main objective is to explore the effectiveness of the attachment process and its impact on the meaning acquisition within a human-robot interaction. We expect that gradually, the communication will be clearer. We setup an indoor ground for navigation task that contains cross points (Fig.3). To pick the right behavior, the participant is instructed by the robot. We asked the participant to talk to ROBOMO with simple words and slowly. 12 participants with age varying in [22−30], take part in 3 sessions. We have chosen several configurations during the 3 sessions to guarantee the diversity of the participant's responses. It helps also to ensure that any successful meaning guessing of ROBOMO's behaviors is not related to the fact that we are using the same configuration but it is related to the social bonding which enhances the participants' adaptation. In our scenario, if the human does not perceive the robot's response, he will repeat his question within a short period for direction's confirmation. In such case, the robot exhibits a body behavior such as pointing to the left or right direction using its upper body part combined with the right inarticulate sound as a response. On the other hand, in the short periods of silence (when the user is not addressing any request), a nodding behavior is displayed. Each student interacts with ROBOMO for 2 minutes and then answers the same 5-Likert Scale questionnaire (13 questions). The table 2 contains the different questions.

Fig. 3. A snapshot of our mobile accompanying robot interacting with a participant during the experiment

Table 2. The questionnaire evaluating the social bonding's five factors

Factors	Code	Questions
Cooperation	Q1	Has ROBOMO tried the best it can to help you?
	Q2	Do you feel that ROBOMO needed your help?
	Q3	Have you wanted to help ROBOMO?
Achievement	Q4	Had you recognized the direction indicated?
	Q5	Can you distinguish ROBOMO behaviors' different meanings?
	Q6	Do you think that you established a good relational contact?
Friendliness	Q7	Can you consider ROBOMO as a friend?
	Q8	Have you felt that ROBOMO was familiar for you?
Stress-Free	Q9	Was it hard for you to understand ROBOMO?
	Q10	Can you get the feeling of ROBOMO?
Adaptability	Q11	Do you think that ROBOMO is a smart robot?
	Q12	Can you feel that ROBOMO showed some animacy?
	Q13	Do you think that ROBOMO behaved like a baby?

Our evaluation of the social bonding process is articulated around five factors: the adaptation, the stress, the friendliness, the cooperation and the achievement. We tried to record on log files the participants' requests and the robot's instructions. We recorded also the interaction videos that helped us to detect the spatial points when the gestures were used.

6   Results

6.1   Questionnaire Based Results

To statistically identify the most ameliorated social bonding factors, we applied ANOVA based on the users' answers. Table 3 exhibits the different p-values and the Fig.4 displays the average mean opinion score (MOS) values of the different subjects per session where the horizontal axis shows the social bonding five factors combined with their related questions during the three sessions and the vertical axis shows the MOS values for 12 subjects. The MOS is the arithmetic mean of all the individual scores, that ranges from 0 (worst) to 5 (best) where a value that is equal to 3 is acceptable. Based on the Figure4, we can see that cooperation, achievement and stress-free factors slightly went up by means of sessions. Table 3 showed that, the questions Q1, Q2 and Q3 which evaluate the cooperation factor were statistically significant with p-values respectively equal to ***p=0.0024<0.005; *p=0.0927<0.1 and *p=0.0993<0.1. The questions evaluating the achievement (Q4, Q5 and Q6) showed also significant results with p-values respectively equal to ***p=0.001<0.005, *p=0.0615<0.1 and **p=0.0137<0.05. Finally, the questions that concern the stress-free (i) Q9: **p=0.0391<0.05 (ii) Q10: **p=0.0185<0.05 showed also that there were statistically significant results. These results suggest that the robot's cooperation capability using the inarticulate sounds and the gestures helped on achieving the task and leaded to stress reduction while interacting with ROBOMO.

Based on the Figure.4, we can see that friendliness and adaptability increase slightly while statistically there was no significant differences between the different sessions with respectively (i) Q7: p=0.2439 (ii) Q8: p=0.1573 for friendliness and (i) Q11: p=0.2038 (ii) Q12: p=0.2875 (iii) Q13: p=0.4785 for adaptability.

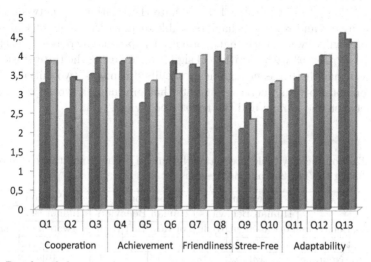

Fig. 4. Results of the average mean opinion score (MOS) based on the 13 questions' answers and for the 3 sessions of the experiment

Table 3. ANOVA evaluation of the questionnaire results

Factors	Code	P-value	Results
Cooperation	Q1	$*p = 0.0927 < 0.1$, d.f=11	significant
	Q2	$* * *p = 0.0024 < 0.005$, d.f=11	significant
	Q3	$*p = 0.0993 < 0.1$, d.f=11	significant
Achievement	Q4	$* * *p = 0.001 < 0.005$, d.f=11	significant
	Q5	$*p = 0.0615 < 0.1$, d.f=11	significant
	Q6	$* * p = 0.0137 < 0.05$, d.f=11	significant
Friendliness	Q7	$p = 0.2439$, d.f=11	not significant
	Q8	$p = 0.1573$, d.f=11	not significant
Stress-Free	Q9	$* * p = 0.0391 < 0.05$, d.f=11	significant
	Q10	$* * p = 0.0185 < 0.05$, d.f=11	significant
Adaptability	Q11	$p = 0.2038$, d.f=11	not significant
	Q12	$p = 0.2875$, d.f=11	not significant
	Q13	$p = 0.4785$, d.f=11	not significant

We asked from people to write down their opinions before and after experiment. We analyzed the participants' different subjective answers and we found out that users confirm that it is easy to adapt with ROBOMO. They found its friendly and cute before even starting the experiment. Thus, the robot's appearance played a key role to reduce the adaptation gap and to give a good first impression.

6.2   Real Time Interaction Results

Based on the stored log files of the speech recognition system and the recorded videos, we counted the user's picked directions based on the robot's indications and the related robot's behaviors (getures, nodding, inarticulate sounds) (table 4) We used the data of the table 4 to evaluate the relationship between participants' behaviors and robot's behaviors. Table 5 shows the different Chi-square test's results where we can see that gradually the p-value increases by means of sessions: $p1 < p2 < p3$ with a statistical significance during the third session. We noticed also that there was no significant results during the two initial sessions. This incremental p-value increase suggests that gradually a strong relationship evolves between the human and the robot's behaviors.

Table 4. The contingency table integrating the human behavior and the related robot's behavior during the 1st, 2nd and 3rd sessions

	Session 1			Session 2			Session 3		
	Human Behaviors			Human Behaviors			Human Behaviors		
Robot's Behaviors	Forward	Left	Right	Forward	Left	Right	Forward	Left	Right
Inarticulate Sounds	9	13	12	13	20	32	9	12	27
Nodding	13	12	18	14	7	11	16	12	13
Gestures	12	6	21	11	11	10	7	17	11

Table 5. Chi-Square test of independency and the corresponding P-values evaluating the relationship between the human behaviors and the robot's behaviors during the different sessions of the experiment

Sessions	Chi-Square Values	P-Values	Results
Session 1	χ^2=5.21, dof=4	$p = 0.266$	not significant
Session 2	χ^2=7.53, dof=4	$p = 0.110$	not significant
Session 3	χ^2=12.2, dof=4	$p = 0.016 < 0.05$	significant

6.3   Correspondence Analysis Results

In order to visualize the relationship between the robot and the users' behaviors, we used a visual approach which is the correspondence analysis. The bi-dimensional map exposed the relationship among categories spatially on empirically derived dimensions. The frequency for each category (*forward, right, left*) and for each variable (nodding, inarticulate sounds (IS) or gestures) is considered in order to expose the Euclidean distance in two dimensions. Figure 5 depicts the associations between categories of robot's behaviors and participants' picked directions during the three trials. The red triangles represent the participants' chosen directions and the blue dots represent the robot's behaviors. Considering the first trial's correspondence analysis Fig.5 (left), we can see that there was no clear relationship between the robot's behaviors and the human's chosen directions. By analyzing the second session results Fig.5 (center), we can see that the robot's behaviors starts to be mapped with the human chosen directions. In fact, there is a tendency to attribute the nodding behavior with the *left* direction, the inarticulate sounds with the *right* direction while the gestures were associated with the *forward* direction. During the final session Fig.5 (right), the Euclidean distance between the robot's behaviors and the human chosen directions becomes shorter and the tendency to associate for each direction a specific robot's behavior becomes clearer. In fact, human turning *right* behavior was related to inarticulate sounds, turning *left* was associated with the nodding, while going *forward* occurred when the robot exposes gestures.

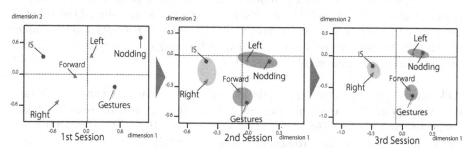

Fig. 5. The correspondence analysis of the trial 1 (left), trial 2 (center) and trial 3 depicting the association between the robot's behaviors (inarticulate sound and gestures) and the directions (forward, right, left)

7   Discussion

Based on the questionnaire results (Fig.4 and table3), we noticed a gradual amelioration on the human's attachment process. The stress was decreasing during the interaction (Fig.4) which explains the different significant p-values ($p=0.0391$, $p=0.0185$). Cooperation had also significant values with $p=0.0024$, $p=0.0927$ while achievement $p=0.001$, $p=0.0615$. This highlights the effectiveness of using inarticulate sounds and iconic gestures to decrease the stress, encourage the human to cooperate with the robot in order to achieve the task and thus helps on creating a social bonding during the human-robot interaction which may facilitate the meaning's acquisition. In fact, we remarked a common interest on finding the frequent successful patterns combining for each robot's behavior a particular direction. Based on the table 5, we remarked that there was an increasing tendency to associate the robot's behaviors with the available directions during the navigation task ($p1 < p2 < p3$). The incremental formation of attuned patterns which maps the robot's behaviors with the human's chosen direction was clearer during the sessions 2 and 3 as the Fig.5 shows. Our experiment leads us to the conclusion that our accompanying mobile robot succeeded in eliciting positive and affectionate behavior from participants. We conclude then that the inarticulate sounds and gestures that were used by ROBOMO during this dyadic interaction appeared sufficient for the attachment evolvement and helped on acquiring the meaning of the robot's behaviors.

8   Conclusion

Our study explored the human's attachment toward our accompanying robot. ROBOMO used inarticulate sound and iconic gestures in order to help people navigating in a block-based environment. It was surprising to see no anxiety-avoidance type of attachment existing in the participants towards ROBOMO which helped to decrease the stress and strengthens the human-robot cooperation in order to achieve the task. The results showed that inarticulate sounds and iconic gestures helped on grounding the attachment process during the experiment and that the participants gradually acquire the meaning of the robot's behaviors. In our future work, we intend to integrate in ROBOMO a self-learning mechanism to improve its adaptation capability and measure the attachment process during the human-robot interaction.

Acknowledgments. Authors would like to express their gratitude to robotics designing group at Kyoto University of Art and Design. This research has been supported by Grant-in-Aid for scientific research of KIBAN-B (26280102) from the Japan Society for the Promotion of science (JSPS).

References

1. Hirschi, T.: Causes of Delinquency. University of California Press, Berkeley (1969)
2. Aitken, K., Trevarthen, C.: Self/other organization in human psychological development. In: Developmental Psychopathology, pp. 653–677 (1997)
3. Custodero, L., Fenichel, E.: The Musical Lives of Babies and Families. Zero to three (2002)
4. Papousek, M.: Intuitive parenting: a hidden source of musical stimulation in infancy. In: Musical Beginnings: Origins and Development of Musical Competence (1996)
5. Beebe, B., Lachmann, F.M.: Infant Research and Adult Treatment: Co-Constructing Interactions. Analytic Press (2005)
6. June, M., Gilbey, A., Rennie, A., Ahmedzai, S., Dono, J., Ormerod, E.: Pet ownership and human health: a brief review of evidence and issues In: BMJ (2005)
7. Kidd, A.H., Kidd, R.M.: Seeking a theory of the human/companion animal bond. In: Anthrozoos, pp. 140–157 (1987)
8. Melson, G.F.: A multidisciplinary journal of the interactions of people and animals. Berg Journals, 45–52 (1987)
9. Melson, G.F.: Children's attachment to their pets: Links to socio-emotional development. Children's Environments Quarterly, 55–65 (1991)
10. Forlizzi, J., DiSalvo, C.: Service robots in the domestic environment: A study of the roomba vacuum in the home. In: ACM SIGCHI/SIGART Conference on Human-robot Interaction, pp. 258–265 (2006)
11. Sung, J., Guo, L., Grinter, R.E., Christensen, H.I.: my roomba is rambo": Intimate home appliances. In: Sung, J., Guo, L., Grinter, R.E., Christensen, H.I. (eds.) Proceedings of the 9th International Conference on Ubiquitous Computing, pp. 145–162 (2007)
12. Jones, K.S., Schmidlin, E.A.: Human-robot interaction toward usable personal service robots. Reviews of Human Factors and Ergonomics, 100–148 (2011)
13. Samani, H.A., Cheok, A.D., Foo, W.N., Nagpal, A., Mingde, Q.: Towards a formulation of love in human - robot interaction. In: RO-MAN, pp. 94–99 (2010)
14. Hiolle, A., Canamero, L., Davila-Ross, M., Bard, K.A.: Eliciting caregiving behavior in dyadic human-robot attachment-like interactions. Interactive Intelligent Systems (2012)
15. Kina, N., Tanaka, D., Ohshima, N., De Silva, R., Okada, M.: Culot: Sociable creature for child's playground. In: Human-robot Interaction, pp. 407–408 (2013)
16. Matsumoto, N., Fujii, H., Okada, M.: Minimal design for human agent communication. In: Artificial Life and Robotics, pp. 49–54 (2006)
17. Cassell, J., Steedman, M., Badler, N., Pelachaud, C., Stone, M., Douville, B., Prevost, S.: Modeling the interaction between speech and gesture (1994)
18. Kanda, T., Ishiguro, H., Ono, T., Imai, M., Nakatsu, R.: Development and evaluation of an interactive humanoid robot "robovie". In: ICRA, pp. 1848–1855 (2002)
19. Sidner, C.L., Lee, C., Kidd, C.D., Lesh, N., Rich, C.: Explorations in engagement for humans and robots. In: Artificial Intelligence, pp. 140–164 (2005)
20. Moon, A.J., Parker, C., Croft, E., der Loos, H.F.: Did you see it hesitate? - empirically grounded design of hesitation trajectories for collaborative robots. In: IROS, pp. 1994–1999 (2011)
21. Noriko, S., Yugo, T., Kazuo, I., Okada, M.: Effects of echoic mimicry using hummed sounds on human computer interaction. Speech Communication, 559–573 (2003)

Sociable Dining Table:
Incremental Meaning Acquisition
Based on Mutual Adaptation Process

Khaoula Youssef, P. Ravindra S. De Silva, and Michio Okada

Interaction and Communication Design Lab, Toyohashi University of Technology
1-1 Hibarigaoka, Tempaku, Toyohashi, Aichi
youssef@icd.cs.tut.ac.jp, {ravi,okada}@tut.jp
www.icd.cs.tut.ac.jp

Abstract. Our main goal is to explore how social interaction can evolve incrementally and be materialized in a protocol of communication. We intend to study how the human establishes a protocol of communication in a context that requires mutual adaptation. Sociable Dining Table (SDT) integrates a dish robot put on the table and behaves according to the knocks that the human emits. To achieve our goal, we conducted two experiments: a human-controller experiment (Wizard-of-Oz) and a human-robot interaction (HRI) experiment. The aim of the first experiment is to understand how people are building a protocol of communication. We suggest an actor-critic architecture that simulates in an open ended way the adaptive behavior that we have seen in the first experiment. We show in a human-robot interaction (HRI) experiment that our method enables the adaptation to the individual preferences in order to get a personalized protocol of communication.

Keywords: Mutual Adaptation, Communication Protocol, Actor-Critic.

1 Introduction

Developing robots with mutual adaptation skills and understanding the meaning acquisition process in the human-human interaction is a cornerstone to build robots that can work alongside humans and learn swiftly from intuitive interaction. By using the natural ability of humans to adapt to other artifacts, the robots can be capable of adapting to humans. Such an adaptation process would commonly be observed in a pair who can communicate smoothly, such as a child and a caregiver. Understanding how the caregiver behaves with the child affords many ideas to design intuitive robots facilitating the communication with people [1]. In fact, the caregiver's voice and physical contact lead to a mutual interest in communication. As a response the child generates some movements and utterances transferring his own assumptions to the caregiver. Incrementally, mutual adaptation evolved since both parties are trying to find the common successful patterns of communication which we name a communication protocol [2]. Our main goal is to explore how a communication protocol is established during the

M. Beetz et al. (Eds.): ICSR 2014, LNAI 8755, pp. 206–216, 2014.

mutual adaptation process in a human-human context and a human-robot context. We intend to develop a computational architecture that helps to simulate the human's adaptive capability using the Sociable Dining Table (SDT). SDT affords the possibility to interact with the humans by displaying its behaviors while the human can interact through a knocking sound with the robot (Fig.1). Knocking is the only channel of communication used in our study that helps to draw a minimalistic scenario similar to the child-caregiver interaction's scenario. It requires mutual adaptation from both parties in order to master and mirror the different most successful knocking and robot's behaviors combinations [3].

Fig. 1. A participant interacts with the sociable dining table

2   Background

To enable the robot to learn flexible mapping relations when interacting with humans in daily life, many studies point out to the mutual adaptation as a very promising solution [4][5]. Mutual adaptation guarantees that if the human proposes new behaviors during the HRI, the robot will try to adapt and acquire the meaning of these new behaviors. Meanwhile, humans also will try to adapt to the robot if it proposed new behaviors [5]. The concept of adaptation was explored in many HRI studies [6][7]. Thomaz et al [8] used the active learning to adapt the robot's knowledge. The robot addresses multiple types of explicit queries to learn the new concepts. Subramanian et al [9] used the explicit answer of the Pacman's users concerning the best interactive options to propose a convenient adaptive Pacman agent that can learn from users. These studies explore the one-sided explicit adaptation (the artifact's adaptation) while a mutual adaptive behavior has to exploit two levels of adaptation to evolve a flexible communication protocol. They also depend on explicit meaning affordance to teach the robot while the meaning can be inferred implicitly in the behavioral interaction between the human and the others. As an example, one can refer to the implicit communication between the caregiver and the child when they autonomously create their own meaning structure through a series of implicit interaction. Our work focuses on the implicit meaning's acquisition and the incremental communication protocol formation through mirroring the patterns of each others' behaviors to guarantee that double sided adaptation emerges.

3   Experiment 1: Human-Controller Experiment

3.1   Experimental Protocol

We conducted a Wizard-Of-Oz experiment that aims to ground the interaction between the human and the controller. 32 participants were grouped into 16 pairs (controller that controls the robot and a user that emits the knocking patterns) in order to lead the robot to the different checkpoints (Fig.2). To avoid the distraction by other sensory channels, the controller is located in another room, ignores the goal, check points and refers only to the knocks. The user knows about the different checkpoints and has to lead the robot through knocking to the final goal after passing by the different checkpoints. The robot uses 5 reflectors [10] to avoid falling from the table. There are 3 trials where in the 1st and the last one we have chosen several configurations by proposing different checkpoints coordinates to guarantee the diversity of the patterns suggested by the participants (Fig.2). Both parties were informed that during the 1st trial the robot can operate only two behaviors (*right, forward*). Since we hypothesized that the pairs will try to build together a communication protocol, we chose 2 behaviors for the 1st trial in order to facilitate finding the successful patterns of communication. In the trial 2 and 3, we increase the degree of difficulty. We told the pairs that the robot can execute 4 behaviors (*right, left, back, forward*). Trial2 is a transitional stress-free session without any checkpoints which we believe that it can enhance the mutual understandability between the two parties. We informed the knocker and the controller during the trial2 that there were no specific trajectories nor checkpoints that the robot has to land on. By changing the configurations and the sessions' conditions, we aim at verifying whether the pairs human-controller can always mutually adapt to each others' behaviors.

Fig. 2. In the 1st trial (left), each participant has to move the creature into 5 places (start, 1, 2, 3, goal) by knocking using 2 behaviors (right, forward). The 2nd trial (center), is a stress-free session where we do not assign any configuration. In the 3rd trial (right), we changed the place of the former points, and then the user has to guide the robot into the new points using 4 behaviors (right, forward, left, back).

4   Experiment 1: Results and Discussion

4.1   Behavior Adaptation Process

Although, we set up 20 minutes as a time limitation to achieve the task, all the participants reached the different checkpoints in less than 15 minutes. Thus,

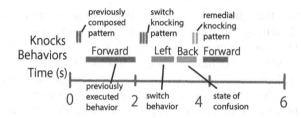

Fig. 3. A scenario showing examples of switch knocking pattern, switch behavior, state of confusion and remedial knocking pattern

to study the incremental adaptation to each others' behaviors, we calculated the number of switch knocking patterns, switch behaviors, states of confusion and the remedial knocking. Figure 3 helps to understand the meaning of these four concepts. As you may see in the Figure 3, the robot executed initially the *forward* behavior and when the controller detected that he received the switch knocking pattern (3 knocks in red), he picked *left* as a new behavior which we call according to this scenario a switch behavior. Thus, we call a switch knocking pattern a new received pattern that is different from the previous received one and a switch behavior the controller's picked behavior as a response for the received switch knocking pattern. Within few milliseconds, we can see that again the controller changes the behavior to *back*. We call such situation a state of confusion since the controller changes the behavior without being prompt by any knocking. As a response the knocker, composed 2 knocks (in orange) as a remedial knocking pattern for the controller's state of confusion. If for each switch knocking pattern, we have systematically a switch behavior then we may conclude that the controller is trying to adapt to the knocker's patterns of knocking. The presence of states of confusion indicate that the controller is trying to establish the rules of communication but may go through some confusing states. Consequently, the knocker also tries to adapt to the controller's state of confusion by composing a remedial knocking pattern and thus the existence of mutual adaptation can be proved. We calculated the test of independence between the switch knocking patterns and the switch behaviors. Table 1 exhibits the Chi-square test results and Cramer V values. A Cramer V value ranging from 0,15 to 0,20 showed that a minimally acceptable dependence exists between the two measured variables while a value ranging from 0,20 and 0,25 showed that we have a moderate dependence and finally a value ranging between 0,35 and 0,41 showed that a very strong relationship exists between the two variables.

Table 1 revealed that during the trial 1, there was no statistically significant relationship between the knocker's switch knocking and the controller's switch behaviors. However, during the trial 2 and 3 we had significant values with p-values respectively equal to 0,036 and 0,0001. By comparing the two Cramer's V values of trial 2 and trial 3, we have $Cramer's V_{trial2} = 0,170 \leq Cramer's V_{trial3} = 0,245$ showing that the dependency between the two variables is becoming gradually larger. This proves that there was incrementally an attempt to combine each pattern to a robot's behavior.

Table 2 revealed that during the trial 1, there was no statistically significant relationship while during the trial 2 and 3 the p-values were respectively equal to 0,019 and 0,004 were significant. By comparing the two Cramer's V values of trial 2 and trial 3, we have $Cramer'sV_{trial2} = 0,260 \leq Cramer'sV_{trial3} = 0,279$ showing that the dependency between the two variables is becoming gradually larger. This proves that the controller was trying to adapt himself and thinking about the best behavior that may correspond to the heard patterns.

We calculated the test of independence between the states of confusion and the remedial knocking. Table 3 exhibits the Chi-square test results and Cramer's V values.

Finally, the Table 3 revealed that during the trial 1, there was no statistically significant relationship. However, during the trial 2 and 3 the p-values were significant with values respectively equal to 0,043 and 0,001. By comparing the two Cramer's V values of trial 2 and trial 3, we have $Cramer'sV_{trial2} = 0,316 \leq Crame'sV_{trial3} = 0,410$ showing that the dependency between the two variables is becoming gradually larger. This proves that the knocker was adapting himself in order to afford for the controller the suitable pattern so he can find his way to the correct behavior. Consequently, based on the 3 tables we can confirm that a double sided adaptation emerges.

4.2   Interaction Smoothness

It is generally assumed that almost any human behavior that involves information processing and decision-making tends to increase the reaction time. We

Table 1. The test of independence between the switch knocking patterns and the switch behaviors as well as the Cramer's V values by means of trial

Trial	χ^2 value	P-value and significancy	Cramer's V (CV)
Trial1	$\chi^2 = 1,112$;df=4	P-value=0,892 at $\alpha = 0.05$ not significant	no significance
Trial2	$\chi^2 = 22,104$;df=12	P-value=0,036 at $\alpha = 0.05$ significant	CV=0,170
Trial3	$\chi^2 = 42,987$; df=12	P-value=0,0001 at $\alpha = 0.05$ significant	CV=0,245

Table 2. The test of independence between the switch knocking patterns and the states of confusion as well as the Cramer's V values by means of trial

Trial	χ^2value	P-value and significancy	Cramer's V (CV)
Trial1	$\chi^2 = 2,334$;df=4	P-value=0,675 at $\alpha = 0.05$ not significant	no significance
Trial2	$\chi^2 = 24,16$;df=12	P-value=0,019 at $\alpha = 0.05$ significant	CV=0,260
Trial3	$\chi^2 = 28,787$;df=12	P-value=0,004 at $\alpha = 0.05$ significant	CV=0,279

Table 3. The test of independence between the states of confusion and the remedial knocking by means of trial as well as the Cramer's V values

Trial	χ^2value	P-value and significancy	Cramer's V (CV)
Trial1	$\chi^2 = 2,635$;df=4	P-value=0,621 at $\alpha = 0.05$ not significant	not significance
Trial2	$\chi^2 = 4,505$;df=12	P-value=0,043 at $\alpha = 0.05$ significant	CV=0,316
Trial3	$\chi^2 = 33,227$;df=12	P-value=0,001 at $\alpha = 0.05$ significant	CV=0,410

wanted to verify whether the controller's response time[1] changes by means of trial (Fig.4). If the response time becomes shorter, we conclude that an adaptation process has facilitated the decision making. The results showed that 75% of the reaction time is in the range of [2-4] seconds . Kruskal-Wallis test proved that there were statistical differences concerning the controller's reaction time during the different 3 trials with(K (Observed value)=13.835; df=2; p-value (Two-tailed)=0.001; alpha=0.1). The multiple pair wise comparisons using the Steel-Dwass-Critchlow-Fligner test showed that there were significant differences between the trial 1 and 2, the trial 3 and 1 but there was no significant differences between the trial 3 and 2. Figure 4 depicts the average reaction time by means of trial for each one of the 16 pairs (knocker-controller) where blue color corresponds to trial 1, red to trial 2 and green to trial 3. During the trial 2 and 3 that involves a higher degree of difficulty, the reaction time decreases slightly in comparison to the trial 1 when the pairs were trying to adapt with a lower task difficulty (2 behaviors). Consequently, even if the complexity of the task increases, the pairs were more engaged during the 2 last trials to acquire incrementally the communication protocol and the decision making becomes easier.

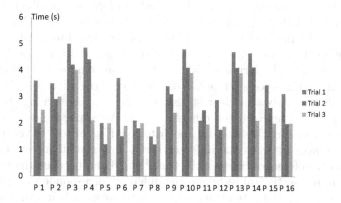

Fig. 4. The response time during the three trials

4.3 Visualization of the Incremental Acquisition of the Protocol of Communication

Using a visual approach which is the correspondence analysis, we succeed in representing the protocol of communication that can be defined as a map which represents the different pairs' knocking patterns and the robot's behaviors. The frequency for each behavior (*forward, right, left, back*) and for each knocking pattern (e;g: 2 knocks, 3 knocks) is considered in order to expose the Euclidean distance in two dimensions. Figure 5 depicts the correspondence analysis for the pair 15 during 3 trials. The red triangles represent the robot's behaviors and

---

[1] It is the time between the onset of the knocking pattern and the time of the 1st response of the controller regardless of whether it was correct or not.

the blue circles represent the knocking patterns. During the trial 1 (Fig.5 (left)), the *right* behavior was associated with 4 and 2 Knocks and *forward* with 3 and 6 knocks. During the trial 2 (Fig.5 (center)), the behavior *back* was associated with 3 knocks, *left* and *forward* with 1 knock while *right* with 2 and 4 knocks. Finally, during the trial 3 (Fig.5 (right)), the pair successfully distinguished the different combinations where 4 knocks was associated with *back*, 2 knocks with *right*, 1 knock with *left* and 3 knocks with *forward*. The different correspondence analysis results proved that the pairs try to establish a communication protocol incrementally.

4.4   The Convergence to a Protocol of Communication

We wanted to explore statistically the differences between the participants' communication protocols during the 3 trials. For this issue, based on the correspondence analysis results, we calculated the euclidean distance between each of the robot's behaviors (red triangles as presented in the Fig.5) and the different patterns (blue circles as presented in the Fig.5). After, we picked for each behavior the minimum distance. We sum up the 4 minimum distances[2] and the resultant value which we call convergence metric, affords an information about the minimum distance that the pair knocker-controller reached to form stable rules. We repeated the same procedure for the 16 pairs and for the three trials.

To verify whether there was statistically convergence differences during the three trials, we used the Kruskal-Wallis test. As the computed p-value=0,01 is lower than the significance level alpha=0,1, we accept the alternative hypothesis confirming that there was a clear statistical difference concerning the convergence to a protocol between the different trials. We applied the multiple pair wise comparisons using the Steel-Dwass-Critchlow-Fligner test to verify the significant differences between the different trials. The statistical results showed that there were differences between the trial 2 and 3 and between the trial 1 and 3. Combining the statistical tests and the different correspondence analysis, we conclude that there was a tendency to associate for each behavior a knocking pattern especially during the trial 3.

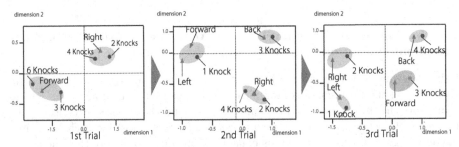

Fig. 5. The correspondence analysis representing the communication protocol during the trial 1 (left), the trial 2 (center) and the trial 3 (right)

---

[2] Each minimum distance is associated with one behavior.

5   Actor-Critic Architecture

Through the 1st experiment, we noticed that incrementally people use in a trial-error process the different successful combinations of (knocking pattern/ robot's behavior) to establish the rules of communication. We proposed a similar trial and error method that is based on the reinforcement learning. Our solution consists on an actor-critic architecture which we expected that it will help to establish a communication protocol.

5.1   Actor Learning

Each knocking pattern has its own distribution $X(S_t) = N(\mu_{X(S_t)}, \sigma_{X(S_t)})$ where $X(S_t)$ is defined as the knocking pattern, $\mu_{X(S_t)}$ and $\sigma_{X(S_t)}$ are the mean value and the variance. We chose 2 s as a threshold for the user's reaction time based on the human-controller experiment. In fact, the results showed that the reaction time is in the range of [2-4] seconds (s) and thus we assumed that we assumed that we have a disagreement state if the human interrupted the robot when it is executing the chosen behavior within 2s. When the robot observes the state S_t that is materialized by a knocking pattern, the behavior is picked according to the probabilistic policy $\Pi(s_t)_{nbknocks}$. If within 2s there was no knocking pattern, we suppose that the robot has succeeded by choosing the right behavior and the critic reinforces the value of the executed behavior in the state S_t to increase its chances to be picked in the future if the robot receives the same knocking pattern. Finally the system will switch to the state S_{t+1}. But if a new knocking pattern is composed before that 2s elapsed, the state of the interaction changes to the state S_{t+1} indicating that the knocker disagrees about the behavior that was executed, the probabilistic policy failed to propose the correct behavior. The critic updates thus the value function before choosing any new behavior. As long as the knocker is interrupting the robot's behavior before that 2 seconds elapsed, the actor chooses the action henceforth by pure exploration (until we meet an agreement state: no knocking during 2 seconds) based on (1). The random values vary between $0 \leq rnd1$, and $3 \leq rnd2$ the above range was decided to bring the values of the action (1) between 0 and 3 (corresponding to the behaviors' (forward, right, back, left) numerical codes). We assume in such case that the knocker will randomly compose the patterns just to switch desperately the robot's behavior.

$$A(S_t) = \mu_{X(S_t)} + \sigma_{X(S_t)} * \sqrt{-2 * log(rnd_1)} * Sin(2\Pi * rnd_2) \tag{1}$$

5.2   Critic Learning

The critic calculates the TD error δ_t as the reinforcement signal for the critic and the actor according to Equ.2

$$\delta_t = r_t + \gamma V(s_{t+1}) - V(s_t) \tag{2}$$

with γ is the discount rate and $0 \leq \gamma \leq 1$. According to the TD error, the critic updates the state value function $V(s_t)$ based on (3).

$$V(S_t) = V(S_t) + \alpha * \delta_t \tag{3}$$

where $0 \leq \alpha \leq 1$ is the learning rate. As long as the knocker disagrees about the executed behavior before 2 s elapsed, we refine the distribution $N(\mu_{X(S_t)}, \sigma_{X(S_t)})$ which helps us to choose the action according to (1). The distribution update consists on computing (4) and (5).

$$\mu_{X(S_t)} = \frac{\mu_{X(S_t)} + A_{S_t}}{2} \tag{4}$$

$$\sigma_{X(S_t)} = \frac{\sigma_{X(S_t)} + |A_{S_t} - \mu_{X(S_t)}|}{2} \tag{5}$$

6   Experiment 2: the Human-Robot Interaction

6.1   Experimental Setup

A second experiment HRI was conducted to show that our architecture learns in real time how to establish the protocol of communication based on the knocking patterns. In this experiment, 10 participants accomplish the same task as in the 1st experiment with two different configurations for the two trials that are also different from those used in the trial 1 and 3 of the experiment 1 (Fig.2) .

6.2   Visualization of the Incremental Acquisition of the Protocol of Communication

We remarked that the human-robot pairs were able to establish communication protocols that allowed the robot to reach the different checkpoints. As in the first experiment, we applied the correspondence analysis for all the participants' interaction data to visualize the communication protocol. Figure 6 exhibits respectively the results of the 1st (left) and the 2nd (right) trial. Figure 6 (left) shows that there was some tendency to attribute for the behaviors different patterns. *Right* was combined with 1 knock, *forward* with 2 knocks with some confusion for the *left* behavior (1 and 4 knocks). During the 2nd trial (Fig.6(right)), the Euclidean distance between *forward* and the pattern 2 knocks decreases, *right* was combined with 1 knock and *left* with 3 knocks.

6.3   The Convergence to a Protocol of Communication

As in the 1st experiment, we calculated for the two trials, the convergence metric values of the 10 participants based on the correspondence analysis results. To verify whether there was statistically some convergence differences during the 2 trials, we used the Mann-Whitney two-tailed test. As the computed p-value=0.027 is lower than the significance level alpha=0.05, we accept the alternative hypothesis confirming that there were a clear differences concerning the

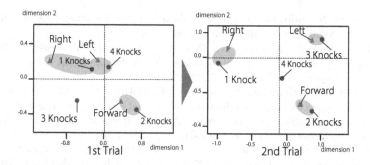

Fig. 6. The correspondence analysis displaying the communication protocol during the trial 1 (left) and the trial 2 (right)

convergence between the trial 1 and 2. As a conclusion, we acknowledge that each participant is collaborating with the robot in order to find out the common best practices associating each behavior with the most convenient generated knocking pattern exactly as in the human-controller experiment.

7 Conclusion

The results showed that the WOZ experiment helps to explore how mutual adaptation evolves between the controller and the knocker and how a protocol of communication can emerge incrementally. The 2nd experiment indicates that there was an incremental formation of a protocol of communication as in the 1st experiment. Although the promising results that we gathered, we have seen that in some cases there are some participants that have slowed adaptation in comparison to others which can be justified by the fact that there are some people that gets along with a different kind of learning. In our future work, we intend to elaborate a learning method that helps to boost the convergence to a communication protocol using inarticulate sounds.

Acknowledgments. This research is supported by Grant-in-Aid for scientific research of KIBAN-B (26280102) from the Japan Society for the Promotion of science (JSPS).

References

1. Michaud, F., Laplante, J., Larouche, H., Duquette, A., Caron, S., Letourneau, D., Masson, P.: Autonomous spherical mobile robotic to study child development. In: IEEE International Conference on Systems, Man, and Cybernetics, vol. 4, pp. 1–10 (2005)
2. Condon, W.S., Sander, L.W.: Neonate movement is synchronized with adult speech:interactional participation and language acquisition. Science 183, 99–101 (1974)

3. Matsumoto, N., Fujii, H., Okada, M.: Minimal design for human agent communication. In: Artificial Life and Robotics, pp. 49–54 (2006)
4. Okada, Y., Ueda, S., Komatsu, K., Takeshi, O., Kamei, K., Yasuyuki, S., Nishida, T.: Formation conditions of mutual adaptation in human-agent collaborative interaction. Applied Intelligence, 208–228 (2012)
5. Xu, Y., Ueda, K., Komatsu, T., Okadome, T., Hattori, T., Sumi, Y., Nishida, T.: Woz experiments for understanding mutual adaptation. AI Society, 201–212 (2008)
6. Thomaz, A.L., Breazeal, C.: Teachable robots: Understanding human teaching behavior to build more effective robot learners. Artificial Intelligence, 716–737 (2000)
7. Mitsunaga, N., Smith, C., Kanda, T., Ishiguro, H., Hagita, N.: Robot behavior adaptation for human-robot interaction based on policy gradient reinforcement learning. In: Intelligent Robots and Systems, pp. 218–225 (2005)
8. Crystal, C., Cakmak, M., Thomaz, A.L.: Transparent active learning for robots. In: Human-Robot Interaction, pp. 317–324 (2010)
9. Subramanian, A., Charles, K., Isbell, L., Thomaz, A.L.: Learning options through human interaction. In: Agents Learning Interactively from Human Teachers, pp. 208–228 (2011)
10. Kado, Y., Kamoda, T., Yoshiike, Y., De Silva, P.R.S., Okada, M.: Reciprocal-adaptation in a creature-based futuristic sociable dining table. In: 2010 IEEE RO-MAN, pp. 803–808 (2010)

A Sociological Contribution to Understanding the Use of Robots in Schools: The Thymio Robot

Sabine Kradolfer[1], Simon Dubois[1], Fanny Riedo[2],
Francesco Mondada[2], and Farinaz Fassa[1]

[1] Faculté des Sciences Sociales et Politiques,
Université de Lausanne, Switzerland
[2] Laboratoire de Systèmes Robotiques,
Ecole Polytechnique Fédérale de Lausanne, Switzerland

Abstract. The Thymio II robot was designed to be used by teachers in their classrooms for a wide range of activities and at all levels of the curriculum, from very young children to the end of high school. Although the educationally oriented design of this innovative robot was successful and made it possible to distribute more than 800 Thymio robots in schools with a large majority in the French-speaking part of Switzerland, it was not sufficient to significantly raise the number of teachers using robot technology in their teaching after three years of commercialization. After an introduction and a first section on the design of this educational robot, this paper presents some results of a sociological analysis of the benefits and blockages identified by teachers in using robots, or not, with their pupils.

Keywords: Educational robot, Thymio II, curriculum, school, sociology.

1 Introduction

Youngsters find mobile robots fascinating, as some of their features, like movement and interaction with the environment, are similar to those of living beings [1]. Moreover, they have been widely presented in literature, movies and the media as machines that are intelligent and even have feelings (see C-3PO and R2-D2 in the movie Star Wars, for example). But they also have a strong link with the real world because of their various applications, ranging from manufacturing to medicine, from rescue to environmental monitoring.

This wide spectrum of applications allows to use robots in courses on robotics but also in many other robotic-related fields. According to Barreto [2], 80% of studies on robots in education address fields linked to math or physics. But they can also be used as educational tools for other disciplines, such as geography, history, languages, or the arts. The recent trend to move from STEM (Sciences, Technologies, Engineering and Mathematics) to STEAM (adding an A for Arts) education [3] illustrates this interesting possibility. Moreover, educational robots can be used in formal educational environments such as schools, or in informal education, for instance in festivals or similar outreach events.

M. Beetz et al. (Eds.): ICSR 2014, LNAI 8755, pp. 217–228, 2014.

There are few robots widely used in education. Even fewer are designed only for education. Most of them are toys or hobbyist products used for education. One of the best-known robots used in schools is the LEGO Mindstorms, in its various versions, the latest being the EV3. This robot is developed from the famous LEGO bricks, integrating sensors, actuators and a power/computational brick. These additional elements are very classical ones, and the main added value for education comes from the adaptation of the LabVIEW graphical programming environment for programming them. The target market is not only schools, but also the general public, as educational toys. The LEGO Mindstorms is a very flexible tool, making it possible to study programming as well as mechanics, and it can be used to approach robotics and extra-robotic activities. A cheaper version, focused more on the educational market, is the LEGO WeDo kit [4], simpler than the Mindstorms and only allowing one sensor and one actuator to be connected. Also focusing on schools and targeting a low price, TTS produces a robot for primary schools called Bee-Bot [5]. This robot has very simple and classic hardware, making it possible to program a displacement by a sequence of steps and turns of the robot. Also in this case the educational effort has not been focused on the robot hardware, but on the mats on which the robot moves. Such mats make it possible to train abilities that are not directly related to robotics, such as reading, doing simple math, recognizing colors and improving laterality for very young children. While the Bee-Bot looks like a little animal and not a robot, there are a set of kits looking very technical and focused on advanced electronic, mechanical or computer science skills. Those kits can have a visible and modifiable electronics, real mechanical parts such as screws and bolts and classical programming environments based on C++ or Python. The electronics is emphasized in kits based on Arduino [6] or the Raspberry PI boards, for instance. Mechanics is emphasized in kits such as those sold by VEX. In all these kits, the focus of the hardware is on technology and not on educational support. The price of a robot based on this technology is normally above 200$.

There is also a set of educational platforms in the research community that is not spread commercially. We can mention Cubelets[7], Play-i[8], Linkbot[9] or the soft robots by the group of Iida[10]. All these platforms bring to various ages of learners some very interesting concepts developed in robotics research, such as modularity, softness or control concepts, but none digs into the human-robot interface targeting specifically formal or informal education. We therefore decided to start the multidisciplinary design of an educational platform to be validated by a large usage in the general public and schools.

In this paper, we will first present the principles we considered when designing the Thymio robot for pedagogical use inside classrooms, and then the results of a sociological analysis on the use and acceptance of robots by teachers in classrooms in the French-speaking area of Switzerland (cantons of Geneva and Vaud). The team of Luc Bergeron, Ecole Cantonale d'Art de Lausanne, provided the product design support. The sociological study was carried out by sociologists of the University of Lausanne with the support of the Swiss National Center of Competence in Research "Robotics".

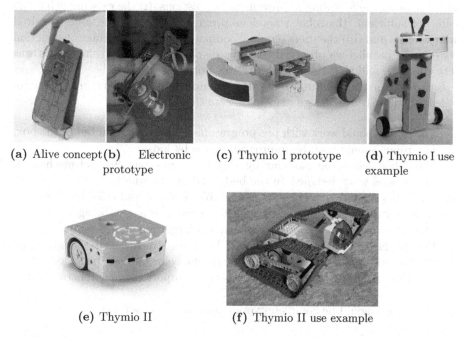

(a) Alive concept **(b)** Electronic prototype   **(c)** Thymio I prototype   **(d)** Thymio I use example

(e) Thymio II   **(f)** Thymio II use example

Fig. 1. Steps in design from the initial concept (a) to the first working prototype (b) to the first version of Thymio with mechanical modularity (c) and an example of use (d) to the final version of Thymio II (e) and an example of use (f)

2   The Design, from the Concept to Thymio II

The initial concept of this project came from the designers Julien Ayer and Nicolas Le Moigne during a workshop held at the Ecole Cantonale d'Art of Lausanne, Switzerland, in 2010. Their idea was to have sensor and actuator modules to robotize any object, as illustrated in Figure 1a with a cardboard structure. The goal was to enable children to develop their creativity.

This concept has been implemented in some raw prototypes (see figure 1b on a potato) that have been tested with children during several workshops. The success of these workshops led us to develop a more finalized version of this system, called Thymio, illustrated in Figure 1c. Figure 1d shows an example of use. The main change between this and the previous concept is that, instead of starting from modules that can be assembled into a robot, which proved to be a difficult task, users start with a working robot that can be disassembled into modules. It is then possible to re-assemble it in new constructions. The appearance was very neutral (white), allowing the children to customize it. The robot had three pre-programmed behaviors. Their respective working principle was illustrated by animated color LEDs. One thousand of these robots were produced and sold at several workshops. This ensured a large set of users that could be asked for feedback. Such feedback was collected from 70 families [11] and generated the following improvement suggestions: (i) the existing behaviors were not sufficient, the robot should be programmable, (ii) the users required more sensors and (iii) the users required compatibility with existing construction systems. In parallel we observed that the display of behavior using LEDs was very effective, and that the modularity was not really used.

Based on this experience, we decided to build the actual version of Thymio, called Thymio II. The design principles were the following:

1. The robot should work with pre-programmed behaviors right out of the box.
2. The robot should then be re-programmable by kids.
3. The robot should be modular by allowing extensions. Several mechanical connections were designed on the body and on the wheels.
4. The robot should be as neutral as possible in shape and color to encourage creativity and not to appeal to a specific gender or a specific age.
5. The robot has been equipped with a much larger number of sensors.
6. The display of functions and behavior has been strengthened, introducing a specific display for each sensor and a specific color for each behavior.
7. The production price could be increased to support the new functionalities, but still keeping it cheap, i.e. below 45$.
8. The robot should be completely open hardware and software.

These design choices have been complemented by a participative wiki website describing the robot, educational material and videos. More than 5000 robot have been produced and sold worldwide, 800 being in schools. Several training sessions have been organized for teachers in the French-speaking part of Switzerland.

Despite the new design, its effectiveness [12–14] and after three years of dissemination work accomplished in schools, robots, and among them Thymio II, are seldom used in education. This situation was confirmed by our sociological research. As we shall see, according to the informants of the research, this situation is not mainly due to the relevance of robots in education but results from the poor capacity of state subsidized schools to adapt to new technological devices. The same kind of blockages could be observed 25 years ago with the first computers [15] and now applies to tablets. In a period where any expense in education is harshly discussed, these blockages are explained mainly by the financial investment required to introduce new IT devices in schools. Whereas large experimentations should be done to evaluate what education could gain from a regular use of IT and/or robots, the development of IT in education raised intense political debates and were presented as unaffordable for state schools whereas it is not the case in private education [16].

3   Analysis of Diffusion in Schools

A one-year pilot study was carried between August 2013 and July 2014 to better understand how robots are used by teachers in classrooms. We were particularly interested in gathering information on (1) the acceptability of and interest in robots as socio-technical educational tools and (2) the resistance and/or acceptance expressed by teachers toward this new type of device. As robots are not common in education and for educational goals, we decided to concentrate our study on the views of the teachers who use robots in education and to investigate more deeply the general disposition to bring such apparatus into the classroom.

We first present the advantages of a wide use of robots in education according to the teachers we met and then describe the kind of blockages they face when they try to integrate robotics into their routine as teachers.

3.1 Methodology and Population

As the aim of our research was to tackle the benefits and the blockages identified by teachers in using robots with their pupils, we chose a qualitative comprehensive research design which allows the actors' subjectivity to emerge and their good reasons to do what they do to be given. Following the principles of the comprehensive apprehension of social determinants proposed by Max Weber [17] at the beginning of the 20th century and developed since then by numerous sociologists (for more information see [18]), we consider in line with socio-constructivist positioning that an appropriate explanation of a social fact only can be achieved through two preliminary stages: the comprehension and the interpretation of social action as they are given by the actors themselves. The comprehension of the sense and the motivations, in our case for teachers to use robots, was undertaken during the semi-oriented interviews that followed a structured interview guide constructed in line with our research questions. All interviews were recorded and transcribed before undertaking a discourse analysis [19] that isolated the different themes organizing the discourses of the teachers we met in order to highlights shared points of view and therefore to clarify what is at the roots of social actions and in this case at the roots of pedagogical activities and choices. Compared to quantitative methods, which use statistics to identify trends in human actions through deduction, this qualitative approach is inductive. Therefore the opinions and discourses of individuals have an explanatory nature that is central for understanding their actions as the reasons people give to their actions contribute to construct the social realm in which there are enacting [20]. As our informants were mainly (14 out of 15 persons) recruited from the population of teachers that had participated in an event organized by the Laboratoire de Systèmes Robotiques, they all show a particular interest in IT and/or robotics. We tried to get information from people opposed to using robots in classrooms but they were almost impossible to identify. Although our informants regularly stated they were facing reluctance to the introduction of robots by some of their colleagues, they didn't convey us the contacts of such of their colleagues when asked to provide us with their details. Several renditions could be given to this situation : one can consider that the robots interested teachers were unable to cooperate with us because, despite they claiming to face opposition from their colleagues, they were actually isolated, their interest in robotics not being neither shared nor discussed by their colleagues. One can also argue that their reluctant colleagues might not been interested to answer a sociological research that raise issues they were opposed to and/or that they didn't wish to give voice to opponents of a deep involvement with this technology such as theirs.

Taking these points into consideration, we nevertheless decided to analyze the discourse of this group of teachers considering it as a specific point of view on robots in education. It is interesting to understand what are the reasons of these committed teachers to introduce robots in education (should it be regular or aimed at children afflicted with cognitive difficulties) and what are the main blockages they say to face when they want to innovate their pedagogical activities thanks to robots. Therefore, the analyses we propose here on the acceptability of robots in education reflect mainly on one hand the understanding of this group of pioneers and on the other the institutional understanding that prevails in the French speaking part of Switzerland, where our research took place.

We carried out 14 interviews with 15 teachers; 7 out of them were men and 8 women, 11 worked in public schools, 3 in private ones and one person was specialized in giving workshops in robotics. They belong to all levels of education, starting with kindergarten (4 year old children) through high school level (18 year old pupils): 8 of our informants work at the pre-primary and the primary level, 2 at the secondary level, 4 at the post compulsory school and 1 who intervenes at different levels. 8 worked either as ICT officer/manager or as PRessMITIC in their schools. This means that this last group of teachers has been specially trained in IT support to help their colleagues or to intervene with pupils on specific topics regarding IT and therefore - sometimes - also robotics. Seven of the 11 users of robots in classroom use Thymio II.

3.2   Robots and the School Curriculum

The period during which our study was carried out is particularly interesting. The organization of compulsory education has been radically changed since a new educational program for all schools of the French-speaking part of Switzerland "Plan d'études romand" - PER) began to be implemented in 2011/2012. Due to the federalist structure of Switzerland, previously all the regions ("cantons") were independent in their choice of educational policies and curricula. The need to ensure more convergence and more coherence among the different cantons led to an inter-cantonal reorganization of the educational curricula through the PER. This school curriculum does not mention robots directly but one of its topics for the general education of pupils throughout their school years is dedicated to the study of MITIC; PRessMITIC training was created to give support to teachers in these domains. It aims to educate children in ICT and Media tools. Although this domain is not precisely defined, it allows the use of robots for different kind of purposes ranging from raising awareness of technology to a learning of language with the help of robots. This means that the decision to use robots in classes depends mainly upon the teachers' willingness and we could observe that educational robots mainly appear in extracurricular activities such as workshops.

3.3   Benefits

Robots Disrupt the Traditional Teaching Styles. Robots present great advantages, as several teachers observed, because they disrupt the classical school order, which Vincent et al. [21] call "the scholastic form":

> "they [the pupils of a pre-school class] tried them out, they found some of the functions, but not all of them, for example those where had to clap your hands, they never found them, because they were told: "Be careful with it, handle it gently, this equipment has been lent to us," so they never imagined you could touch them to make something happen [she laughs], it goes against a principle we try to teach them from the outset [...] They really liked the fact that they could work on the table and it doesn't fall off, whereas at first they were immediately afraid that it would. They hadn't realized at the start that it wasn't going to fall, so that was a feature they really liked." (101).

The same teacher observed that the benefits of robots for her pupils greatly outweigh the time investment she had to make before being able to start to use robots in her classroom. The break with usual school knowledge was also identified in the application of theoretical knowledge:

> "The Thymio robot uses event algorithms, so you have to change their [the pupils'] point of view somewhat, and that's interesting, yes, suddenly you have to... you apply things that you've... that the pupil has learned and then she has to apply it slightly differently. You have to move away a bit from the scholastic application of things, and that's no easy step, but a very useful one if you manage it and especially if you can get the pupils to do it." (106).

Thus, because robots provide new teaching tools, they induce and allow innovative pedagogical practices likely to challenge the dominance of written text as a means of access to knowledge:

> "I think it shows them that they understand certain things. But that it isn't math or French. In the class of eleven year olds, who has repeated their year [...] there was a child who went up to his teacher and said, "Today, I feel I'm living again." " (107).

For this very reason, robots are also welcome in working with children who face major difficulties when a normal curriculum is applied:

> "And I found that the workshop was a good way to bring together pupils who were performing poorly but who had real abilities, others who found it hard to conceptualize things, math and even physics and chemistry, they can be very abstract sometimes, and pupils who weren't integrated, and so I ran that workshop for pupils aged 11 to 15. I was a bit worried about the age difference, but I wanted to try out a kind of... group spirit." (102).

Although robots may help children to understand concretely what they are doing and enable them to progress, they also break with theoretical knowledge and traditional teaching styles that are based on the practice of written language according to the work of Vincent et al. [21]. In line with their findings, we can conclude that in so doing, robots are also likely to destabilize the balance of power that formal education reproduces according to Bourdieu [22].

Which Robots for Which Children?. In our interviews, we could identify that different robots are used at different ages. Although Thymio II was designed for types of pupils from pre-school classes up to university, it is often integrated into education after an introduction to BeeBots (from the first to the sixth school year for children aged 4 to 10 and before or in parallel with the Arduino and the Lego Mindstorms, which are mostly used in senior high school (pupils aged 15 and over). One of our interviewees considered that:

"for the youngest, Beebot is almost better [than Thymio II], because they can really control the movements" (101),

and another one saw Thymio as useful for an introduction:

"As I see it, at whatever age, from 11 to 18, even if they of course don't approach it in the same way, Thymio is an introduction to robotics. It's... a way in. It seems to me to be the simplest." (102).

Two different types of practice were described by the teachers we met: robots can be used either (1) as tools to enhance various types of learning (including languages), or (2) as technical devices to study robotics and the disciplines that are at its roots (computer science, math, physics, electronics, etc.). Activities in the first category are mainly mentioned in primary schooling. For these pupils, there are linked to spatial awareness, which can easily be done with BeeBot or Thymio II:

"I think that with children who have a lot of problems with spatial orientation you could... yes, stimulate them by getting the robot to move around, follow a route [...]. And they could say "Now I'm upside down, now I'm the right way up". Starting from an early age. I think that would help them a lot." (111).

A seminal workshop to learn German is often cited as an example that could be followed to use robots in an innovative way at the secondary level, but apart from the teacher who developed this lesson, very few people seem to have endorsed it.

We nevertheless observed a situation in which these devices are very seldom present in the classrooms: between the use of robots with youngsters in primary school and their use in post-compulsory education. The reasons for this are complex and they relate in good part to the school curriculum in secondary classes and to the school organization.

"There's the world of infants' and primary schooling, where the teacher is in practice relatively free to choose what he does, the activities, and there, working with Thymio is great. And then you enter the world of the secondary school, where the French teacher is there to do French dictations and spelling and then comes the math teacher who has to do math theorems, and... and... there's not much room for robots, unless you create specific options." (105).

3.4   Blockages

As the PER does not explicitly mention robots, even where teachers work in subjects related to robotics, they have to face barriers to their proper use.

Money. The main difficulty stems from the fact that robots are still an expensive technology. The high cost of Lego Mindstorms was often mentioned by our interviewees, but even with less expensive products such as Thymio II, the question of very limited school budgets appears to be central:

> "We are told, "Oscilloscopes are a good thing," because there's an official text that says that every school teaching for that qualification must have an oscilloscope. No room for argument there. You go to the principal and say "It's not my idea, it says so there." But it doesn't say that every school must have a robot." (105).

or

> "I had to beg for funds... I bought two [Thymio] robots at the robotics fair, with my own money. And then I asked if it might be possible to reimburse me, and that was allowed, and then I asked if I could buy two more." (111)

And as the same interviewee explains a little later, even when she was allowed the money to buy robots, she finds it difficult to order them. She told us how a colleague was interested in using robots with his pupils and he asked her to order them, but she could not do so:

> "That colleague could help me make the missing link for those who would like to do some robotics in the school. And then you come up against the CADEV catalogue (the central purchasing department of the State of Vaud), which doesn't offer everything! [...] So I phone Mr. X. and say, "You see, X, my colleague Y wants to do some robotics and he's familiar with the Raspberry Pi: what did you do to order them for your school, because I can't get them through that CADEV catalogue?" " (111).

Private schools are in a rather better situation in that regard. Although they seem more open to working with robots on pedagogical grounds, financial questions also play a large role. In contrast to the subsidized state schools, using robots in private education can even be viewed as a means to generate a benefit for everyone. Although the optional workshops offered to the pupils (on computer science) require an additional payment from the parents, the cost they entail are no longer a barrier and this innovation can be brought into the school as a useful extra educational contribution to the children's future:

> "This is a private school, no problem there, the parents are very interested, the principal is very interested, he bills it as an extra at... incredible prices for the parents. The parents are happy, I'm happy, and the kids are happy because they each have their robot. That's not the problem. It's more a question of: "But what place does that have in a lesson?" " (102).

Time, Training and Curriculum. The current poor integration of robots into education has the consequence that teachers don't view the immediate utility of robots and that they approach activities with this socio-technical device as particularly time-consuming. While the most convinced among them overcome this difficulty, it may create barriers for the beginner. Due to the lack of time to invest in creating pedagogical sequences including robots, they wish to get ideas and ready for use instructions to carry out a proper lesson.

> "If we really wanted to develop good material for the Thymios, we'd need a great amount of time, and some reduction in our teaching load to do it. [...] But given the duties assigned to me, I can't really do it." (111)

Robots in classrooms being a pioneers' activity, the pedagogical equipment is not clearly identified, even if teachers working with Lego Mindstorms did not mention it as often as the ones who explore Thymio II. So setting up activities with robots is seen as time-consuming, and teachers sometimes feel lonely in this odyssey:

> "I'm all alone in this [...]. I set up the workshop on my own. Defining the educational aims and objectives was all down to me, sorting out software bugs in the evening or early morning, running the workshop, all that is my work. I have no team, no support, and in particular no colleagues who are interested." (102)

Finally, even when teachers are interested in using robots, either because they know robots from previous experiments outside school with this kind of device, or, having attended a workshop offered by the HEP-VD or robot specialists from the EPFL, they might be willing to introduce robots into their teaching activities, they are confronted to budget constraints. But they also have to face difficulties due to the fact that the school curriculum does not really allow them to teach this kind of knowledge:

> "So I wanted to run a robotics course [but due to the curriculum spelling] I realized I couldn't then integrate it into my lessons, whether in math, or physics and chemistry, or electricity, which are the subjects I am spread over [...]. I had to call it a workshop [i.e. optional], because I wasn't allowed to call it a lesson (laughs)." (102).

4   Conclusion: Using Robots in Classrooms without Institutional Injunctions?

The Thymio project was started to bring technology education to a large number of youngsters. We started this project with mechatronics, product and interaction design, targeting the best learning experience, and we had considerable success in three years of informal education events. In formal education (schools) Thymio achieved a similar acceptance and diffusion to those of other tools such as the BeeBot or Mindstorms robots, which have fewer education-oriented technical features, are not open-source, have fewer sensors and are not gender or age neutral. After three years of sales and with more than 800 Thymios being mainly used in Swiss schools, the sociological study presented enables to better understand the perceived benefits but also the factors blocking a wider diffusion.

Among the broader benefits, it has been observed that robots such as Thymio break with the classical school order and can facilitate the education of children who face difficulties when following a normal curriculum. This social aspect of the use of robots was observed in several teaching disciplines.

The observed blocking factors often come from the school structure. Although money can be a problem and Thymio brings a solution with its lower price and broad use across ages, this is not the only problem. The lack of injunctions in favor of robotics in educational policies or from local authorities (school area directors) is also at the basis of a lack of pedagogical research on the benefits that robots could offer to education. Therefore, working with these devices implies a commitment that is difficult to fulfill for regular teachers. Not having a proper training and not having activities at hand that they can offer to their pupils, planning activities with robots seems so time-consuming that a great proportion of them give up before trying.

Therefore it is crucial to better spread knowledge on the possible benefits of using robots in education and to develop research on that field [23]. This requires a real backing from educational policies and a larger involvement of entities that are in charge of teacher training or are references in the use of technology. We have already tested, for one year, the organization of regular training sessions for teachers, with a big attendance despite the fact that the sessions were during the teachers' free time. We will continue this effort and are starting to develop ready-to-use material fitting the standard school programs in math, science and other disciplines.

To conclude, introducing a robot into the educational ecosystem of a school requires a strong interdisciplinary effort involving technology, sociology, pedagogy and politics. We hope that this study will encourage other interdisciplinary efforts in this critical domain.

Acknowledgments. This research was supported by the Swiss National Center of Competence in Research "Robotics".

References

1. Turkle, S.: The Second Self. Simon and Schuster New York (1984)
2. Benitti, F.B.V.: Exploring the educational potential of robotics in schools: A systematic review. Computers & Education 58(3), 978–988 (2012)
3. Jin, Y.-G., Chong, L., Cho, H.-K.: Designing a robotics-enhanced learning content for steam education. In: 2012 9th International Conference on Ubiquitous Robots and Ambient Intelligence (URAI), pp. 433–436 (November 2012)
4. LEGO, "Lego wedo" (2013), http://www.legoeducation.us/eng/categories/products/elementary/lego-education-wedo

5. TTS Group: Bee-Bot (2011), http://www.beebot.org.uk
6. Balogh, R.: Educational robotic platform based on arduino. In: Proceedings of the 1st International Conference on Robotics in Education, RiE2010. FEI STU, Slovakia, pp. 119–122 (2010)
7. Gross, M.D., Veitch, C.: Beyond top down: Designing with cubelets. Tecnologias, Sociedade e Conhecimento 1(1), 150–164 (2013)
8. Play-i, "Play-i: Delightful robots for children to program" (2014), https://www.play-i.com
9. Barobo, "Barobo, inc. - educational robotics." (2014), http://www.barobo.com
10. Yu, X., Assaf, D., Wang, L., Iida, F.: Robotics education: A case study in soft-bodied locomotion. In: 2013 IEEE Workshop on Advanced Robotics and its Social Impacts (ARSO), pp. 194–199 (November 2013)
11. Riedo, F., Rétornaz, P., Bergeron, L., Nyffeler, N., Mondada, F.: A two years informal learning experience using the thymio robot. In: Rueckert, U., Joaquin, S., Felix, W. (eds.) Advances in Autonomous Mini Robots, vol. 101, pp. 37–48. Springer, Heidelberg (2012)
12. Magnenat, S., Shin, J., Riedo, F., Siegwart, R., Ben-ari, M.: Teaching a Core CS Concept through Robotics. In: 19th Annual Conference on Innovation and Technology in Computer Science Education, ITiCSE (2014)
13. Magnenat, S., Riedo, F., Bonani, M., Mondada, F.: "A Programming Workshop using the Robot "thymio ii": The Effect on the Understanding by Children. In: Proceedings of the IEEE International Workshop on Advanced Robotics and its Social Impacts, pp. 24–29. IEEE (2012)
14. Riedo, F., Chevalier, M.S.D., Magnenat, S., Mondada, F.: Thymio II, a robot that grows wiser with children. In: 2013 IEEE Workshop on Advanced Robotics and its Social Impacts (ARSO), pp. 187–193. IEEE (2013)
15. Fassa, F.: Société en mutation, école en transformation: le récit des ordinateurs. Éd. Payot, Lausanne, Switzerland (2005)
16. Burgeois, L.: Une cole vaudoise mise sur la tablette. 24 Heures (December 2, 2013)
17. Weber, M.: From Max Weber: essays in sociology. Routledge (2009)
18. Kaufmann, J.-C.: L'entretien compréhensif. Armand Colin (2011)
19. Beaud, S.: L'usage de l'entretien en sciences sociales. plaidoyer pour l'"entretien ethnographique. Politix 9(35), 226–257 (1996)
20. Berger, P.L., Luckmann, T.: The Social Construction of Reality. Doubleday, New York (1966)
21. Vincent, G.: L'éducation prisonnière de la forme scolaire? Scolarisation et sociali-sation dans les sociétés industrielles. Presses Universitaires, Lyon (1994)
22. Bourdieu, P., Passeron, J.C.: Reproduction in Education, Society and Culture, vol. 4. Sage (1990)
23. Han, J., Kim, D.: r-learning services for elementary school students with a teaching assistant robot. In: 2009 4th ACM/IEEE International Conference on Human-Robot Interaction (HRI), pp. 255–256. IEEE (2009)

We, Robots: Correlated Behaviour
as Observed by Humans

Christian Kroos[1] and Damith C. Herath[2]

[1] Alternate Anatomies Lab, School of Design and Art, Curtin University,
Perth, Australia
christian.kroos@curtin.edu.au
[2] Robological, Sydney, Australia
damith@robological.com

Abstract. In this study participants judged on the relationship between
two interacting robots, one of them a mobile robot, the other one a stationary, robot arm-based artistic installation with a high flexibility in
orienting its anthropomorphic face. The robots' behaviour was either
(1) weakly correlated through a loose tracking function, (2) independently random, or (3) independently random, but constrained to the
same closely limited area. It was found that the true degree of coupling
was reflected on average in the rating responses but that pseudo-random
behaviour of one of the robots was judged less random if a relationship
between the two robots was present. We argue that such robot-robot
interaction experiments hold great value for social robotics as the interaction parameters are under complete control of the researchers.

Keywords: Robot-robot interaction, behaviour coupling, agency,
Articulated Head.

1 Background

When thinking about social robotics and social robots, we appear to have primarily their interaction with humans in mind and pay rarely attention to Robot-
Robot Interaction (RRI). Given that the purpose of social robots lies in the
interaction with humans, it does not surprise, although with social robots becoming more common place in the near future, it can be expected that they
will have to interact with each other and the outcomes of these interactions
might have consequences for their human owners. This fact on its own would
make experiments in RRI worthwhile, but it is not the primary reason why
we are intrigued by RRI. At the centre of our interest is a more fundamental question: Why shouldn't robots socialise with each other in the first place?
Social interactions were arguably responsible for the rapid and unique cognitive development of the human species [13,4] and in the same way social robots
could be become more sophisticated (e.g., along the lines described in [12]).
While an autonomous robotic evolution is still more or less in the realm of science fiction, a more methodologically orientated motivation is already applicable

M. Beetz et al. (Eds.): ICSR 2014, LNAI 8755, pp. 229–238, 2014.

and - we would argue - advisable: Human-robot interaction experiments include by definition a human participant who reacts to the robot and can adjust to interactive shortcomings of the robot, possibly even without being aware of the changes in behaviour. Thus, the combined human-robot entity is studied, usually intentionally so, but posing the difficulty to identify clearly the robot's part in the success or failure of the interaction.

Robot-robot interaction is largely under the control of the researcher. The drawback of RRI experiments on the other hand is their low explanatory value with respect to human-robot interaction, though probabilistic methods and machine learning would allow for very valuable insights to be gained nevertheless: If the experiment is situated in an ecologically valid environment, say, a supermarket, the interaction is influenced by the environment and the resulting behaviour of the robots is likely to become complex. The results move from an area of easy predictability to uncharted territory, even if the adaptation ability of the robot control system is severely limited and despite that no human participant is directly involved in the interaction. The emphasis in the previous sentence must be on 'directly' since the two or more robots would act in a environment shaped by humans and intended for human use. Therein lies the specific value of such experiments. In a thought experiment one could have a robotic shopping assistance in a supermarket and the customer to be helped would be a robot, too. As pointed out before, it would be less motivated by the expectation of this being a likely scenario any time soon but more by the advantage of complete control over the parameters that drive the robots. Obviously, the experiment does not need to be a thought experiment; it could be done in reality right now with the current state of art of robots.

Whether or not this line of research belongs in a social robotics conference is an open question. We believe it should. The current study uses an RRI experiment in this way, but with a slightly different focus: it investigates human behaviour relative to robots by removing the human participants from the experiment and making them observers. Our previous work on robotic agency [8] raised questions about the recognition of agency by human interaction partners. We concluded that - at least within the specific work of art at the centre of this research - the impression of agency originated from the human-robot interaction itself, that it was largely attributed due to specific behaviour of the robot that evoked the impression of agency.

The inner workings of this process, however, could not be clarified. One of the simplest assumptions is that any clear 'wilful' relationship between the behaviour of two interacting parties would lead to the impression of agency. For trivial reasons, direct dependency can be ruled out as a plenitude of physical phenomena fall in this category due to cause-effect relationships. Thus, less than perfect correlations become a candidate, e.g., one interaction partner tracking the movements of the other though not constantly, but rather with substantial deviations. The control condition would be random instead of pursuit movements. But would human participants be able to identify loose couplings when looking at machines with doubtful capability for intentional agency and given a tendency to emphasise seemingly meaningful over random behaviour [5]?

Our primary robotic platform, the Articulated Head (see below) has a pursuit behaviour that is driven by an underlying attention model. It exhibits the property of a loose coupling to a moving person or a robot very remotely resembling a human (our secondary robot, a PeopleBot) in its vicinity. Our hypothesis was that observers would notice the relationship despite being rather loose and be able to distinguish it from random movements. Alternatively, participants might consider any random behaviour of the two robots connected because of short accidental movement similarities. As a third hypothesis it could be assumed that a spatial constraint on the movement would be enough to elicit the impression of mutually influenced behaviour. We were then further interested in knowing whether the relationship would influence randomness judgements with respect to either robot.

We included three conditions in the experiment of this study to be able to distinguish between the alternative explanations. In the first condition, the Articulated Head was steered by the attention model (THAMBS, see below) in normal mode. In the second condition, all sensing was switched off and the Articulated Head performed random idle movements as it would normally do if for some time its environment is void of any stimulus able to attract its attention. To create the impression of naturalness, the target values for each joint for these movements are drawn from normal or log-nomal probability distributions following a few simple rules [7]. In the third condition, the Articulated Head was driven by the simulated input of a single (small) person performing a constrained random walk within the area that was in the real world dedicated to the second robot, the PeopleBot. As a consequence, the motion of the Articulated Head was in the height range of the PeopleBot and the orientation of its end effector (the LCD monitor) constrained to point to the area in which the PeopleBot was moving while not following it or only accidentally so.

We recorded the interaction between the remotely controlled PeopleBot (PB) and the Articulated Head (AH) on video and the clips were judged by human participants off-line in the lab. The independent variable was the movement control condition of the Articulated Head with three levels: fully random (FR), constrained random (CR) and tracking with the attention model applied (TR). With regard to the latter see also section 2.1 for more details. The dependent variables were five ratings by the participants. We predicted a linear trend with the order $FR < CR < TR$ of the judgements referring to perceived regularities (correlations) between the behaviour of the two robots (first rating). With regard to two complementary questions, asking about whether the PeopleBot (second rating) or the Articulated Head (third rating) was leading and the respective other robot was following, we expected a linear trend with the order $FR < CR < TR$ for the first question (PB leading) and no statistically significant differences between conditions for the second (AH leading), but a significant deviation from the midpoint of the scale towards ascribing no leading role to the Articulated Head. For the last two questions, asking whether the behaviour of the PeopleBot

Fig. 1. The Articulated Head in the Powerhouse Museum, Sydney, Australia

(fourth rating) or the Articulated Head (fifth rating) was considered random, we predicted no significant trend for the first question (PB random) and a linear trend with the order $FR > CR > TR$ for the second (AH random).

2   Method

2.1   Materials

Articulated Head. The Articulated Head is an interactive robot as a work of art designed by the Australian performance artist Stelarc [8]. It was realised by a small team of engineers and cognitive scientists within the Thinking Head project [1] and was displayed for two years in the Powerhouse Museum, Sydney, Australia. It consists of an LCD monitor mounted as the end effector of an industrial Fanuc LR Mate 200iC robot arm (see Figure 1). On the monitor, an animated virtual talking head is shown. For safety reasons the Articulated Head had to be within an enclosure. Multiple input sensors and associated software provided the Articulated Head with situational awareness among them a stereo camera with associated tracking software following people in the vicinity of the Articulated Head (PeopleTracker).

The artistic goal was to create a robotic system that would have a physical, sculptural presence and that would be recognised as a conscious and even intelligent being despite not resembling a human and clearly announcing its machine character. The non-verbal behaviour (the movements of the robot arm) were considered crucial in achieving this goal. Human participants have to be found to attribute animacy, agency, and intentionality to objects dependent on their motion pattern alone [11] and HRI studies confirm that robots are no exceptions though differences remain if compared for instance to the treatment of motor actions of other humans [2,9].

PeopleBot. The PeopleBot is a differential drive research robot platform from Adept MobileRobots with a height of 112 cm at the top base where an LCD touch screen is mounted. Its upright slender build adds a certain anthropomorphic quality to it.

THAMBS. The high-level processing of the sensing information and the behaviour control of the Articulated Head, in particular the motor response to visitor movements, is accomplished by the Thinking Head Attention Model and Behavioural System (THAMBS). A detailed description of THAMBS can be found in [7,8].

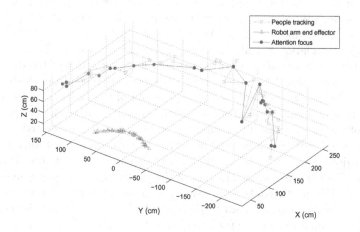

Fig. 2. Example of an episode with a person interacting with the Articulated Head. The person walks around the enclosure with varying speed. The input from the People Tracker as received by THAMBS is shown (orange line, square markers), the spatial location of the attention focus over time (green line, circular markers) and the location of the robot arm end effector (light blue line, triangular markers).

Figure 2 depicts a typical instance with tracking results, location of the attention focus and the location of the robot arm end effector shown. As can be seen, the shifting attention focus follows the indicated location of the person, however, not 'slavishly' but with some room to manoeuvre. Often the differences are delays, THAMBS intentionally not reacting to fast, but relatively small movements. This is caused by the previous attention focus being still dominant over other appearing candidates and not decaying rapidly enough. It does not save

THAMBS, however, from paying attention to one of the failures in the tracking, an error in the determination of the person's height.

2.2   Stimuli

The stimuli consisted of video clips of the two robots (Articulated Head and PeopleBot) in close proximity at the location of the exhibition of the Articulated Head in the Powerhouse Museum. An area of roughly four by three meters extending from one side of the triangular enclosure of the Articulated Head was used as the movement space for the PeopleBot (see Figure 3). The Articulated Head was set to one of the three conditions and the PeopleBot brought into its dedicated area. Since the PeopleBot moved in the public space of the museum, its movements were for safety reasons controlled by a human operator (none of the authors) who could not see the Articulated Head and was not aware of the tracking condition it was in. The robot operator was instructed to cover the whole area with the movements of the PeobleBot without pursuing any further aims. A Sony HDRFX100 camera was used to record the movements of the Articulated Head and the PeopleBot from a single viewpoint, slightly to the left of a frontal position to minimise occlusions and allow better depth perception with respect to the motion of the Articulated Head.

With the PeopleBot already moving, the camera was started and the scene recorded for 3 min and 10 s. To obtain the final stimuli, clips of the duration of 60 s were excised using Adobe Premiere. The clips were taken from the beginning of the recording or immediately following the end of the previously excised clip, resulting in nine items altogether. No attention was paid to the orientation or location of the two robots. The sound tracks of the clips were erased.

The video stimuli were presented using the experiment control software Alvin [6] and projected on a wall in the HRI lab at the MARCS Institute, University of Western Sydney, Australia. The size of the projection was approximately 210 by 170 cm. The participants were seated in a distance of approximately 320 cm from the projection. A small table in front of them provided the necessary support for a computer mouse used to obtain their responses.

2.3   Participants and Procedure

Twenty-four graduate students and members of the lab (16 female) aged 24-57 (mean: 33.83) participated. They were not familiar with the aim of the study and only 3 participants were accustomed to robotic research. They were instructed to watch the video clip and then respond to five statements about the actions of the robots in the clip by clicking with the computer mouse on the labelled buttons below the area where the video clip was shown. After each clip the text of the first statement appeared on the left hand side and the corresponding set of buttons was activated on the right hand side. After the participant selected a response the next statement and the next set of buttons appeared below. This continued until all five statement were answered and clicking on one of the

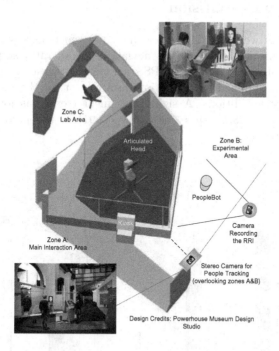

Fig. 3. The Articluated Head environment in the Powerhouse Museum and the set-up for the experiment

response buttons of the last statement triggered a new screen and the next video clip to be shown.

The statements were:

[CONNECT] 'There was a connection between the behaviour of the two robots.'
[PB_LEAD] 'The PeopleBot was leading, the Articulated Head was following.'
[AH_LEAD] 'The Articulated Head was leading, the PeopleBot was following.'
[PB_RAND] 'The movements of the PeopleBot appeared random to me.'
[AH_RAND] 'The movements of the Articulated Head appeared random to me.'

The response buttons were labelled:

- 'Strongly disagree' coded as 1,
- 'Disagree' coded as 2,
- 'Undecided' coded as 3,
- 'Agree' coded as 4, and
- 'Strongly agree', coded as 5,

implementing a five point Likert scale [10,3].

The nine video clips were repeated three times resulting in 27 clips to be rated by each participant. Thus, altogether 72 ratings for each rating statement were obtained. The order of presentation was fully (pseudo-)randomised.

3   Results and Discussion

The rating results were averaged over the three repetitions per participants. Figure 4 shows means and standard deviations of all ratings split by condition. Using the statistics software SPSS a repeated-measures General Linear Model (GLM) was applied to the rating data to test for the predicted trends across the motion conditions. A significant linear trend was found for CONNECT ($F(1, 23) = 223.19$; $p = .000$; $\eta_p^2 = .91$) in the predicted direction ($FR < CR < TR$).

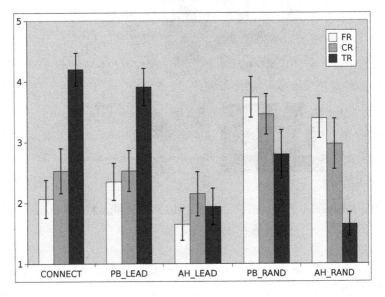

Fig. 4. Rating means split by movement condition. FR (fully random): yellow bars; CR (constrained random): green bars; TR (THAMBS tracking): red bars. Error bars denote one standard deviation.

As hypothesised, the participants rated the degree of connectivity between the behaviour of the two robots according to the real relationship if one considers being confined in orientation and location to the same area as a weak but yet existing link. The mean of 2.53 in the CR condition close to the midpoint of the scale (at 3) indicates that the participant detected, maybe subliminal, some relationship but were not confident about it. The difference between FR and CR is - though statistically significant (post-hoc comparisons as part of the GLM model - FR vs CR: diff = .46; $F(1, 23) = 5.62$; $p = .026$; $\eta_p^2 = .20$) - very small compared to the difference between CR and TR which is at 2.13 and thus substantial given a 5 point scale.

A significant trend in the predicted direction ($FR < CR < TR$) was also detected for PB_LEAD ($F(1, 23) = 94.89$; $p = .000$; $\eta_p^2 = .81$). Contrary to our predictions a significant trend in the same direction was found for AH_LEAD

$(F(1, 23) = 4.99; p = .035; \eta_p^2 = .18)$. Thus, unexpectedly, both trends for a leading role of any of the robots reached significance. However, when considering ratings in each condition separately and comparing Articulated Head and PeopleBot, the PeobleBot was rated significantly higher as the leader in the FR and TR condition (paired-sample t-test; FR: $t(23) = 4.59$; $p = .000$; TR: $t(23) = 8.76$; $p = .000$) but not in the CR condition ($t(23) = 1.98$, $p = .60$). The difference is very pronounced in the TR condition which is of course in line with the factual circumstances in this condition.

In line with expectations a significant trend in the predicted direction ($FR > CR > TR$) was confirmed for AH_RAND ($F(1, 23) = 113.48$; $p = .000$; $\eta_p^2 = .83$). Contrary to expectations a significant linear trend ($FR > CR > TR$) for PB_RAND was also attested ($F(1, 23) = 30.30$; $p = .000$; $\eta_p^2 = .57$).

The degree of randomness in the behaviour of the Articulated Head is reflected in the trend found for AH_RAND. For FR and CR the ratings are distributed close to the midpoint of the scale indicating that the participants were on average not sure whether or not the behaviour of the Articulated Head was random. This points toward the tendency mentioned in the Section 1 to mistake randomness for intentional behaviour. In the TR condition, however, the rating result clearly indicates that the participant were confident about the lack of randomness in the behaviour of the Articulated Head. The difference between CR and TR is strong (1.31). Note that the difference can only originate from the recognition of the stronger coupling between the two robots' behaviours, since it was the only pronounced discrepancy between the CR and TR condition.

The trend found for PB_RAND is indeed surprising since there was no change in the behaviour of the PeopleBot in the three conditions. Post-hoc comparisons as part of the GLM model revealed that the trend is primarily due to condition TR (FR vs CR: diff = .28; $F(1, 23) = 2.75$; $p = .11$; $\eta_p^2 = .11$; CR vs TR: diff = .65; $F(1, 23) = 13.84$; $p = .001$; $\eta_p^2 = .38$). It suggests that if a relationship between the behaviour of two robots is present, it biases the perception of both robots towards attesting meaningful behaviour to both of them. Indeed in an unexpected way, this is in line with the original conjecture that the impression of agency arises - at least partially - from the interaction itself.

Given that humans of course see themselves always as intentional agents, there might be a tendency to attribute agency to any entity that shows some loose coupling of its own behaviour to the one of the human. It can be speculated that it does not require a robot for this impression to be evoked let alone a humanoid robot but rather that it applies to all kinds of physical phenomena unless an effect-cause relationship can be established. One's own shadow, for instance, is typically characterised through a tight coupling that makes it easy to assert a cause-effect relationship. If the correlation of the movement of the shadow to the movements of the body, however, is lowered, the cause-effect explanation might become doubtful and the impression of an uncanny agency might arise. In the same way some of what is typically considered by cultural believes to fall into the realm of magic and the paranormal might be the impression of agency without a home, that is, without a proper explanation. Social robots could both profit from the phenomenon and suffer in acceptance - dependent on circumstances. Obviously, much more research is needed.

Acknowledgements. The authors wish to thank Zhengzhi Zhang for his help with driving the robot and administering the rating experiment, Stelarc for conceiving and initiating the Articulated Head and the MARCS Institute of the University of Western Sydney for their support in the form of the HRI lab at the Bankstown campus. The authors wish to express their gratitude to the Powerhouse Museum, Sydney.

References

1. Burnham, D., Abrahamyan, A., Cavedon, L., Davis, C., Hodgins, A., Kim, J., Kroos, C., Kuratate, T., Lewis, T., Luerssen, M., Paine, G., Powers, D., Riley, M., Stelarc, S., Stevens, K.: From talking to thinking heads: report 2008. In: International Conference on Auditory-Visual Speech Processing 2008., Moreton Island, Queensland, Australia, pp. 127–130 (2008)
2. Castiello, U.: Understanding other people's actions: Intention and attention. Journal of Experimental Psychology: Human Perception and Performance 29(2), 416–430 (2003)
3. Cox III, E.P.: The optimal number of response alternatives for a scale: A review. Journal of Marketing Research 17(4), 407–422 (1980)
4. Dunbar, R.I.: The social brain: mind, language, and society in evolutionary perspective. Annual Review of Anthropology 32, 163–181 (2003)
5. Ebert, J.P., Wegner, D.M.: Mistaking randomness for free will. Consciousness and Cognition 20(3), 965–971 (2011)
6. Hillenbrand, J.M., Gayvert, R.T.: Open source software for experiment design and control. Journal of Speech, Language, and Hearing Research 48(1), 45–60 (2005)
7. Kroos, C., Herath, D.C.: Stelarc: From robot arm to intentional agent: The Articulated Head. In: Goto, S. (ed.) Advances in Robotics, Automation and Control, pp. 215–240. InTech (2011)
8. Kroos, C., Herath, D.C.: Stelarc: Evoking agency: Attention model and behaviour control in a robotic art installation. Leonardo 45(5), 133–161 (2012)
9. Liepelt, R., Prinz, W., Brass, M.: When do we simulate non-human agents? Dissociating communicative and non-communicative actions. Cognition 115(3), 426–434 (2010)
10. Likert, R.: A technique for the measurement of attitudes. Archives of Psychology 22(140), 44–53 (1932)
11. Scholl, B.J., Tremoulet, P.D.: Perceptual causality and animacy. Trends in Cognitive Sciences 4(8), 299–309 (2000)
12. Steels, L.: Experiments in cultural language evolution. John Benjamins Publishing (2012)
13. Tomasello, M.: The cultural origins of human cognition. Harvard University Press, Cambridge (1999)

An Approach to Socially Compliant Leader Following for Mobile Robots

Markus Kuderer and Wolfram Burgard

Department of Computer Science, University of Freiburg

Abstract. Mobile robots are envisioned to provide more and more services in a shared environment with humans. A wide range of such tasks demand that the robot follows a human leader, including robotic co-workers in factories, autonomous shopping carts or robotic wheelchairs that autonomously navigate next to an accompanying pedestrian. Many authors proposed follow-the-leader approaches for mobile robots, which have also been applied to the problem of following pedestrians. Most of these approaches use local control methods to keep the robot at the desired position. However, they typically do not incorporate information about the natural navigation behavior of humans, who strongly interact with their environment. In this paper, we propose a learned, predictive model of interactive navigation behavior that enables a mobile robot to predict the trajectory of its leader and to compute a far-sighted plan that keeps the robot at its desired relative position. Extensive experiments in simulation as well as with a real robotic wheelchair suggest that our method outperforms state-of-the-art methods for following a human leader in wide variety of situations.

1 Introduction

There is a wide range of applications for mobile robots for which it is desirable that the robot follows a human leader. For example a robotic co-worker that provides tools to a human in a factory needs to stay in a position where the human can reach the robot. Similarly, a mobile shopping cart should always stay in a position where the human is able to place objects into it. A further application is a robotic wheelchair that stays side by side to an accompanying pedestrian, allowing interaction with the pedestrian during the navigation task.

When following a human leader, it is beneficial for the robot to reason about the natural navigation behavior of pedestrians. During navigation, pedestrians interact with their environment, which includes obstacles, other nearby humans and also the robot itself. A robot that has a better understanding of this interactive behavior is able to fulfill its task in a socially compliant way, i.e., in a way that does not unnecessarily hinder nearby pedestrians. Such a robot is able to predict the behavior of the humans and to plan far-sighted trajectories that keep the robot close to its desired position in the long run.

There has been a wide range of research on controlling a group of robots in formation, which have, to some extent, also been applied in the context of social

M. Beetz et al. (Eds.): ICSR 2014, LNAI 8755, pp. 239–248, 2014.

robotics [12, 14]. Many of these approaches utilize control-theoretic methods to steer the robot towards a virtual target that moves along with the leader [5, 14]. However, these methods mostly neglect information about the more complex navigation behavior of pedestrians that strongly depends on the environment.

In this paper, we propose to utilize a feature-based model of human navigation behavior to predict the path of the leading pedestrian [9]. This model accounts for the intention of a human to reach a certain goal while keeping a comfortable velocity, avoiding strong accelerations and to stay clear of obstacles. The individual characteristics of different pedestrians, or distinct behavior in different environments can be learned from observation.

The contribution of this paper is a method that simultaneously predicts the most likely trajectory of the pedestrian and computes the trajectory for the robot that minimizes the distance to its desired relative position along the whole trajectory in a forward-looking manner. Such a predictive planning method leads to a socially more compliant behavior of the robot. In addition, planning long-term trajectories mitigates the problem of local minima in a local control function, especially in the presence of arbitrary, non-convex obstacles in the environment. We conducted a simulated comparison of our method to related approaches as well as experiments with a real robot that show the applicability of the proposed approach to navigate a robotic wheelchair next to an accompanying pedestrian.

2 Related Work

In the past, many authors proposed methods to navigate a group of robots in formation. Liu et al. [11] cast the joint path planning task of a robot formation as a linear programming problem. Similar to our approach, they plan the trajectories to the target position of each robot. However, Liu et al. control the group of robots in a central manner and each robot executes the optimal trajectory. Balch and Hybinette [1] propose to use social potential fields that pull the robots towards attraction points to achieve a certain formation. Our experiments include a comparison to a social potential-based approach.

A different problem arises when the task of the robot is to follow a leader whose goal is unknown. Chiem and Cervera [4] and Huang et al. [8] propose to compute a cubic Beziér curve between the leading robot and the follower. The follower then navigates along this trajectory, using a velocity controller. In addition, if the robots task is not only to follow the same path but to stay in a certain formation, they propose to compute virtual targets for each of the robots and compute Beziér curves to these target positions. However, they follow the leading robot without active obstacle avoidance. Desai et al. [6] and Das et al. [5] use control theoretic approaches to keep each robot close to its designated position within the formation, also considering obstacles in the environment. If the desired shape of the formation changes, they introduce control graphs to assign the robots to their new position in the formation. Qin et al. [15] use artificial forces to navigate each robot close to the desired position in a formation. Similarly, Tanner and Kumar [16] propose to use navigation functions to keep a group of robots in a certain formation. Navigation functions also lead the robot along the gradient of a smooth function, similar to artificial forces, but there are no local minima allowed, except of the target position. In general, however, it is difficult to design such a function for arbitrary environments [10]. Chen and Wang [3] provide a survey on different approaches to robot formation control.

The abovementioned approaches use local control methods to steer the robot either directly to the desired position in the formation, or to some local virtual target position. In contrast, we predict the trajectory of the leader based on its current state and the state of the environment. At the same time, we compute the trajectory that minimizes the distance to the desired relative position along this trajectory while satisfying further constraints. This prevents the robot to get stuck in local minima of the cost function and allows it to adapt the planned trajectories to the environment early on.

Similar methods have also been used to enable a robot to follow a human leader. Pradhan et al. [13] utilize a navigation function method and set the tracked positions of the pedestrian as virtual target positions. Therefore, the robot is only able to follow the person, but not to stay at a fixed relative position. Prassler et al. [14] aim at coordinating the motion of a human and a robot and also apply it to a robotic wheelchair. They propose to use the velocity obstacles approach [7] to guide the robot to a local virtual target. We compare our method to a similar approach in our experimental section. Most similar to our approach is the work of Morales et al. [12]. They optimize a utility that encodes the desired relative position as well as the walking comfort of the pedestrian. However, they optimize the planned trajectory locally, whereas we optimize future trajectories to a distant subgoal, which allows the robot to adapt its behavior to the environment in a predictive manner.

3   A Socially Compliant Follow the Leader Approach

A better understanding of the natural navigation behavior of pedestrians enables a mobile robot to follow a human leader in a socially more compliant way. In this section, we first formalize the problem of following a leader. We consider the navigation task to stay close to a fixed relative position with respect to its leader. To solve this task, we propose an approach that predicts the trajectory of the pedestrian and at the same time computes a forward-looking trajectory that minimizes the deviation to the desired position.

3.1   Problem Definition

In this work, we consider the 2D navigation behavior of a mobile robot and a leading pedestrian. A trajectory τ_h of the human and τ_r of the robot are mappings $\tau : \mathbb{R} \to \mathbb{R}^2$ from time to a 2D position. The position of the robot, or the pedestrian, respectively, at time t is thus given by $\tau(t)$ and their velocity by $\dot{\tau}(t)$. We assume a mobile robot with a differential drive that is always oriented in driving direction. Similarly, we assume that the pedestrian is always headed

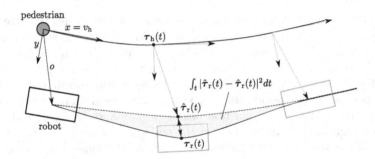

Fig. 1. The desired position of the robot is a fixed location in the local coordinate system of the pedestrian. The dashed line illustrates the desired trajectory of the robot $\hat{\tau}_r$ given the predicted trajectory $\hat{\tau}_h$ of the pedestrian. Deviation from the desired trajectory yields an additional cost integrated along the trajectory, as illustrated by the shaded area.

in walking direction. Thus, the orientation $\theta(t)$ at time t is the direction of the vector $\dot{\tau}(t)$.

We define the desired position of the robot by a fixed position $\mathbf{o} = (o_x, o_y)$ in the local coordinate system of the pedestrian, i.e., the robot is supposed to always maintain the same position relative to the human. Given the trajectory $\tau_h(t)$ of the human, we can compute the desired trajectory of the robot

$$\hat{\tau}_r(t) = \tau_h(t) + q(\theta_h(t))\mathbf{o}, \tag{1}$$

where $q(\theta_h(t))$ is the rotation of the human at time t. In practice the robot cannot always follow this desired trajectory due to obstacles in the environment, or other dynamic constraints. We cast the resulting navigation goal in a utility-optimizing manner, where the cost function is a linear combination of the squared norm of the deviation from the desired trajectory and an additional term $g_{nav}(\tau, t)$ that comprises acceleration and velocity bounds and clearance to obstacles. Therefore, the desired trajectory minimizes the navigation cost function

$$c(\tau_r) = \int_{t=0}^{T} \left(\theta_1 |\tau_r(t) - \hat{\tau}_r(t)|^2 + \theta_2 g_{nav}(\tau_r, t) \right) dt, \tag{2}$$

where the weights θ_1 and θ_2 are model parameters to adjust the behavior to the given application. Fig. 1 illustrates the predicted trajectory of the pedestrian, the offset in the local reference frame of the pedestrian and the resulting desired trajectory of the robot. The challenge of this approach is to predict the trajectory of the human, which determines the desired trajectory of the robot $\hat{\tau}_r(t)$. To this end we utilize a predictive model of natural human navigation behavior, which we shortly recap in the following.

3.2   Modeling Human Navigation Behavior

Our approach relies on an accurate model of human navigation behavior that allows the robot to predict the movements of the leading pedestrian. To achieve socially compliant behavior of the robot, we want to explicitly model the fact that the human is also aware of the robot and reacts to the actions of the robot.

Kretzschmar et al. [9] describe a probabilistic model of such an interactive navigation behavior. For given start and goal positions, the proposed model yields a distribution over the joint space of the trajectories of each agent involved in the navigation process. This probability distribution depends on a weighted sum of features $\mathbf{f}$ that capture important properties of human navigation behavior. Each feature is a function that maps a composite trajectory, i.e., the set of trajectories for all agents, to a real value. Kretzschmar et al. propose features that describe the individual properties of each trajectory, such as the integrated velocity and acceleration along the trajectory, and the time to reach the target. In addition, they propose features that describe interaction between the agents, such as their mutual distance. A weight vector $\boldsymbol{\theta}$ parameterizes the model and describes the importance of each feature in the feature vector $\mathbf{f}$.

In the special case of two agents h and r, the model yields the distribution

$$p_{\boldsymbol{\theta}}(\boldsymbol{\tau}_h, \boldsymbol{\tau}_r) \propto \exp(-\boldsymbol{\theta}^T \mathbf{f}(\boldsymbol{\tau}_h, \boldsymbol{\tau}_r)), \tag{3}$$

where $\boldsymbol{\tau}_h$ and $\boldsymbol{\tau}_r$ are the trajectories of the two agents, as introduced in the previous section. One can interpret $\boldsymbol{\theta}^T \mathbf{f}(\boldsymbol{\tau}_h, \boldsymbol{\tau}_r)$ as a cost function. The agents are thus exponentially more likely to select a trajectory with lower cost. To adapt the model to the individual navigation behavior of different pedestrians or to a certain environment, we can learn the feature weights $\boldsymbol{\theta}$ from observed data, such that the predicted trajectories accurately resemble the navigation behavior of real humans in the designated environment. Find details on the proposed learning approach as well as a description of features that capture important properties of natural navigation behavior in Kretzschmar et al. [9].

3.3   Unifying Prediction and Planning

We utilize the model proposed by Kretzschmar et al. [9] to predict the trajectory of the pedestrian, and to plan a trajectory for the robot simultaneously. In particular, we adopt the proposed features that capture accelerations, velocities, distances to obstacles and the time to reach the target to predict the natural navigation behavior of the pedestrian. In addition, we introduce the feature

$$f_d(\boldsymbol{\tau}_h, \boldsymbol{\tau}_r) = \int_{t=0}^{T} |\boldsymbol{\tau}_r(t) - \hat{\boldsymbol{\tau}}_r(t)|^2 dt, \tag{4}$$

that describes the squared deviation from the desired position of the robot along the trajectory, and

$$f_n(\boldsymbol{\tau}_h, \boldsymbol{\tau}_r) = \int_{t=0}^{T} g_{\text{nav}}(\boldsymbol{\tau}_r, t) dt, \tag{5}$$

to account for further navigation constraints of the robot, as described in Sec. 3.1. During navigation, we compute the most likely composite trajectory $(\boldsymbol{\tau}_h, \boldsymbol{\tau}_r)$ with respect to the probability distribution given by Eq. (3). Due to the additional features f_d and f_n, this most likely composite trajectory not only predicts

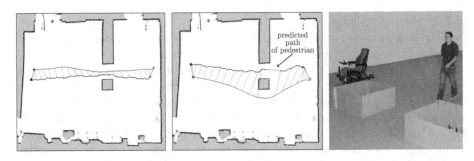

Fig. 2. Left: observed trajectories of the robot (red) and the human (blue) during navigation. The robot falls back behind the pedestrian in the narrow passage. Middle: observed trajectories in an experiment where the robot bypasses the obstacle on the lower side to meet the pedestrian after the passage. Right: Experimental setup.

the trajectory of the pedestrian but also computes the trajectory of the robot that minimizes the navigation cost function of the robot (Eq. (2)). In particular, this method accounts for the effect that the pedestrian interacts with the robot, i.e., that the pedestrian behaves cooperatively and navigates in a way that helps the joint navigation goal. By adjusting the weights of the features we can adapt the level of cooperative behavior that we ascribe to the human. Fig. 3 and Fig. 4 illustrate the predicted trajectory of the pedestrian and the planned trajectory for the robot in two different scenarios.

In addition, the predictive model is beneficial in situations where the leading pedestrian is not in the field of view of the robot's sensors for some time. Instead of stopping the navigation task, the robot is able to predict the trajectory of the pedestrian and to continue its plan. When the human reappears in the observation of the robot, the people tracker can use the prediction to solve the data association problem, i.e., to select the correct pedestrian as leader.

The predictive model yields predictions of trajectories to known target positions. However, the final target position of the pedestrian is not known in general. In our experiments, we interpolate the observed trajectory of the pedestrian to estimate its target position. In environments where prior information of the typical paths of pedestrians is available, we can also use more sophisticated methods to estimate their target position [2, 18].

4   Experiments

In this section, we describe a set of experiments using a real robotic wheelchair that suggest that our method is applicable to successfully navigate alongside an accompanying pedestrian in the presence of obstacles. Furthermore, we compare our approach in simulation to two related methods. These experiments intend to show the advantages of our predictive planning approach over local control methods, especially in situations where the environmental conditions hinder the robot to remain at its desired position. During the navigation task, our method

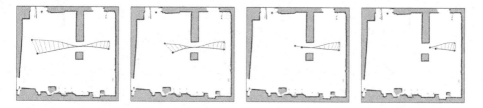

Fig. 3. Predictions computed by the wheelchair at four successive time steps. The robot predicts the human to pass the passage. Since the passage is too narrow for the robot (red) and the human (blue) the robot leaves its desired position and let the human pass first. After the passage, the robot resumes its desired position.

continuously computes the most likely composite trajectory by optimizing its probability at a rate of 5 hz.

4.1   Real Robot Experiments

In the following experiments, we use the method proposed in this paper to navigate a robotic wheelchair next to a pedestrian at a distance of 1 m. The robot relies on on-board sensors only. It localizes itself in the environment using Monte Carlo localization [17] and tracks the pedestrian using a laser based people tracker. Fig. 2 shows the paths of the wheelchair and the pedestrian as observed by the wheelchair in two different scenarios.

In the first run (Fig. 2 left), the robot's desired position is on the left of the pedestrian. It starts moving alongside the pedestrian, falls back behind the pedestrian during passing the passage and catches up afterwards. Fig. 3 shows the predictions of the wheelchair during the navigation task in the same run. As soon as the pedestrian starts to move, the robot computes the most likely composite trajectory of the robot and the pedestrian. It predicts that the pedestrian walks through the passage and that the robot itself stays behind and regains the position to the left of the pedestrian afterwards.

In the second run (Fig. 2 middle), the robot is supposed to keep its position on the right hand side of the pedestrian. Since there is enough space on the lower side of the obstacle, the robot decides to pass the obstacle on a this side, which allows the robot to stay at the human's side as long as possible. While the pedestrian is in the passage, the obstacle blocks the laser scanner and the robot cannot observe the pedestrian. However, since the robot maintains predictions about the movement of the pedestrian, it is able to follow its planned path and join the pedestrian after it is tracked again. Fig. 4 shows the predictions of the robot during this second run. First, the estimated target is still on the left side of the obstacles due to the low velocity of the pedestrian. However, as soon as the pedestrian proceeds to its goal position, the robot predicts that the pedestrian moves through the passage and plans to pass the obstacle on the other side. While the obstacle occludes the pedestrian, the robot updates its beliefs based on the current prediction of the pedestrian's position.

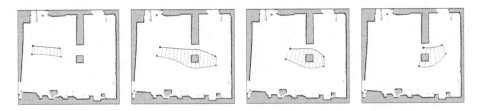

Fig. 4. Prediction computed by the wheelchair at four successive time steps. In this experiment, the desired position of the robot is on the right hand side of the human. The robot stays at the human's side as long as possible. It then evades the obstacle on the right side and continues to move to its desired position relative to the human.

4.2   Comparison in Simulation

Fig. 5 and Fig. 6 show a comparison of our method in simulation to a social forces (SF) based approach [1] and a velocity obstacles (VO) approach, similar to the method proposed by Prassler et al. [14]. To allow for a fair comparison of the methods, we scripted the pedestrian's path on a rectangular path with a velocity of $0.5\,\mathrm{ms}^{-1}$. The desired position of the robot is $1\,\mathrm{m}$ to the left of the pedestrian for all experiments. We set the parameters of all approaches such that the robot always kept a safety distance of at least $0.25\,\mathrm{m}$ to the pedestrian, as well as to obstacles in the environment.

Both, SF as well as VO compute control commands towards a virtual target position. To compute this position, we adopt the method proposed by Prassler et al. [14]. They linearly extrapolate the current velocity of the pedestrian in a small time horizon Δt to avoid that the robot lags behind the desired position. We adjust Δt for both methods such that the robot converges to the desired position when the pedestrian moves on a straight line with $0.5\,\mathrm{ms}^{-1}$.

In the test environments, the challenge for the robot is to catch up to the desired position after the pedestrian takes turns on its path. Furthermore, there is a narrow passage in which the robot cannot keep its desired position. Fig. 5 shows that all methods manage to pass the passage. However, the bar plot on the right shows that our method is able to stay closer to the desired position on average along the trajectory. This is due to the fact that our method predicts the trajectory of the pedestrian and computes the trajectory of the robot that minimizes the deviation along the whole path, while also incorporating properties of the robot, such as limited acceleration or velocity constraints. Such long term planning is better suited to accomplish the navigation task compared to greedily approaching the desired position.

Fig. 6 shows a similar experiment with an additional obstacle that resembles an open door in a typical indoor environment. The first image shows that our approach is able to negotiate the passage in a similar way as in the first setup. The robot falls back behind the pedestrian and catches up afterwards. Both SF as well as VO, however, get stuck behind the open door, since there is a local minimum in their local control functions. The bar plot reflects the advantage of the predictive planning in this experiment. Whereas our method shows a similar

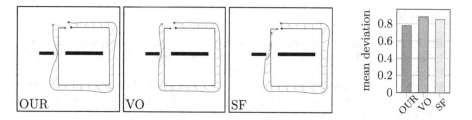

Fig. 5. Comparing our method to velocity obstacles (VO) and social forces (SF). The desired position of the robot (red) is one meter to the left side of the human (blue). The bar plot shows that our method stays closer to the desired position on average.

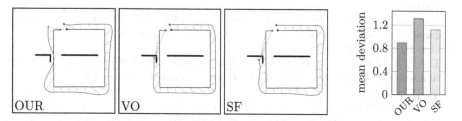

Fig. 6. Comparison to VO and SF that illustrates the advantages of our method over local control methods. Both VO and SF get stuck in the non-convex obstacle.

mean deviation from the desired position as in the first experiment, SF and VO gain a higher deviation whilst stuck in the local minima.

5   Conclusion

In this paper, we presented a novel method that allows a robot to follow a leading person in a socially compliant way. Our approach uses a feature-based model of natural navigation behavior to predict the trajectory of the leading human. In contrast to previous approaches, our method allows the robot to compute far-sighted plans that minimize the long-term deviation from the desired trajectory. In addition to features that describe natural intents of navigating pedestrians, our method uses features that capture the navigation goals of the robot. The resulting model thus unifies prediction of the human's behavior, and path planning of the robot. In several experiments also carried out with a robotic wheelchair we demonstrated that the proposed model is applicable to real world scenarios such as navigating alongside an accompanying person in the presence of obstacles. A comparison in simulation suggests that our method outperforms previous models that rely on local control strategies.

Acknowledgements. This work has been partially supported by the EC under contract numbers ERC-267686-LifeNav, and FP7-610603-EUROPA2.

References

[1] Balch, T., Hybinette, M.: Social potentials for scalable multi-robot formations. In: Proceedings of the IEEE International Conference on Robotics and Automation, ICRA 2000, vol. 1, pp. 73–80. IEEE (2000)

[2] Bennewitz, M., Burgard, W., Czielniak, G., Thrun, S.: Learning motion patterns of people for compliant robot motion. Int. Journal of Robotics Research (IJRR) 24(1), 31–48 (2005)

[3] Chen, Y.Q., Wang, Z.: Formation control: a review and a new consideration. In: 2005 IEEE/RSJ International Conference on Intelligent Robots and Systems (IROS 2005), pp. 3181–3186 (2005)

[4] Chiem, S.Y., Cervera, E.: Vision-based robot formations with bezier trajectories (2004)

[5] Das, A., Fierro, R., Kumar, V., Ostrowski, J., Spletzer, J., Taylor, C.: A vision-based formation control framework. IEEE Transactions on Robotics and Automation 18(5), 813–825 (2002)

[6] Desai, J., Ostrowski, J., Kumar, V.: Modeling and control of formations of non-holonomic mobile robots. IEEE Transactions on Robotics and Automation 17(6), 905–908 (2001)

[7] Fiorini, P., Shillert, Z.: Motion planning in dynamic environments using velocity obstacles. Int. Journal of Robotics Research (IJRR) 17, 760–772 (1998)

[8] Huang, J., Farritor, S., Qadi, A., Goddard, S.: Localization and follow-the-leader control of a heterogeneous group of mobile robots. IEEE/ASME Transactions on Mechatronics 11(2) (2006)

[9] Kretzschmar, H., Kuderer, M., Burgard, W.: Learning to predict trajectories of cooperatively navigating agents. In: IEEE Int. Conf. on Robotics and Automation (2014)

[10] LaValle, S.M.: Planning algorithms. Cambridge University Press (2006)

[11] Liu, S., Sun, D., Zhu, C.: Coordinated motion planning of multiple mobile robots in formation. In: 2010 8th World Congress on Intelligent Control and Automation (WCICA), pp. 1806–1811 (2010)

[12] Morales Saiki, Y., Satake, S., Huq, R., Glas, D., Kanda, T., Hagita, N.: How do people walk side-by-side?: using a computational model of human behavior for a social robot. In: Proceedings of the Seventh Annual ACM/IEEE International Conference on Human-Robot Interaction, pp. 301–308. ACM (2012)

[13] Pradhan, N., Burg, T., Birchfield, S., Hasirci, U.: Indoor navigation for mobile robots using predictive fields. In: American Control Conference, pp. 3237–3241 (2013)

[14] Prassler, E., Bank, D., Kluge, B., Hagele, M.: Key technologies in robot assistants: Motion coordination between a human and a mobile robot. Transactions on Control, Automation and Systems Engineering 4(1), 56–61 (2002)

[15] Qin, L., Zha, Y., Yin, Q., Peng, Y.: Formation control of robotic swarm using bounded artificial forces. The Scientific World Journal 2013 (2013)

[16] Tanner, H.G., Kumar, A.: Formation stabilization of multiple agents using decentralized navigation functions. In: Robotics: Science and Systems, pp. 49–56 (2005)

[17] Thrun, S., Fox, D., Burgard, W., Dellaert, F.: Robust monte carlo localization for mobile robots. Artificial Intelligence 128(1-2), 99–141 (2000)

[18] Ziebart, B., Ratliff, N., Gallagher, G., Mertz, C., Peterson, K., Bagnell, J., Hebert, M., Dey, A., Srinivasa, S.: Planning-based prediction for pedestrians. In: Proc. of the IEEE/RSJ International Conference on Intelligent Robots and Systems, IROS (2009)

Application of Fuzzy Techniques
in Human-Robot Interaction - A Review

I-Han Kuo and Chandimal Jayawardena

Department of Computing, Unitec Institute of Technology, New Zealand
{ikuo,cjayawardena}@unitec.ac.nz

Abstract. Targeting research challenges in Socially Assistive Robotics (SAR), this paper provides a review of previous work that describe robot or non-robot systems that use fuzzy logic to infer high-level human intention or activities. In comparison to statistical and probabilistic approaches which are very popular in SAR and Human-Robot Interaction (HRI), this review focuses on fuzzy logic-based systems. As fuzzy logic has already been widely used in almost all research areas in robotics, this review does not consider systems that uses fuzzy logic for sensing, modelling or planning tasks except for inferencing or reasoning tasks. From this review, it was found minimal research has been done in this special research niche and is deemed to gain more attention as the research communities shifts from sensing toward modelling and inferencing in the loop of Sense-Model-Plan-Act or Sense-Plan-Act.

1 Introduction

Since research in the area of Social Assistive Robotics (SAR) took off in the last two decades, a variety of challenges have been identified. Due to the difficulty of understanding and modelling human behaviour or intelligent thinking, several focus areas of research were deemed critical for the advances in this research area. Of all the possible applications, Tapus in [15] identified elderly care, physical rehabilitation and training and care for people with cognitive disabilities as the three main target areas in SAR. To realise these applications, she further pointed out questions that need to be answered through research including what roles does physical embodiment play in social interaction, how to display empathy and emotional intelligence for enhancing therapeutic outcomes and how to better engage users through eye gaze and awareness of human presence [15]

Motivated by goals in SAR and other application areas, there have been many concrete advancements in the area of signal processing, and knowledge modelling. Signal processing techniques for detection algorithm (e.g. human body detection, face detection) and recognition algorithms (e.g. face recognition, palm vein recognition) have seen maturity and even break-through. Improvements in sensor hardware in terms of cost and efficiency have also added to the research momentum. Good examples include Microsoft Kinect sensor, stereo-vision camera and 3D accelerometer. Features that can be detected and recognised through these algorithms and hardwares include human face, human body shape

M. Beetz et al. (Eds.): ICSR 2014, LNAI 8755, pp. 249–255, 2014.
© Springer International Publishing Switzerland 2014

(2D or 3D), spatial relationship between human and surrounding objects, or even various human activity (through worn 3D accelerometer). Research has also tried to model these features or combine this data (data fusion) together to form a better understanding of the human user of the technology. For example, 3D shapes can be modelled through point clouds, or volume pixels (voxel) [1] and dialogue between human and robot can also be modelled through speech text in which human language provides a model for semantic understanding or intention inference. This area of work overlaps with signal processing and is being explored in many research endeavours.

According to the authors, the body of research work in SAR and human robot interaction (HRI) and can be dissected and viewed in the following way (See Fig. 1)

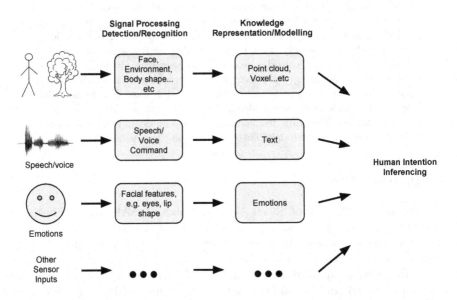

Fig. 1. An overview of the research areas for realising SAR

Due to the success of research in signal processing, and knowledge modelling, some research focus is gradually advancing to human intention inference; using modelled or represented information about human features to reason, interpretate their intention or their activities. To inform and contribute to the SAR research community, this paper focuses on the application of fuzzy logic in the area of human intention and activity inferencing. The differentiation of human intention and activities through their natural (and fuzzy) speech, emotion, body motion is particularly of interest in this review.

In the rest of the paper, the authors aims to summarise the existing literature in three different main topics that are relevant to SAR; namely speech, emotion and body motion.

2   Speech

Natural language is the most familiar and convenient method of communication for humans. Therefore, it is very convenient if a robot has an interface that can comprehend what people say, carry out their orders, and answer them in a peer-like manner. To achieve this goal, just voice recognition is not sufficient; instead, a robot should be able to understand the underline meaning of a user utterance. Although there are several speech recognition systems that can provide satisfactory performance, it is necessary to use different strategies to decode the intended information contained in natural language expressions for true natural language controlled systems. Some notable contributions are summarized below.

The adaptive fuzzy command acquisition network (AFCAN) proposed by Lin and Kan [9], was able to acquire fuzzy commands via on-line learning and accepting criticism from a user. In their method, acquiring commands such as "move forward very fast" was studied. Here, "move forward" represents the action to be performed while "very fast" represents the fuzzy linguistic information. For machine or robot control, interpreting this kind of subjective commands is useful.

Pulasinghe et al. [12] applied a similar method for controlling a mobile robot. They proposed that the significance of action modification words changes contextually and implemented a command interpretation strategy based on a fuzzy neural network. A fuzzy neural network was used for evaluating the meaning of fuzzy terms.

Jayawardena et al. studied using natural language instructions for controlling robotic manipulators. In their work, in addition to simple motion commands, posture control commands such as "bend forward little," "bend backward very little" etc. were used [8][7]. A modified version of a probabilistic neural network and fuzzy inferencing system was used for interpreting the meaning of fuzzy terms.

Jayasekara et al. proposed a method for interpretation of fuzzy voice commands based on the vocal cues [6]. In their work, the fuzzy linguistic terms included in natural language instructions were interpreted as modifying the robots environment based on vocal cues received from the user. A vocal cue evaluation system based on fuzzy techniques was developed to evaluate vocal cues.

3   Emotions

Emotion is a key part to human behaviour and even the way human think [10]. In order to build an intelligent and believable agent, investigation of how human emotion changes in reaction to events and modelling internal emotion states and generating outward expressions are key research challenges. Based on the event-appraisal model, El-Nase applied fuzzy logic in building an emotion framework [3,4]. Different to previous attempts including the OZ project [13] which used thresholding technique to determine the desirability of an event in a binary manner, this framework named FLAME (fuzzy Logic Adaptive Model of Emotions) uses fuzzy variables to represent intermediate states within variables such as desirability. In the framework, El-Nasr used 2 fuzzy variables to represent 1) the impact of an event to a goal and 2) the importance of the goal . A set of fuzzy rules was then used to determine the desirability of an event, which is another fuzzy variable. For example, a goal can be {HighlyUndesired, SlightlyUndesired, Neutral, SlightlyDesired, and HighlyDesired}.

Once the desirability of an event is determined, Ortony's model is used again to trigger emotion states. It is worth mentioning that a total of 14 emotions are modelled in the framework including joy, sad, disappointment, relief, hope and fear. This framework incorporates a few other components to compute other parameters including expectation of an event and the occurence of an event which are also used in the forumlae for triggering emotions. After determining emotion states, the framework further uses a fuzzy variable to represent the intensity of emotions {HighIntensity, LowIntensity or MediumIntensity} triggered for selecting the behaviour of an agent. As a case study, El-Nasr chose a pet dog (PETEEI) as the agent. PETEEI is an 2D animated dog in a GUI environment. In the first experiment, FLAME with fuzzy logic produced better behavior for the pet dog PETEEI when compared to the same system using thresholding technique (or interval mapping technique) or the baseline which randomises emotions and behaviours [4]. In [3], a learning component was introduced into FLAME to include reinforced learning for associating emotions with objects, learning patterns of past events and etc. Another experiment was done and the result shows that the learning component further enhances the system's believability and intelligence. While fuzzy technique provides an easy way to model the pet's emotions, fuzzy technique in this second experiment did not show significant benefit when compared to the thresholding technique.

In a similar approach, Mobahi in [11] explored a reactive approach to construct a believable and emotional robot head with physically actuated facial expressions. At perception level, the robot head (Aryan) senses the distance and the speed of the object moving in front of it and represent them in 2 fuzzy variables. For example, the object can be very near, near or far. These two fuzzy variables are then used in a set of fuzzy rules to determine the robot's emotion. In the implementation, 3 emotions (angry, surprised and fear in 3 different intensities are determined. Emotions are mapped to each of the degree of freedoms (DOFs) in the robot's face. A simple experiment carried out by the authors showed that the fuzzy system produced smoother transition between emotions and better believability.

Friberg in [5] described a fuzzy logic-based tool for analysing emotions from cues extracted from music audio and body motions. The aim of the work is to recognise its user's (a dancer, a music performance or a conductor) emotions through simple cues from body movements including quantity of motions, heights and width of their movements and cues from music including sound level, tempo and articulation. Each cue was modelled as a fuzzy variable with 3 levels {-,0,+}. A total of six fuzzy variables were used to represent all cues (three music audio cues and three body motion cues). Emotions are recognised based on these fuzzy variables. In application, recognised emotions are further used to control an artistic performance through either synthesizer or visual effects. In this work, Friberg utilised qualitative data from previous research that investigated emotional expressions in music and body motions to define the membership functions of the fuzzy variables. This is different to data-driven approaches such as Neural network and Hidden Markov Models.

The work was demonstrated through an visual application that visualises a ball in different sizes, shapes, colors and positions according to a combination of the cues and emotions of the music. It was also evaluated in a collaborative game which considers both body motions and music cues.

4 Body Motion and Activity

With the motivation to reduce falls among the older people, Anderson in [2,1] explains a two-level fuzzy logic inferencing system to infer an elder person's daily activities from camera images. Similar to [11], fuzzy logic was applied in the most of the system. In the first level of inferencing, a person is detected, recognised and modelled through a 3D discrete model called "Voxel person" which is constructed through images from two cameras. The system reasons about a person's state {Upright, On-the-ground, In-between} based on signal inputs including the person's centroid, height, major body orientation, similarity of the body orientation with the ground plane normal. In this level of inferencing, 24 fuzzy rules were determined empirically with the help of nurses. The output is combined with time information which is also fuzzified through a fuzzy variable {brief, short, moderate, long} to produce a linguistic and temporal summary of a person's state. Exemplar outputs include "John is upright in the toilet for a moderate amount of time" and "John is on-the-ground in the living room for a brief amount of time." The effectiveness and usefulness of this summarisation lie in the ability to generate daily activity report succinct enough for caregivers and nurses to review at the end of the day.

In order to complete the system from signal processing, to modelling, through alert generation. Anderson further uses a second inference system (second level) to determine a person's activity in general but with a focus on falls. Average state membership, time duration, confidence in a quick change in Voxel person's average speed ...etc were used for in the fuzzy rules for characterising a falls. Experiments were carried out with videos of actors mimicking different type of falls. The result was promising; the system detected all 14 falls recorded in the videos with 6% false positive rate. It proved its claims on its advantage from using fuzzy logic in rejection of false positive detection, and modelling special cases/activities in comparison to HMM approach or its variants.

It is worth noting that the fuzzy rules (with linguistic variables) are easy add, delete or modify. Because of the ease to understand the rules, the rules were a point of collaboration between the engineers and the nurses. In other words, expert advises from the nurses feed into the system through the definition of the rules.

5   Discussion and Conclusion

In the traditional research fields like robot navigation and signal processing, fuzzy logic has been successfully applied and shown to be very effective in 1) handling uncertainties in sensor data, 2) provide an alternative when the targeted features or environment can not be accurately or easily modelled and 3) achieving real-time operation [14]. At the other end of the loop of Sense-Plan-Act (SPA) or Sense-Model-Plan-Act (SMPA), fuzzy logic can further be used by robots to infer, reason about or make meanings out of perceived features about human who is the target user in SAR applications before actioning. Research reviewed has shown that fuzzy logic can be used to smoothen transition of robot's outward expressions such as emotions, therefore creating more life-like and believable characters e.g. [3,4,11,5]. Fuzzy rules allow actions to be mapped to sensor inputs when the model is impossible or hard to obtain (due to non-linearity) and is computationally expensive to run. The nature of fuzzy rules being linguistic, easy to understand and modify provide a good method to incorporate domain expert's inputs in the design of a robot's behaviour [2]. This is particularly useful in the area of SAR which is inherently inter-disciplinary and involves domain experts including nurses and psychologist. Critical parameters in the system for robot behaviour can be easily adjusted through IF-THEN rules without any previous training. This is very advantageous when compared to HMM-based approaches which require adjustments of likelihood values (hard to understand) and ad-hoc training of different models required [2].

References

1. Anderson, D., Luke, R., Keller, J., Skubic, M., Rantz, M., Aud, M.: Modeling human activity from voxel person using fuzzy logic. IEEE Transactions on Fuzzy Systems 17(1), 39–49 (2009)
2. Anderson, D., Luke, R.H., Keller, J.M., Skubic, M., Rantz, M., Aud, M.: Linguistic summarization of video for fall detection using voxel person and fuzzy logic. Comput. Vis. Image Underst. 113(1), 80–89 (2009), http://dx.doi.org/10.1016/j.cviu.2008.07.006
3. El-Nasr, M., Yen, J., Ioerger, T.: Flame - fuzzy logic adaptive model of emotions. Autonomous Agents and Multi-Agent Systems 3(3), 219–257 (2000)
4. El-Nasr, M., Yen, J.: Agents, emotional intelligence and fuzzy logic. In: 1998 Conference of the North American Fuzzy Information Processing Society - NAFIPS, pp. 301–305 (1998)
5. Friberg, A.: A fuzzy analyzer of emotional expression in music performance and body motion. In: Proceedings of Music and Music Science, Stockholm 2005 (2004)
6. Jayasekara, A.G.B.P., Watanabe, K., Kiguchi, K., Izumi, K.: Interpretation of fuzzy voice commands for robots based on vocal cues guided by user's willingness. In: 2010 IEEE/RSJ International Conference on Intelligent Robots and Systems (IROS), pp. 778–783 (2010)
7. Jayawardena, C., Watanabe, K., Izumi, K.: Controlling a robot manipulator with fuzzy voice commands using a probabilistic neural network. Neural Computing Applications 16(2), 155–166 (2007)

8. Jayawardena, C., Watanabe, K., Izumi, K.: Posture control of robot manipulators with fuzzy voice commands using a fuzzy coach player system. Advanced Robotics 21, 293–328 (2007)
9. Lin, C.T., Kan, M.C.: Adaptive fuzzy command acquisition with reinforcement learning. IEEE Transactions on Fuzzy Systems 6(1), 102–121 (1998)
10. Minsky, M.: The emotion machine: Commonsense thinking, artificial intelligence, and the future of the human mind. Simon and Schuster (2007)
11. Mobahi, H., Ansari, S.: Fuzzy perception, emotion and expression for interactive robots. In: IEEE International Conference on Systems, Man and Cybernetics, vol. 4, pp. 3918–3923. IEEE (2003)
12. Pulasinghe, K., Watanabe, K., Izumi, K., Kiguchi, K.: Modular fuzzy-neuro controller driven by spoken language commands. IEEE Transactions on Systems, Man, and Cybernetics, Part B: Cybernetics 34(1), 293–302 (2004)
13. Reily, W.: Believable social and emotional agents(ph. d. thesis) (1996)
14. Saffiotti, A.: The uses of fuzzy logic in autonomous robot navigation. Soft Computing 1(4), 180–197 (1997), http://www.aass.oru.se/~{}asaffio/
15. Tapus, A., Maja, M., Brian, Scassellatti, o.: The grand challenges in socially assistive robotics. IEEE Robotics and Automation Magazine 14(1) (2007)

The Uncanny Valley:
A Focus on Misaligned Cues

Lianne F.S. Meah and Roger K. Moore

Department of Computer Science, University of Sheffield,
United Kingdom
{lfsmeah1,r.k.moore}@sheffield.ac.uk

Abstract. Increasingly, humanoid robots and androids are easing into society for a wide variety of different uses. Previous research has shown that careful design of such robots is crucial as subtle flaws in their appearance, vocals and movement can give rise to feelings of unease in those interacting with them. Recently, the Bayesian model for the uncanny has suggested that conflicting or misaligned cues at category boundaries may be the main attributing factor of this phenomenon. The results from this study imply that this is indeed the case and serve as empirical evidence for the Bayesian theory.

Keywords: Uncanny valley, social robotics, human-robot interaction.

1 Introduction

Although the phenomenon of the uncanny valley was first proposed by Mashiro Mori [7], the concept of the uncanny can be traced back as far as 1906. In his essay, psychiatrist Ernst Jentsch described the uncanny as *intellectual uncertainty* [4], and several years later it was revisited by Sigmund Freud, who described it as something which seems familiar and yet foreign simultaneously [2]. In his report, Mashiro theorized that an object that is more humanlike in appearance will seem more familiar with an observer.

For example, a robotic arm used in industry may be seen as less familiar than a humanoid robot, as it is visually far less humanoid. This is depicted in Fig. 1, where industrial robots are placed near the origin of the graph with low familiarity and low human likeness. Humanoid robots are placed just before the peak in familiarity. It might then be expected that robots that look *especially* human will continue the trend in the graph, however, they instead fall into the uncanny where their familiarity ratings are akin to those of zombies or corpses. With this drop in familiarity comes an increase in eeriness, which manifests as a feeling of unease or repulsion in observers.

Before proceeding, it is important to clarify what is meant by the terms *robot*, *humanoid robot* and *android* in this study. The term robot shall refer to a programmable machine, or automaton, that bears little to no resemblance of a human being. A humanoid robot, then, is a robot which is humanlike in some sense (it may possess a humanoid body or face) but can visibly be distinguished

M. Beetz et al. (Eds.): ICSR 2014, LNAI 8755, pp. 256–265, 2014.

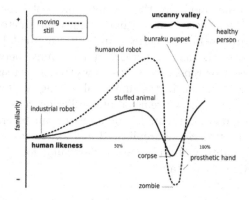

Fig. 1. The uncanny valley diagram. [7]

from a human being, in other words, it is easy to classify as a robot. The term android refers to a humanoid robot with an added layer of complexity; androids are designed to pass as human beings and will own more intricate assets such as artificial skin, hair and so on. They are visually almost human, to the point where they fall into the uncanny.

As the original illustration of the uncanny valley depicts familiarity against increasing visual human likeness, many studies have been carried out with a focus on the visual domain. However, the uncanny valley has also been shown to exist in the audio continuum [3]. As such, it can be suggested that a person's response to a stimulus can be altered by changing either the visuals, audio or both. Indeed, the link between a character's voice and face has already been investigated [8], and a mismatch in these features can induce the uncanny valley effect. For example, pairing a human voice with a robotic, mechanical face produces feelings of unease in observers [8], suggesting that a person or robot's voice and face play a major role in communication. In particular, the eyes are thought to provide a multitude of cues. Abnormal alterations of the eyes alone is enough to produce the uncanny effect [5], [6].

More recently, a Bayesian explanation of the uncanny has been suggested [9]. Based on the categorical perception model of Feldman, Griffiths and Morgan [1], the model of the uncanny proposes that stimuli containing conflicting cues cause 'differential perceptual distortion' which in turn induces perceptual tension. It is suggested that this tension manifests as feelings of eeriness. The key to perceptual distortion is categorization; the uncanny is predicted to manifest from observing androids as they contain multiple conflicting perceptual cues, some of which cause a greater amount of uncertainty regarding their category membership, thus giving rise to perceptual tension (also see [11]). Androids cannot easily be classified into the human or robot category; they lie within or near a category boundary (see [9] for example illustrations), and we find that the *"inability to categorize will then lead to a state of dissonance"* [10].

To date, the Bayesian model of the uncanny has not been documented in an empirical study. We investigated to what extent contradictory or misaligned cues contribute towards feelings of eeriness in both a unimodal and multimodal setting. The model suggests that an increase in uncertainty between cues results in an increase in perceptual tension, thus it follows that a decrease in uncertainty will reduce perceptual tension. In the unimodal setting, we examined the role of an android's eyes and how the removal of conflicting cues from them might alter an observer's response. In the multimodal setting, an experiment performed originally by Mitchell et al [8] was replicated and extended to include a wider range of visual and auditory stimuli, with a particular focus on the degree of conflicting cues they might contain.

2   Materials and Methods

We performed two experiments, one with a focus on unimodal cues and the other focusing on multimodal cues. In both experiments, volunteers were asked to watch several videos and then provide feedback both qualitatively and quantitatively by filling out a questionnaire. Upon watching a video, a participant was required to give ratings for four different attributes of the subject in the video: humanness, eeriness, familiarity and appeal (it should be noted that only the eeriness attribute will be discussed in the results). The ratings were on a Likert scale between 1 and 5. A listening booth was provided by the University Speech and Hearing Lab, where participants could sit at a desk within a quiet environment with the videos being displayed on a computer monitor. Footage of three androids and one humanoid robot were obtained for use in both experiments: the *Geminoid DK*, *'Jules'*, the *Repliee Q* and the *iCub*, respectively. In addition, for the second experiment, a video of a human male was recorded using an HD camcorder. See Fig. 2 for all the visual stimuli.

Fig. 2. All visual stimuli used in the experiments, composed of *a*: one humanoid robot, *b-d*: three androids and *e*: one human. Subjects *a*, *b*, *c* and *d* were used in the first experiment, subjects *a*, *d* and *e* were used in the second. Additionally, audio was recorded from *e* for the second experiment. Images are a single frame taken from each video.

2.1   Experiment One

The primary goal of the first experiment was to investigate the impact of unimodal cues in the visual domain, as such the videos were not combined with any auditory cues. We investigated the impact of an android's eyes and hypothesized that the removal of misaligned cues from them would significantly decrease the eeriness felt in an observer. We also investigated the impact of a humanoid robot's eyes and predicted that, since robots typically do not fall into the uncanny (although this is dependent on design), removal of cues from the eyes would not have the same effect.

To carry out this study, three videos of different androids and one video of a humanoid robot were shown to participants. The original videos were edited only to control the length of time that each video ran for and also to mute the audio. In addition, four other videos were created where the cues from the eyes were blocked by a rectangular black box, which was placed just above the lower lid and beneath the eyebrows, thus covering the eyes. In the final video reel, the videos were paired such that a 'covered' video followed after its 'uncovered' counterpart and vice versa. To summarize, there were eight videos in total, four pairs of 'covered' and 'uncovered' clips.

2.2 Experiment Two

In order to confirm that a mismatch in voice and face induces the uncanny effect, in the second experiment the focus changed from unimodal to multimodal cues and audio was combined with the visual stimuli. For this experiment, we extended a recent study on the uncanny [8]. In the original study, Mitchell et al combined the face and voice of a human with the face of a robot and a synthetic voice in order to create 'matched' and 'mismatched' stimuli. They theorized that matched stimuli (aligned cues) would be significantly less eerie than mismatched stimuli (misaligned cues). For example, it was shown that participants are comfortable in viewing a video of a human face combined with a human voice, but not so comfortable if the human face was paired with a synthetic voice. We extended this experiment to include the visuals of an android and dual-pitched audio, both of which should be regarded as particularly eerie by observers as they are both almost human in their respective domains, thus near category boundaries.

To create dual-pitch voices we recorded audio from a human male (aged in his late thirties) and ran it through a dual-pitch voice changer, developed in Pure Data. This particular method of voice changing gives the impression that two voices are being spoken at once, one of which differs in pitch, and serves as a way of constructing a robotic-sounding voice without disruptions in sentence flow, as is often heard in other text-to-speech voices.

We theorized that the android, despite being visually very close to human, would still be judged as a robot, and that the dual-pitch voices, although derived from and close to the original human audio, would still be judged as robotic. As such, for the android visuals, the synthetic and dual-pitch voices were hypothesized to be the matching audio, with the human voice acting as the mismatching audio. The same was hypothesized for the robot. For the human visuals, we predicted that the human voice would serve as the matching audio, with the synthetic and dual-pitch voices acting as mismatching audio. Additionally, for the human visuals, we hypothesized that a dual-pitch mismatch would be significantly eerier than the synthetic mismatch, as the dual-pitch audio clips are closer to the original audio recorded from the human, and would thus cause a greater amount of perceptual tension.

In the experiment, footage of one android and one humanoid robot were used. Additionally, both video and audio of a human male were recorded speaking the neutral phrase *'a goal is a dream with a deadline'*. The original human audio was run through the voice changer and shifted by three different frequencies, 50Hz, 150Hz and 250Hz, in order to create three different dual-pitch stimuli. Furthermore, a text-to-speech (TTS) synthesizer was used to create a synthetic voice which spoke the same phrase. Upon completion of gathering the required audio, the voices were then overlaid onto the videos. Full lip syncing was not possible, except for the human visuals as they were recorded at the same time as the audio. As there were three different visuals (humanoid robot, android, human) and five different voice conditions (human voice, dual-pitch 50Hz, dual-pitch 150Hz, dual-pitch 250Hz and synthetic), there were fifteen clips overall for the second experiment.

Table 1 gives a summary of the stimuli used for second experiment. In the interest of clarity, hereafter the stimuli will be referred to using *visual-audio* notation, where visual refers to one of the visual categories (human, android, robot) and audio refers to one of the auditory categories (human, 50Hz dual-pitch, 150Hz dual-pitch, 250Hz dual-pitch, synthetic). For example, *Robot-Synthetic* refers to the robot face combined with the synthetic, text-to-speech voice, *Android-50Hz* refers to the android face combined with the dual-pitch voice that has been shifted by 50Hz, and so on.

Upon completion of all 23 videos, they were all combined into one single reel which was presented to the participant. Before the beginning of each experiment, a black screen would be presented with text in white, reading as 'Experiment One' or 'Experiment Two'. The respective stimuli would then follow, in a randomized order unknown to the participant. Each participant thus took part in both experiments and completed experiment one first.

Table 1. Summary of stimuli for the second experiment

	human	android	robot
human voice	match	mismatch	mismatch
50Hz shift	mismatch	match	match
150Hz shift	mismatch	match	match
250Hz shift	mismatch	match	match
TTS (synthetic)	mismatch	match	match

3   Results

The study was conducted over three weeks in March. For both experiments there were 40 volunteers of varying disciplines within the University of Sheffield,

14 female and 26 male, with a mean age of 25.8. Data analysis was conducted using the matched-pairs t-test.

3.1 Experiment One: Unimodal Cues

The full result set for the eeriness ratings is shown in Fig. 3. The pairs of stimuli have been plotted in terms of their humanness and eeriness; the far left denoting lower humanness ratings. As expected, the humanoid robot was rated lowest in terms of humanness whilst the Geminoid DK android was given the highest ratings.

We found that blocking the eyes of the Geminoid DK, thereby decreasing perceptual tension, did indeed have a positive effect on an observer and significantly decreased its average eeriness rating. We theorize that it is because the Geminoid DK is the most humanlike of the androids that the blocking had the most impact; in the middle range of the humanness scale, the ratings for eeriness were not significantly impacted by blocking the eyes. However, on the far left of the humanness scale, covering the eyes of the humanoid robot resulted in an enhanced *negative* response from viewers and significantly increased its eeriness rating. It could be suggested, then, that the less human a robot visually appears to be, the less the covering of the eyes will impact an observer's responses in a positive way.

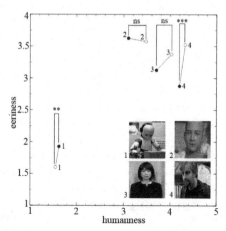

Fig. 3. Mean eeriness ratings from the first experiment. *Filled circles* denote videos where eyes were covered. *Open circles* denote videos were eyes were shown. Average eeriness ratings (from left to right): 1.600, 1.925, 3.625, 3.575, 3.125, 3.375, 2.875 and 3.575. ** denotes $p \leqslant 0.01$, *** denotes $p \leqslant 0.001$.

3.2 Experiment One Discussion

These results agree with the Bayesian model; the eyes of an android contain conflicting cues which give rise to uncertainty and perceptual tension. Covering the eyes, thereby removing the conflicting cues, decreases perceptual tension and thus decreases the eeriness felt in viewers. The impact of cue removal depends on where the subject sits on the humanness scale, or rather, how close to the category boundary it is. The model predicts that removal of cues from an object rated lower in humanness (a humanoid robot) should instead increase eeriness, which is indeed what has been found here.

3.3 Experiment Two: Multimodal Cues

There were two aims of this experiment. The first was to test whether a dual-pitch voice, combined with mismatching visual stimuli would be regarded as significantly eerier than a synthetic voice mismatch. The second was to repeat and extend a recent study on the uncanny, with the additional android footage (the Geminoid DK) and the dual-pitch voices to bring more dimensions to the experiment and test the Bayesian model in a multimodal setting.

The average eeriness ratings for this experiment are given in Fig. 4a and Fig. 4b. The lowest rating of eeriness was given to the Human-Human stimulus and the highest was given to the Android-50Hz stimulus. Generally, stimuli using the android face were given the highest eeriness ratings in each voice condition. Additionally, stimuli using the 50Hz dual-pitch voice were also given the highest eeriness ratings in each visual condition.

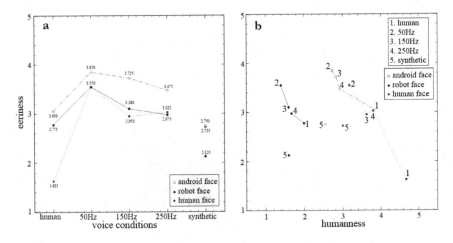

Fig. 4. Average eeriness ratings plotted against **a**: voice conditions, and **b**: average humanness ratings

3.4 Matched and Mismatched Comparisons

For the human visuals, all mismatched stimuli (Human-50Hz, Human-150Hz, Human-250Hz, Human-Synthetic) were significantly eerier than the matched stimulus (Human-Human). Since the Human-Human stimulus is a 'matched' combination and it thus follows that it received the lowest ratings of eeriness. Furthermore, we can conclude that a dual-pitch voice is indeed not judged as human.

For the robot visuals, the synthetic and dual-pitch voices were theorized to be the matching auditory stimuli. However, this was not the case. The Robot-Human combination (mismatch) was significantly eerier than the Robot-Synthetic (match) combination, which was expected. However, the dual-pitch voices were also given significantly higher eeriness ratings than the Robot-Synthetic stimulus, suggesting that the dual-pitch voices are also mismatching stimuli.

We predicted that the android would be judged as a robot; thus the mismatching auditory stimulus for this visual category was proposed to be the human voice, and the matching stimuli were proposed to be the dual-pitch and synthetic voices. Generally, participants gave higher eeriness ratings for the Android-Human stimulus than the Android-Synthetic stimulus, suggesting that it was indeed a mismatched video. Additionally, the Android-Synthetic combination generated the lowest eeriness ratings for the android visuals. Here, it can be suggested that the android was being perceived as as robot. However, this implies there to be a significant increase in eeriness from the Android-Human to the Android-Synthetic stimuli. Statistically however, there was no difference between the two voice conditions. Furthermore, the dual-pitch combinations (Android-50Hz, Android-150Hz, Android-250Hz) were also seen as significantly eerier than the Android-Synthetic stimulus, suggesting that they were also mismatching audio. The full results are given in Fig. 5.

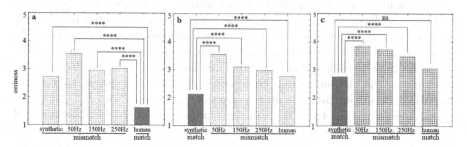

Fig. 5. Average eeriness ratings **a:** for the human, **b:** for the humanoid robot, **c:** for the android. Although initially thought to be matching stimuli, the dual-pitch voices for both the robot and android visuals are instead mismatching stimuli. The average eeriness rating for the Android-Human stimulus is statistically the same as the Android-Synthetic stimulus, highlighting confusion about the android's category membership. ******** denotes $p \leqslant 0.0001$.

3.5 Experiment Two Discussion

This experiment serves as further evidence to support the Bayesian model of the uncanny valley. Here, we have shown that eeriness can be induced by mismatching stimuli, using a variety of different combinations. For a mechanical, 'obvious' humanoid robot that is far away from a category boundary, an 'obvious' synthetic voice is most suited to it. On the other end of the humanness scale, a human face, which is also far away from a category boundary, is best matched with a human voice.

For the android visuals, however, conclusions are a little more difficult to draw. The Android-Human stimulus received a higher rating of eeriness than the Android-Synthetic combination, though not significantly so. However, the Android-Human stimulus received significantly higher ratings of familiarity and appeal (data not shown), which contradicts what should happen in the presence of increased eeriness. It is also theorized that there is confusion about where the android sits in terms of categorical definition, thus why there is no statistical difference between the Android-Human and Android-Synthetic stimuli. Possibly, a dual-nature is being perceived due to there being misaligned cues at category membership.

It was already predicted by the uncanny valley model for visuals of a mechanical humanoid robot, such as the iCub, to be perceived as less eerie than the android, so it follows that generally, the videos of the humanoid robot are rated as less eerie than the videos of the android. However, the introduction of audio implements another layer of complexity to the problem. Multimodal cues are indeed influencing participant judgment, as the eeriness of a certain visual was also dependent on the voice it was combined with. Fig. 4b shows that the Android-Human combination is less eerie than the Robot-50Hz combination, and that the Human-50Hz combination is regarded as eerier than the Android-Human combination. In these cases the audio alone has reversed the uncanny effect, such that a human or humanoid robot is regarded as stranger than an android.

The dips in eeriness in Fig. 4b are hypothesized to be caused by stimuli that can be easily classified, for example, a mechanical humanoid robot paired with a synthetic voice which sits within the non-human category. On the far right of the graph, the Human-Human combination is also well defined in category. The peaks in eeriness may then be explained as the result of misaligned cues. For example, the android visuals combined with dual-pitch voices, that sound almost human, are stimuli that may be regarded as eerie in both the visual and auditory domain. Thus there is an increase in perceptual tension; the face and voice combined give rise to an enhanced peak in eeriness.

4   Conclusions

In this study, two experiments were conducted to investigate the impact of conflicting cues from visual and auditory stimuli. We have shown that removal of unimodal, misaligned cues from the eyes of an android can significantly decrease the eeriness felt in observers and that the impact of cue removal is dependent on where the android sits in terms of humanness, or rather, how far it is from a category boundary. Thus, the eyes of an android have a great impact on observers in human-robot interaction. Humanoid robots, with low ratings of humanness, are seen as eerier when cues from the eyes are removed as they are further away from a category boundary.

We have also replicated and extended an experiment that tests the influence of multimodal cues. Our results agree that the uncanny does indeed exist within the auditory continuum and that visual stimuli regarded as non-eerie can fall into the uncanny with the introduction of audio. Our results also agree that mismatching voices and faces will induce the uncanny. Additionally, we have shown that a dual-pitch voice, derived from a human voice, is regarded as significantly eerier than a text-to-speech synthetic voice when combined with visual stimuli. It is hypothesized that the dual-pitch voices sit near to a category boundary and thus give rise to a greater amount of perceptual tension. Although developed to sound robotic, the dual-pitch voices do not match with robotic faces. Furthermore, we have demonstrated that an android sits near to a categorical boundary which in turn gives rise to perceptual uncertainty about its identity. This results in both the Android-Human and Android-Synthetic combinations being regarded as the same in terms of eeriness ratings.

In both experiments, the results agree with the Bayesian explanation of the uncanny valley and suggest that perceptual distortion, caused by misaligned cues, gives rise to perceptual tension which is felt as unease or eeriness in observers. This study thus serves as empirical evidence for the Bayesian model.

Acknowledgments. The authors would like to thank the Speech and Hearing Lab for providing a booth for use in the experiments.

References

1. Feldman, N.H., Griffiths, T.L., Morgan, J.L.: The influence of categories on perception: explaining the perceptual magnet effect as optimal statistical inference. Psychological Review 116(4), 752 (2009)
2. Freud, S.: The uncanny (j. strachey, trans.). the standard edition of the complete psychological works of sigmund freud, vol. 17 (1919)
3. Grimshaw, M.N.: The audio uncanny valley: Sound, fear and the horror game. Audio Mostly, 21–26 (2009)
4. Jentsch, E.: On the psychology of the uncanny (1906) 1. Angelaki: Journal of the Theoretical Humanities 2(1), 7–16 (1997)
5. Looser, C.E., Wheatley, T.: The tipping point of animacy. How, when, and where we perceive life in a face. Psychological Science 21(12), 1854–1862 (2010)
6. MacDorman, K.F., Green, R.D., Ho, C.C., Koch, C.T.: Too real for comfort? Uncanny responses to computer generated faces. Computers in Human Behavior 25(3), 695–710 (2009)
7. Mashiro, M.: Bukimi no tani (the uncanny valley). Energy 7, 22–35 (1970)
8. Mitchell, W.J., Szerszen Sr, K.A., Lu, A.S., Schermerhorn, P.W., Scheutz, M., MacDorman, K.F.: A mismatch in the human realism of face and voice produces an uncanny valley. i-Perception 2(1), 10 (2011)
9. Moore, R.K.: A Bayesian explanation of the uncanny valley effect and related psychological phenomena. Scientific reports (2012)
10. Pollick, F.E.: In search of the uncanny valley. In: Daras, P., Ibarra, O.M. (eds.) UC-Media 2009. Lecture Notes of the Institute for Computer Sciences, Social Informatics and Telecommunications Engineering, vol. 40, pp. 69–78. Springer, Heidelberg (2010)
11. Ramey, C.H.: The uncanny valley of similarities concerning abortion, baldness, heaps of sand, and humanlike robots. In: Proceedings of Views of the Uncanny Valley Workshop: IEEE-RAS International Conference on Humanoid Robots, pp. 8–13 (2005)

Representation and Execution of Social Plans through Human-Robot Collaboration

Lorenzo Nardi and Luca Iocchi

Dept. of Computer, Control and Management Engineering
Sapienza University of Rome, Italy

Abstract. The use of robots in people daily life and, accordingly, the requirement for a robot to behave in a socially acceptable way are getting more and more attention. However, although many progresses have been done in the last years, robots still have many limitations when they are required to share the environment with humans.

In this paper, we define the concept of *social plans* combining two main ideas: the definition of social behaviors for enabling a robot to live with humans and the establishment of a symbiotic relationship among robots and humans to overcome robots' limitations. Social plans are plans containing both robot and human actions and we provide an execution model for them where human actions are replaced by a human-robot collaboration scheme in which the robot actively drives the interaction with a human in order to obtain the desired effect.

A fully implemented system has been realized following this idea and different examples are provided in order to demonstrate the effectiveness of the approach.

1 Introduction

Recently, advancement of robotics encouraged a gradual move of robots from laboratories into people daily lives for acting as partners or assistants. In order to achieve this, robots should be able to share their working space with the people inhabiting it. The first requirement for a robot that wants to perform tasks closely to people is that it must be safe to humans; however, in the direction of being perceived by people as an actual partner rather than a solely mechanical tool, also human comfort and social acceptability should be considered.

In the last years, different robot social navigation systems have been developed to enable robots navigating into environments inhabited by people and interacting with them for accomplishing diverse and even complex tasks in a socially-acceptable way. These include systems for robots acting as an interactive museum tour-guide [1], escorting residents in nursing homes [6] and giving directions to the clients in malls [3]. However, most of them are task-specific and typically hard to generalize.

Further, current robotic systems still suffer from significant limitations at perception, reasoning and actuation due to factors like poor accuracy of sensors and actuators, high complexity and costs. These limitations cause several evident difficulties to develop complete and autonomous robotic systems able to perform even simple tasks, such as opening a door or bringing a cup of coffee.

M. Beetz et al. (Eds.): ICSR 2014, LNAI 8755, pp. 266–275, 2014.

In order to overcome robots' limitations, in this paper, we introduce a new general approach based on the idea of symbiotic robotics that allows robots for realizing tasks in a social way. In fact, as asserted by Rosenthal et al. [8], since many of robots' difficulties are easy tasks to humans with whom they share the working space, robots should be able to interact and collaborate with them for receiving the help needed. We design this collaboration through plans represented in Petri Net Plans (PNP) [11] that explicitly take into account human actions. We call these plans *social plans*.

The contributions of this article are in the definition and execution of social plans and in the explicit representation of human-robot collaboration. A full implementation of different services on an actual mobile robot has been realized to show the effectiveness of the presented approach.

2 Related Work

Over the last years, some general systems for robot social navigation have been implemented. They focus not only on paths and trajectories a robot should follow to reach its goal, but also on general norms for appearing as social entities to humans. In this regard, Pacchierotti et al. [7] developed a system to navigate in a hallway in which the Hall's concept of proxemic space was exploited. Kirby extended Pacchierotti's work realizing COMPANION framework [5], [4]: a more general system for realizing navigation tasks, able to take into account a wider variety of social cues. Here, a global optimal planner considers a set of constraints including some general social conventions to produce socially correct robot behaviors.

However, despite these systems allowing robots for a socially-acceptable navigation, current robots have significant limitations that affect the accomplishment of even simple tasks. In order to overcome this issue, Rosenthal et al. [8] introduced symbiotic relationships among robots and humans. Each individual involved in a symbiotic relationship performs distinct asynchronous actions and the results affect all individuals involved. Accordingly, when a human assists a robot, people will receive back the service provided from its task completion. The idea of symbiotic robotics has been further extended considering robots capable to ask for help to the actual occupants of the environment without any supervision, distributing the load of help to all the people living in the environment and obtaining assistance from the largest number of humans available in it [9].

We propose a new general approach to enable robots for accomplishing different social services based on the idea of collaboration among robots and humans who share the working space with them for overcoming their limitations. To this end, we introduce a robotic system that, resuming the key principles of symbiotic robotics and accounting a set of shared social conventions, enables to execute robot plans realizing socially-acceptable behaviours.

Cirillo et al. [2] assert that classical robot planning systems, in which the state of the world is only affected by the robot, are no longer applicable to robots sharing their working space with people. They define a human-aware robot task planning, in which robots consider the forecasted future human activities and adapt their action plans accordingly. Similarly, our approach to robot planning takes into account humans sharing the environment with robot; however, we define social plans in Petri Net Plan in which human actions are explicitly represented to realize social tasks. Then, human actions are transformed into sub-plans of robot operations realizing the collaboration among robots and people and allowing for the accomplishment of diverse and complex tasks.

3 Social Plans

In this article, we propose a new approach based on the idea of human-robot collaboration to enable a robot for accomplishing different social tasks. In particular, we are interested in the definition and execution of social plans and in the explicit representation of human-robot collaboration. In the following, we introduce the concept of social plans describing how they can be defined and executed using the PNP formalism.

3.1 Definition of Social Plans

In robot planning, given an action theory, an initial situation and a goal, a planner computes a plan that will be executed by the robot. However, if a robot is not able to perform some actions necessary to reach the goal, a plan may not exist. For example, it would not be possible for a robot with no arms to find a plan for reaching a goal that would require the opening of a door.

We define the concept of *social plan* as a plan that combines actions performed by robots and by humans. Including human actions, a social plan allows for an explicit representation of human-robot collaboration. Given the specification of robot and human actions, the initial situation and the goal, a social plan can be generated by any planner. Otherwise, these plans can be manually designed by the robot programmer using a suitable formalism. In this paper, we are not interested in how social plans can be generated, but we focus on their representation and execution.

Since a social plan includes actions that must be performed by humans who are not aware of the global plan of the robot, it cannot be directly executed by a robot. In order to address this issue, it is necessary to transform the social plan into a robot action plan. In this transformation, human actions are converted into subplans of robot operations which realize a human-robot collaboration scheme in which the robot actively asks for help to any human passing nearby.

3.2 Human-Robot PNP

In order to represent social plans, in this paper, we decided to use Petri Net Plan (PNP) formalism [11]. The motivations for this choice are: 1) the availability of definition and execution model of multi-agent PNP, that we can similarly apply to humans and robots [10]; 2) the higher expressiveness of the formalism with respect to other languages that allow for representing complex plans and complex forms of human-robot interaction; 3) the implementation[1] of the formalism in Robot Operating System (ROS) that allows for an easy development of actual robotic applications.

Given the high expressiveness of this formalism, no automatic planner is available to generate PNPs. However, the main concepts of social plans and of transformation into executable robot plans could be applied to other languages as well.

A social plan in PNP is defined similarly to a multi-robot PNP in which one of the agents acting in the global plan is a human. This plan is called Human-Robot PNP (*HR-PNP*). In a HR-PNP, human actions are represented in the same way as robot actions and are labeled as *H_X*, where X is the name of the action expected to be executed by a human. For example, *H_Open_door* denotes the action of opening a door by a human.

As discussed in the previous section, a HR-PNP cannot be directly executed by a robot because it includes some actions that have to be performed by a human. Therefore, a transformation of the social plan is needed for enabling a robot to actually execute the corresponding behavior. Since we defined a HR-PNP as a particular multi-robot PNP, this transformation will resume the scheme for transforming a multi-robot PNP into a single-robot PNP [10]. In fact, from a multi-robot PNP it is possible to automatically produce a set of single-robot PNPs by dividing the part of the plan relative to each robot. In the same way, we can transform a HR-PNP into an executable PNP by maintaining robot actions and converting those actions that should be executed by a human into PNP sub-plans composed of robot operations for interacting with humans, preserving the correctness of the whole plan.

Since we are interested in automatic transformation of HR-PNPs into executable PNPs, we have to consider a transformation method that can be generally valid for every human action. Accordingly, we realized a template in PNP composed of robot actions that can be applied for replacing each of the human actions included into a social plan. This transformation makes explicit the collaboration among humans and robot. In particular, the PNP template includes all those actions and conditions that allow a robot for establishing an interaction with people in a socially-acceptable way and receiving from them the help needed to achieve its goal.

The transformation scheme from a human action into a subplan of robot operations is illustrated in Figure 1. Here, it is possible to understand that the PNP template is based on three generic robot actions/subplans that can be implemented in different ways depending on the human action considered: preparation action (*PreA_X*), communication action (*CA_X*) and perception action (*PA_X*), where X is the name of the human action taken into account.

- *PreA_X* is the PNP action/subplan executed by the robot for preparing itself to receive the human help needed to perform its task. Here, the robot waits, detects and approaches humans to establish an interaction with them.
- *CA_X* is the PNP action/subplan that allows a robot to ask humans for the assistance needed. Here, a time-out is necessary to handle failures in the

---

[1] pnp.dis.uniroma1.it

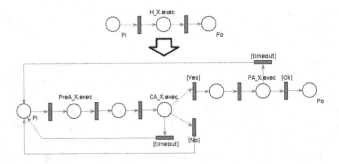

Fig. 1. Template PNP for transforming a human action in robot operations

communication. When a human refuse to help the robot or a failure is verified, the robot interrupts the interaction and tries to approach another person.

- *PA_X* is the PNP action/subplan that allows robot for sensing the space to ensure that the human accomplished the task he has been asked for. This action can be implemented exploiting diverse level of perceptions, from speech recognition to camera/laser data. Even in this case, a time-out is required to handle failures.

As illustrated in Figure 1, it is important to notice that the initial state (P_i) and the final state (P_o) of the original human action exactly coincide with the initial and final states of the template with which it is replaced. In fact, once the subplan defined in the template has been executed, the human action has been actually performed by some human who accepted to help the robot. In this way, human actions can be automatically transformed into subplans of robot operations without affecting the whole plan, given only the specifications of *PreA_X*, *CA_X* and *PA_X* that implement the human-robot collaboration for the action *X*.

Once all human actions of a HR-PNP have been transformed, an action plan that can be executed by a robot for realizing the corresponding behavior is obtained.

4   Implementation

Once the concept of social plan has been defined, we can introduce our system for realizing robot social behaviors describing its information workflow and providing an example of how does it work.

4.1   System Architecture

Currently, most of the systems demonstrating robot social tasks are specific for that particular task. Typically, these systems are effective but each of them is able to face just a few situations. Conversely, in this paper, we introduce a general

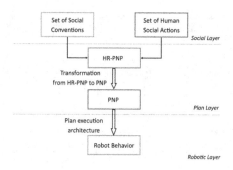

Fig. 2. Information workflow of our system

scheme that can be adopted and reproduced for easily realizing a large variety of robot social behaviors taking advantage from the collaboration with humans.

The information workflow of our robotic system is illustrated in Figure 2. It presents a layered architecture composed of three different levels.

- *Social Layer* includes a set of human actions, which defines the human actions that can be employed in designing social plans, and a set of social conventions, which includes different social norms shared among humans that range from navigation in public spaces to communication and allow a robot for interacting with humans in a socially-acceptable way.
- *Plan Layer* is the level in which a social plan for the desired service is designed in PNP considering the human actions defined in the social layer. Here, once a HR-PNP has been defined, the transformation into an executable PNP described in Section 3 takes place.
- *Robotic Layer* is the layer in which all different functionalities of the robot such as motion, speech and perception are combined together through ROS in order to actually execute the behavior corresponding to the PNP obtained from the Plan Layer.

4.2   An Example of Transformation of a HR-PNP

In order to provide an example of how our system works, we consider the simple HR-PNP illustrated in Figure 3a. Here, it is assumed to work with a mobile robot with no arms that, first, moves in front of a given closed door (*GoTo_door*); then, it is expected that a human opens the door (*H_Open_door*) in such a way the robot can pass through it (*Enter_door*).

Starting from this HR-PNP we want to obtain an equivalent executable PNP exploiting the method introduced in Section 3, i.e. replacing each human action that appears in the HR-PNP with a subplan of robot operations. To this end, we have just to transform *H_Open_door* as described before and to define the corresponding preparation, communication and perception actions that take into account the social conventions defined in the Social Layer of the system. A description of the definition of such actions is given in the following.

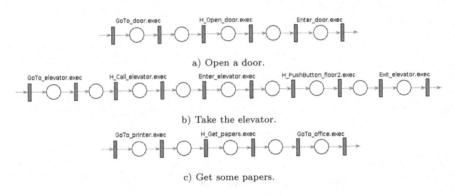

a) Open a door.

b) Take the elevator.

c) Get some papers.

Fig. 3. Examples of HR-PNPs for different robot behaviors

- *PreA_Open_door*: When the robot is in front of the closed door, since it is not able to open it by itself, it should look for help. For example, the robot can passively wait for a person who is passing nearby with whom trying to establish an interaction and to ask for assistance. Once a human has been detected by analysing its laser range finder readings, the robot approaches him greeting and turning in his direction.
- *CA_Open_door*: As a human has been approached, the robot tries to establish a conversation saying "Excuse me. I'm not able to open this door, can you help me?", which realizes the request for help by the robot. If the human refuses to give his help to the robot or the interaction fails for some reasons, it will look for another person who is available to help it.
- *PA_Open_door*: If the human accepts to help the robot, it senses the space through its laser range finder until the door has been opened. As the human opens the door, the robot thanks and greets him. Failure cases are handled by a timeout: if timeout is expired before the door has been opened, robot will restart the subplan asking for assistance to another person.

It is important to notice that we considered in social plans the concept of interrupt of PNP. When there is none in the surrounding of the robot available to help it, interrupt allows for suspending the current task and executing the next one, avoiding that the robot would be stuck forever waiting for assistance.

Table 1. Table for transforming some human actions into subplans of robot operations

Human Actions	Preparation Actions	Communication Actions	Perception Actions
Open a door.	Wait for a human and, when detected, approach him.	"I'm not able to open this door, can you help me?".	Check whether the door has been opened.
Call the elevator.	Wait for a human and, when detected, approach him.	"I'm not able to call the elevator, can you help me?".	Check whether the elevator door is open.
Push the elevator button.	None.	"I would like to go up to the second floor, can you press the button?".	Check whether the elevator door is open.
Bring papers from the printer.	Wait for a human and, when detected, approach him.	"Can you take papers from the printer and put them on my plate?".	Check that there are some papers on the plate.

In Table 1, preparation, communication and perception actions for transforming some human actions we considered in our experiments have been outlined. In particular, the first line reports the transformation we described above.

5   Use cases

We have implemented some use cases on an actual robot to demonstrate the effectiveness of our approach. The working space considered is our Department. Here, a mobile robot with no arms equipped with a laser range finder, a camera, a microphone and a speaker, who knows a priori the map of the environment and some semantic information on it, has performed different social tasks sharing the space with the humans working there. During the tests, some people without being informed a priori have been approached by the robot asking for their help. The videos of the robot accomplishing the diverse use cases are available at: https://sites.google.com/site/robotsocialnavigation/videos.

5.1   Navigation from the Hallway to the Auditorium

In the first use case, the robot should navigate from the hallway to the auditorium passing through two doors. Since the robot is not able to open a door, in order to face those situations in which it may find a closed one along its path, it is necessary to consider human intervention into the social plan (Figure 4a). This is exactly the situation we discussed in the previous section and the first video shows it. Here, a complete human-robot interaction is presented; however, if the human opens the door before the end of the interaction, the robot will be anyway able to take advantage of it.

5.2   Navigation from the First Floor to the Second Floor

In the second use case, the robot should reach the second floor from the first floor taking the elevator. To this end, the robot should call the lift and push the button corresponding to the second floor. However, the robot is not able to perform these two operations by its own and, therefore, the human intervention is needed. This is made explicit into the portion of social plan represented in Figure 3b that includes *H_CallElevator* and *H_PushButton_floor2* actions.

In the second video, it is shown how the action of calling the elevator has been replaced with robot operations outlined in the second line of Table 1. Accordingly, when the robot reaches the elevator, it looks for someone passing there who can help it (*PreA_CallElevator*). As a human has been detected, the robot approaches him greeting and turning in his direction. Then, the human stops and the communication starts with the robot saying "Excuse me, I'm not able to call the elevator, can you help me?" (*CA_CallElevator*). When the human accepts and calls the elevator (Figure 4b), the robot thanks him and waits until the elevator door is open (*PA_CallElevator*). As the elevator arrives, it greets the human and gets in.

Although it is not shown in this video, also *H_PushButton_floor2* can be transformed into a subplan of robot actions (third line of Table 1) to enable the robot for reaching the second floor and completing its task.

a) Open a door.   b) Call the elevator.   c) Get some papers.

Fig. 4. Examples of tasks performed by a human for helping the robot

5.3   Bringing Papers from the Printer to Someone's Office

In the third use case, we foresee the possibility for a user to send a paper to be printed to the robot. Once the robot has received the paper, it chooses the printer and sends there the paper. As the paper has been printed, the robot takes it and brings it to the user's office. The interesting part of this task, corresponding to the plan in Figure 3c, is how the robot can collect the paper from the printer. In fact, since the robot is not able to grasp the paper, the human assistance (H_Get_papers) is necessary to achieve the goal.

In the third video, it is shown how H_Get_papers has been replaced by a subplan of robot actions (forth line of Table 1). Preparation action $PreA\_Get\_papers$ is implemented by looking for and approaching a human. Then, the robot asks him: "Can you take papers from the printer and put them on my plate?" (CA_Get_papers). As the robot detects through the camera that the human have put the paper on its plate (PA_Get_papers) (Figure 4c), it thanks and greets him, and navigates towards the user's office. Once the robot arrives in front of the office door, it sends a Skype message to the user notifying that it is outside his door and he can came to take the paper. When the robot detects that there are no papers on its plate, it greets the user and takes charge of the next service.

6   Conclusions

In this paper we have presented the definition and execution of social plans as plans that combine robot and human actions and that are represented using PNP formalism. In order to be executed, human actions are transformed into PNP subplans explicitly representing a human-robot collaboration scheme driven by the robot. This formalism allows for describing complex tasks and complex human-robot interactions. The implementation of a set of use cases realized on a mobile robot in our Department has been used to demonstrate the effectiveness of the approach.

The proposed method shows that a robot behaving in a socially acceptable way can overcome its limitations by asking for help to humans. In contrast with previous forms of human-robot interaction where the human drives the conversations, in this approach, the robot is actively looking for human collaboration asking people for help to execute a task. We believe that if this request is reasonable with respect to the situation and it is made in a polite and socially acceptable way, people would not deny their help to the robot.

Several further studies should be considered along this line. First, the social norms that drive the robot behaviors may be explicitly represented and used to generate the social plans. At this moment these are considered by the human designer of the system. Second, extensive user studies should be done to evaluate the system when used by non-expert users (e.g., typical visitors of our offices). Third, more complex tasks and increased forms of social interactions that also take into account context may be explored. Although further research is needed in order to fully assess the proposed method, the reported activities show its feasibility and promising results.

References

1. Burgard, W., Cremers, A.B., Fox, D., Hähnel, D., Lakemeyer, G., Schulz, D., Steiner, W., Thrun, S.: Experiences with an interactive museum tour-guide robot. Artificial Intelligence 114(1), 3–55 (1999)
2. Cirillo, M., Karlsson, L., Saffiotti, A.: Human-aware task planning: an application to mobile robots. ACM Transactions on Intelligent Systems and Technology (TIST) 1(2), 15 (2010)
3. Gross, H.M., Boehme, H., Schröter, C., Mueller, S., Koenig, A., Einhorn, E., Martin, C., Merten, M., Bley, A.: Toomas: interactive shopping guide robots in everyday use-final implementation and experiences from long-term field trials. In: IEEE/RSJ International Conference on Intelligent Robots and Systems, IROS 2009, pp. 2005–2012. IEEE (2009)
4. Kirby, R.: Social robot navigation. Ph.D. thesis, Carnegie Mellon University, The Robotics Institute (2010)
5. Kirby, R., Simmons, R., Forlizzi, J.: Companion: A constraint-optimizing method for person-acceptable navigation. In: The 18th IEEE International Symposium on Robot and Human Interactive Communication, RO-MAN 2009, pp. 607–612. IEEE (2009)
6. Montemerlo, M., Pineau, J., Roy, N., Thrun, S., Verma, V.: Experiences with a mobile robotic guide for the elderly. In: AAAI/IAAI, pp. 587–592 (2002)
7. Pacchierotti, E., Christensen, H.I., Jensfelt, P.: Embodied social interaction for service robots in hallway environments. In: Field and Service Robotics, pp. 293–304. Springer (2006)
8. Rosenthal, S., Biswas, J., Veloso, M.: An effective personal mobile robot agent through symbiotic human-robot interaction. In: Proceedings of the 9th International Conference on Autonomous Agents and Multiagent Systems, vol. 1, pp. 915–922. International Foundation for Autonomous Agents and Multiagent Systems (2010)
9. Rosenthal, S., Veloso, M., Dey, A.K.: Is someone in this office available to help me? Journal of Intelligent & Robotic Systems 66(1-2), 205–221 (2012)
10. Ziparo, V., Iocchi, L., Lima, P., Nardi, D., Palamara, P.: Petri Net Plans - A framework for collaboration and coordination in multi-robot systems. Autonomous Agents and Multi-Agent Systems 23(3), 344–383 (2011)
11. Ziparo, V.A., Iocchi, L.: Petri net plans. In: Proceedings of Fourth International Workshop on Modelling of Objects, Components, and Agents (MOCA), pp. 267–290 (2006)

Why Industrial Robots Should Become More Social
On the Design of a Natural Language Interface
for an Interactive Robot Welder

Andreea I. Niculescu, Rafael E. Banchs, and Haizhou Li

Institute for Infocomm Research, Human Language Technology,
1 Fusionopolis Way, 138632 Singapore
{andreea-n,rembanchs,hli}@i2r.a-star.edu.sg

Abstract. The capability of communicating in natural language as well as manifesting responsive behavior is fundamental in social communities. In this extended abstract, we present our work in progress towards creating a natural language interface for a robot welder that will use natural language to manifest responsive behavior in interaction with human workers. Based on the results of two data collections we argue that these robot capabilities can be highly beneficial in industrial environments, enhancing the human-robot communication and contributing to a shift in the interaction paradigm from industrial towards social.

1   Introduction

Nowadays industrial robots are used in multiple areas, such as automotive, constructions, manufacturing, mining, etc. In all these areas human-robot interaction plays an important role during programming, operation and validation of the required tasks. We collected and analyzed data concerning a robotic application meant for welding on shipyards, more specifically targeting jack-up rig constructions. We found that programming and interacting with such a robot require highly skilled and thoroughly trained human operators. On the other side, the welders working on the shipyard do not have such knowledge, nor can they be extensively trained. In order to make the interaction more natural, friendly and efficient, we propose a natural language interface (NLI) that would ease the communication process and enable the robot to behave more responsive towards a human operator.

To the extent of our knowledge, this is the first initiative in Asia to promote the use of NLI in industrial environments with the goal of replacing or complementing system-dependent protocols and computer interfaces. Only two other European projects have addressed this issue in the past in combination with industrial robots: SMERobotics [1] and JAHIR [2]. Apart from trying to solve a challenging communication problem, our attempt is also meant to motivate further initiatives to build more social interactions between humans and industrial robots.

M. Beetz et al. (Eds.): ICSR 2014, LNAI 8755, pp. 276–278, 2014.
© Springer International Publishing Switzerland 2014

2   Data Analysis and Approach

We performed our data collection at two different locations: at the **shipyard** – where the robot is going to be deployed - and at **SIMTech** (Singapore Institute of Manufacturing Technology) – where the robot is currently programmed and tested.

The data collected on the **shipyard** enabled us to determine the common welding routine, as well as several environmental details and workers profiles. Concerning the welding routine four sequences of operations could be defined: 1) surface cleaning, 2) parameter setup, 3) welding operation and 4) feed-back. Regarding the shipyard environment we found it to be very loud for common standards in automatic speech recognition (ASR). Additionally, most of the workers were foreigners and spoke highly accented English, a fact that would pose additional challenges to an ASR trained with Standard English. However, since the robot is going to be used in tele-operation mode the human workers would use a head-set with a built-in microphone. This would reduce considerably the noise impact on the ASR. Further, training the engine with Indian accented speech data – the Indian accent was the most common accent encountered on the shipyard - would considerably reduce the negative effects of the accent on the ASR accuracy.

On the other side, since coordination during work requires communication, we observed workers talking in English with each other and using welding specific terminology learned during the basic training courses they have to attend on the shipyard. This finding was encouraging for our idea of using a natural language interface.

From our data collection at **SIMtech** we observed that the current robot interface lacks intuitiveness and direct feedback while its pre-programming requires extensive knowledge of CAD. During the welding simulation, the robot's behavior was rather passive, i.e. there was no feed-back on the accuracy of its movements or on the setup parameters. The human operator had to check manually each sequence of movements to ensure the robot arm was reaching the intended location and had the right calibration.

For the natural language interface we are currently working on two different input/out modalities: one is using speech, the other one is using text input. Both modalities are to be used in combination with a screen input/output as an additional option to control and monitor the robot. The interaction through natural language is planned to include different communication types mapped along the four operation sequences mentioned above. These types refer to: **chat interactions**, containing a set of greetings pairs; **commands**, simple and specific instructions given by the operator to the system (e.g. *"abort"*, *"stop"*, *"move right/left,"*, *"locate welding position"*); **question-answering (Q&A)**, specific questions about parameter set-up or feedback messages from the robot (e.g. *"What does this error code mean?"*) and **task-oriented dialogue,** a more complex and type of communication involving several rounds of interactions (e.g. **Operator**: *"What voltage are you using?* **Robot**: *"40V"*, **Operator**: *"Make it 60V"*, **Robot**: *"Sorry, this value is out range for the current welding operation. Please set the voltage below 50V."*). The robot's responsive behavior will be expressed through immediate feed-back on completed tasks or wrongly set-up parameters. Additionally, the robot will be able to communicate autonomously visual recognition difficulties, (e.g. if it cannot localize the welding path or there is not enough light available), report welding results, estimate welding quality and take decisions autonomously based on the information exchanged with the human operator. We believe that this behavior would simplify the current human-robot interaction paradigm in welding.

All natural language interface components will be in integrated in APOLLO, a domain independent, spoken dialogue platform developed in our department [3]. As for the dialogue management strategy, we are planning to use a hybrid combination between rule-based and example-based approaches given the size of our current data collection [4].

References

1. http://www.smerobotics.org/
2. http://www6.in.tum.de/Main/ResearchJahir
3. Jiang, R.D., Tan, Y.K., Limbu, D.K., Li, H.: Component pluggable dialogue framework and its application to social robots. In: Proc. Int'l Workshop on Spoken Language Dialog Systems (2012)
4. Banchs, R.E., Jiang, R., Kim, S., Niswar, A., Yeo, K.H.: AIDA: Artificial Intelligent Dialogue Agent. In: Proceedings of the SIGDIAL 2013 Conference, pp. 145–147 (2013)

Classical Conditioning in Social Robots

Rony Novianto[1], Mary-Anne Williams[1],
Peter Gärdenfors[2], and Glenn Wightwick[3]

[1] Quantum Computation and Intelligent Systems,
University of Technology Sydney, Australia
rony@ronynovianto.com, mary-Anne.williams@uts.edu.au
[2] Lund University, Sweden
peter.gardenfors@lucs.lu.se
[3] University of Technology Sydney, Australia
glenn.wightwick@uts.edu.au

Abstract. Classical conditioning is important in humans to learn and predict events in terms of associations between stimuli and to produce responses based on these associations. Social robots that have a classical conditioning skill like humans will have an advantage to interact with people more naturally, socially and effectively. In this paper, we present a novel classical conditioning mechanism and describe its implementation in ASMO cognitive architecture. The capability of this mechanism is demonstrated in the Smokey robot companion experiment. Results show that Smokey can associate stimuli and predict events in its surroundings. ASMO's classical conditioning mechanism can be used in social robots to adapt to the environment and to improve the robots' performances.

Keywords: Classical Conditioning, Maximum Likelihood Estimation, ASMO Cognitive Architecture.

1 Introduction

Classical conditioning is a cognitive skill crucial to learn and predict events in terms of associations between stimuli, and to produce responses based on these associations. People are expected to develop a classical conditioning when a right condition is presented repeatedly. If social robots have a similar cognitive skill to develop a classical conditioning like people, then people will know how they behave and can interact with them more naturally, socially and effectively. Social robots require cognitive skills that support the necessary social intelligence to engage with people and other robots effectively [15].

In this paper, we present a novel classical conditioning mechanism for social robots. This mechanism is implemented in ASMO cognitive architecture [6]. Section 2 first describes a definition of classical conditioning. Section 3 discusses existing computational models of a classical conditioning proposed in the literature and how they are different to this work. Section 4 describes the design and implementation of the classical conditioning mechanism in ASMO cognitive architecture. Section 5 evaluates ASMO's classical conditioning mechanism in the 'Smokey robot companion' experiment and shows that the robot can predict

M. Beetz et al. (Eds.): ICSR 2014, LNAI 8755, pp. 279–289, 2014.

users' requests. Finally, Section 6 summarises the benefit and future work of ASMO's classical conditioning mechanism.

2 Definition of Classical Conditioning

Classical conditioning (or Pavlovian conditioning) [10, p.109–110] [9] is an association of a neutral stimulus that does not elicit a response (called the *conditioned stimulus* or CS) with another stimulus that elicits a response (called the *unconditioned stimulus* or US), such that the presence of the CS will elicit the same response that would be elicited by the US, despite the US not actually being present. For example, if John repeatedly asks Mary to cook him the same dish every time he visits Mary, then Mary may develop the association between his visit and the dish, such that his presence will trigger Mary to *accidentally* start cooking the dish, even though John had asked Mary to go to a restaurant.

A classical conditioning is different to an operant conditioning. They are both a form of associative learning. However, a classical conditioning creates an association between *involuntary* behaviours and a stimulus before the behaviours are performed, whereas an operant conditioning creates an association between *voluntary* behaviours and their consequences after the behaviours are performed [2, pp. 141–142].

3 Existing Computational Models

Computational models of classical conditioning can be divided into models based on neural network and models that are not based on neural network. They can also be divided into trial-level and real-time models [3]. In trial-level models, the association between the stimuli is computed after all relevant stimuli have been observed and terminated. In real-time models, the association between stimuli is computed at every time-frame and the computation can cope with those frames being arbitrarily small.

In Furze's dissertation [3], he has reviewed a large number of trial-level and real-time computational models of classical conditioning (for both neural network and non-neural network models):

- The trial-level neural network models reviewed were the Pearce and Hall model and the Kehoe model.
- The trial-level non-neural network models reviewed were the Stimulus Substitution model, the Rescorla–Wagner model and the Mackintosh's Attention model.
- The real-time neural network models reviewed were the Grossberg model, the Grossberg–Schmajuk (G.S) model, the Klopf model (also called the drive-reinforcement model), the Schmajuk–DiCarlo (S.D) model and the Schmajuk–Lam–Gray (S.L.G) model.
- The real-time non-neural network models reviewed were the Sometimes-Opponent-Process (SOP) model, the Temporal Difference (TD) model and the Sutton–Barto (S.B) model.

In this paper, we focus more details on the real-time non-neural network models: SOP, TD and S.B models. This is because robots are required to operate in real-time. In addition, non-neural network models allow robots to learn without the need to be trained based on some prior input stimuli. Thus, they allow robots to predict stimuli that have not been trained previously.

The SOP model [14] represents a stimulus in one of three states: A1 (high activation), A2 (low activation) or I (inactive). A stimulus in the A1 state will elicit a primary A1 response (observed as an unconditioned response) whereas a stimulus in the A2 state will elicit a secondary A2 response. Two stimuli that are both in the A1 state will become associated and cause the strength of their association to increase. A stimulus that is either in the A1 or A2 state will induce its associated stimuli to enter their A2 states, which will then elicit their A2 responses (observed as conditioned responses). This inducement occurs in proportion to the strength of the association between the two stimuli. This model supports different phenomena of classical conditioning. However, it requires a stimulus to be represented in one of the three states and it is not implemented in robots.

The Temporal Difference (TD) model [13] is an extension of the Sutton–Barto model [12] proposed by the same authors. These two models rely on reinforcement (or rewards) and eligibility to determine the association strength of a stimulus (1). They have the same operations and equations, except that the reinforcement is determined by R_{TD} for the TD model (2) or R_{SB} for the SB model (3). Unconditioned stimuli have a starting association strength of a positive value. Other stimuli have a starting association strength value of zero.

$$\Delta V_t(i) = \beta R \times \alpha(i)\overline{X}(i)$$
$$\overline{X}_{t+1}(i) = \delta \overline{X}_t(i) + (1 - \delta)X_t(i) \tag{1}$$

$$R_{TD} = \lambda_t + \gamma Y_t - Y_{t-1} \tag{2}$$

$$R_{SB} = \dot{Y}_t = Y_t - Y_{t-1} \tag{3}$$

Where:
$R \in R_{TD}, R_{SB}, 0 < \beta < 1, 0 < \alpha < 1$
$V(i)$ and $\Delta V(i)$ are the association strength and the change of the association strength of stimulus i respectively
β and α are the constant reinforcement and eligibility learning rates respectively
$X_t(i)$ and $\overline{X}_t(i)$ are the strength and the weighted average strength (called eligibility trace) of conditioned stimulus i at time t respectively
δ is the decay rate of the eligibility trace
λ_t is the strength of the unconditioned stimulus at time t
γ is the discount factor
Y_t is the prediction made at time t of the unconditioned stimulus being associated

This paper presents the novel ASMO's classical conditioning mechanism based on attention and manipulation of memory. This mechanism differs from previous works in the following: (i) it does not require reinforcement values to learn and does not require specific representations of stimuli and responses, (ii) it is embedded in a cognitive architecture, (iii) it is not based on neural network, (iv) it is a real-time model and (v) it is implemented in a robot.

4   Design and Implementation in ASMO Cognitive Architecture

In this section, we describe the design and implementation of ASMO's classical conditioning mechanism based on the inspiration of human classical conditioning. We first review the overview of ASMO cognitive architecture. We follow by describing the mechanism and how it fits in the architecture.

4.1   Overview of ASMO Cognitive Architecture

ASMO [4,5,7] is a flexible cognitive architecture that orchestrates and integrates a diversity of artificial intelligence components based on bio-inspired model of attention. It can be used to explain and understand human cognition, however it does not aim to *imitate* the human cognitive architecture (i.e. it is bio-inspired rather than biomimetic).

ASMO cognitive architecture contains a set of self-contained, autonomous and independent processes (also called modules) that can run concurrently on separate threads (see Fig.1). Each module requests 'actions' to be performed. An action can be a low-level command to actuators, such as move head to a ball or walk to a specific location, or it can be a high-level function, such as store data to a memory, recognise objects (i.e. percept) or find the shortest path (i.e. plan).

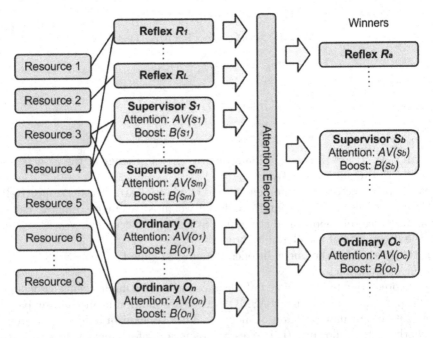

Fig. 1. Attention election in ASMO cognitive architecture

Actions can only be performed if the resources required by the actions are available (e.g. hand, leg, CPU, memory, virtual resource, etc). They can be performed simultaneously when there is no conflict in using the resources. Otherwise, modules with actions that require the same resources have to compete for 'attention' to use the resources. The winner of the competition is chosen based on the modules' types and attention levels.

Modules are divided into few types, including supervisor and ordinary modules. The supervisor and ordinary modules are non-reflex modules that have a 'attention value' attribute and 'boost value' attribute to determine their total attention levels (used to compete for attention). The 'attention value' attribute captures the degree of attention the module seeks based on the demand of the tasks whereas the 'boost value' attribute represents the bias associated with the module as a result of learning [6] or subjective influence [8]. Supervisor and ordinary modules are similar, except that supervisor modules can influence the total attention levels of ordinary modules but not vice versa.

Currently, modules that have the highest total attention levels will win the competition. The total attention level is given by the sum of boost and attention values. Under ordinary operation, attention values are (by convention) bounded between 0.0 and 100.0, equivalent to scaled values between 0.0 and 1.0. Modules with attention values of 0 demand the least attention whereas modules with attention values of 100 demand full or maximum attention. The boost value will bias this demand in the competition.

4.2 ASMO's Classical Conditioning Mechanism

ASMO's classical conditioning mechanism is created to trigger non-reflex modules to propose actions when the *conditional* stimulus is present even though the *unconditional* stimulus is not actually present. This mechanism is implemented in a supervisor module and its algorithm is described in the following five steps:

1. **Capture sequences of stimuli**
   ASMO's classical conditioning mechanism will capture sequences of stimuli. It represents each sequence of stimuli using a Markov chain where each node represents a stimulus.
2. **Calculate probabilities of stimuli will occur given an occurring stimulus**
   For every occurring stimulus, ASMO's classical conditioning mechanism will calculate the probabilities of other stimuli will occur (i.e. called 'candidates') given this occurring stimulus. In other words, it will calculate the probabilities of unconditioned stimuli being associated with a given conditioned stimulus. It calculates these probabilities by using the maximum likelihood estimation algorithm [1, p. 615]. These probabilities represent the strengths (or rather the confidences) of the associations between stimuli.
3. **Pick associated stimuli**
   ASMO's classical conditioning mechanism will pick the candidates that have significant probabilities as the stimuli being associated with the occurring

stimulus (i.e. pick the likely unconditioned stimuli). A candidate is significantly different if its root mean square deviation is above a threshold (4).

$$RMSD(c) = \sqrt{\frac{1}{n}\sum_{i=1}^{n}(P_c - P_i)^2}$$

$$Significance(c) = \begin{cases} True & \text{if } RMSD(c) \geq T_{\text{RMSD}} \\ False & \text{otherwise} \end{cases}$$

(4)

Where:

$Significance(c)$ is the significance function of candidate c

$RMSD(c)$ is the root-mean-square deviation of candidate c

T_{RMSD} is the threshold of a candidate being significant

n is the number of candidates

P_i is the probability of candidate i

4. **Trigger modules to propose actions**

   ASMO's operant conditioning mechanism will add the likely unconditioned stimuli to ASMO's memory as if these stimuli are currently occurring. This addition will cause non-reflex modules to believe that these stimuli are present, despite the fact that these stimuli are not physically present. As a result, it will trigger non-reflex modules to compete and propose actions in order to respond to these stimuli. Hence, the conditioned stimulus has triggered actions that are associated with the unconditioned stimuli without the unconditioned stimuli being physically present. This implementation allows a conditioned stimulus to be paired with a single unconditioned stimulus or multiple unconditioned stimuli.

5. **Repeat step 2 to step 4 for other occurring stimuli**

   ASMO's operant conditioning mechanism will repeat step 2 to step 3 if there are other stimuli that are currently occurring.

The Markov chain model used by ASMO's operant conditioning mechanism may require many observations to provide an accurate estimation of reality. However, many observations are often not available and can be difficult to obtain. Thus, this mechanism uses a smoothing technique, such as the *Laplace smoothing* (also called *additive smoothing*) [11], to smoothen the observations in order to provide a better estimation.

5   Evaluation

ASMO's classical conditioning mechanism is experimented in Smokey robot companion project using a bear-like robot called Smokey [6,8,7]. This project aims to bring Smokey to 'life' and explores the meaning of life by interacting socially with people. It has potential applications in nursing, healthcare and entertainment industries by providing companionship to people with disabilities, people with autism, the elderly and children.

As part of the experiment, in a simplified scenario, Smokey has to accompany or entertain a person (i.e. the target user) while simultaneously regulating the person's rest. Smokey can play either a red ball game or drums to accompany the user. It can also go to sleep to encourage the user to rest (since it will not interact with the user when it is sleeping). When playing, Smokey will also pay attention to any motion in the environment from people other than the user.

Smokey can receive a request from the user through a graphical user interface to either play the red ball game, play the drums or go to sleep. It will consider this request, but does not necessarily have to perform this request. In addition, Smokey is desired to learn to predict the request that the user tends to ask and to perform this request before the user asks (i.e. to be conditioned by the appearance of the user so as to perform his/her request). Conditioning to the appearance of the user is similar to the example of classical conditioning described in Section 2. It will make Smokey more personalised to the user, which results in better companionship.

In summary, our hypothesis in this experiment was that ASMO's classical conditioning mechanism could model a classical conditioning: it could learn the association between the appearance of a user and the user's request. The methodology to validate this hypothesis was to show that after learning Smokey would perform the request that a user tended to ask (if any) when the user was seen. In addition, we would show the probability of the request compared to other requests. This experiment involved five users (i.e. participants) with different requests.

There were four ordinary modules and two supervisor modules created in this experiment to govern Smokey's behaviours:

- **The 'attend_motion' ordinary module**
  The 'attend_motion' module proposed an action when Smokey was not sleeping to look at the fastest motion in the environment. Its attention value was set to the average speed of the motion scaled between 0.0 and 100.0. The faster the motion, the more attention demanded by the module to look at the motion.
- **The 'play_ball' ordinary module**
  The 'play_ball' module proposed an action when Smokey was not sleeping either to track or to search for the ball depending on whether the location of the ball was known or not respectively. Its attention value was set to a constant value of either 60.0 when the user preferred Smokey to play the ball than to do other things, or 50.0 when the user preferred Smokey to do other things than to play the ball.
- **The 'play_drums' ordinary module**
  The 'play_drums' module proposed an action when Smokey was not sleeping either to play, track or search for the drums depending on whether the location of the drums was known and within reach, known but not within reach or unknown respectively. Similar to the 'play_ball' module, its attention value was set to a constant value of either 60.0 when the user preferred Smokey to play drums than to do other things, or 50.0 when the user preferred Smokey to do other things than to play the drums.

- **The 'go_sleep' ordinary module**
  The 'go_sleep' module proposed an action to go to sleep and wake up in every defined period. Its attention value was linearly increased until either it won an attention competition or its attention value reached 100.0 (i.e. maximum value of attention). This module then reset back its attention value to 0.0 after Smokey had enough sleep (i.e. predefined time).
- **The 'attend_request' supervisor module**
  The attend_request module proposed an action to increase the boost value of the play_ball, play_drums or go_sleep module when Smokey was not sleeping and requested to play the red ball game, play the drums or go to sleep respectively. It increased these boost values proportionally to the probability of the request (5). This probability was set to 1.0 when a request was received through a graphical user interface, or set to a value calculated by ASMO's classical conditioning mechanism when an associated stimulus was determined.

  The attend_request module did not require any resource. Its attention value was set to a constant arbitrary value of 10.0. This value does not hold any significant meaning. It does not have to be 10.0 and could be any value between 0.0 to 100.0. The reason is because the attend_request module did not need to compete for attention to gain access to resources since this module did not require any resource. Thus, this module will always be selected regardless of its attention value.

$$BV(pb) = P(b) \times 20.0$$
$$BV(pd) = P(d) \times 20.0 \tag{5}$$
$$BV(gs) = P(s) \times 20.0$$

Where:
$BV(pb)$ is the boost value of the play_ball module
$BV(pd)$ is the boost value of the play_drums module
$BV(gs)$ is the boost value of the go_sleep module
$P(b)$ is the probability that the request to play the ball is received
$P(d)$ is the probability that the request to play the drums is received
$P(s)$ is the probability that the request to go to sleep is received

- **The 'classical_conditioning' supervisor module**
  The classical_conditioning module performed the five steps described in the previous section to learn the associations between the appearance of a user and his/her request. This module was specified by developers to observe users' requests. It calculated the probability of a user requesting Smokey to play the ball, to play the drums and to go to sleep. It determined requests with significant probabilities and added these requests into ASMO's memory every time the user was appear. This addition caused the attend_request module to believe that the user had made a request even though the user did not ask. As a result, the attend_request module increased the boost value of either the play_ball, play_drums or go_sleep module as if the user made an actual request.

Table 1 shows the requests received when interacting with the five users where -, b, d and s denote no request, play the ball request, play the drums request and go to sleep request respectively. Table 2 shows the probability of each request that might be asked by each user given when the user was seen. These probabilities were calculated based on the users' requests in Table 1 using the expectation maximization algorithm and Laplace smoothing with k of 1.0. Note that the probability of the request (or no request) that a user tended to ask was higher than other requests.

Table 1. Users' Requests

User	Requests
Anshar	b,s,d,d
Ben	d,d,d,d,d
Evelyn	-,-,-
Michelle	b
Xun	s,d,b,-,s

Table 2. Probability of Requests Asked by Users

User	Probability of Request Given User is Seen			
	Play Ball	Play Drums	Go to Sleep	No Request
Anshar	0.25	0.375	0.25	0.125
Ben	0.1111	0.6666	0.1111	0.1111
Evelyn	0.1429	0.1429	0.1429	0.5714
Michelle	0.4	0.2	0.2	0.2
Xun	0.2222	0.2222	0.3333	0.2222

Figure 2 shows the result of the experiment without and with ASMO's classical conditioning learning mechanism when Smokey was interacting with Evelyn and then replaced by Ben. Both Evelyn and Ben preferred Smokey to play the ball rather than the drums. Thus, the total attention level of the play_ball module was initially higher than the total attention level of the play_drums module.

Without ASMO's classical conditioning mechanism, the total attention level of the play_drums module did not change when Smokey saw Ben. Thus, Smokey still chose to play the ball instead of the drums when interacting with Ben (i.e. no change of behaviour).

With ASMO's classical conditioning mechanism, the total attention level of the play_drums module was increased when Smokey saw Ben. This increase caused the total attention level of the play_drums module to be higher than the total attention level of the play_ball module. Thus, Smokey chose to play drums instead of the ball when interacting with Ben (i.e. change of behaviour). This change of behaviour showed that Smokey was classically conditioned to the appearance of Ben: it could learn the association between Ben's appearance and his requests.

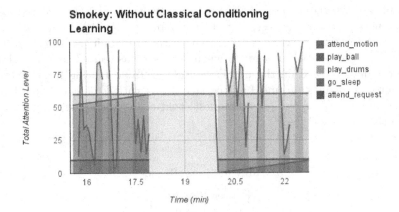

(a) Without Classical Conditioning Learning

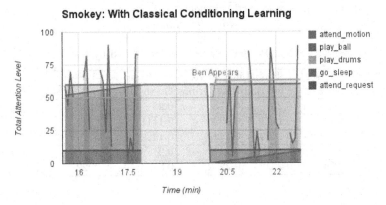

(b) With Classical Conditioning Learning

Fig. 2. Smokey's Classical Conditioning Learning

6 Conclusion

This paper has demonstrated the capability of ASMO's classical conditioning mechanism to learn in real-time in a physical robot without requiring reinforcement values. This mechanism is not based on neural network and has been embedded in ASMO cognitive architecture. It allows social robots to learn and predict events in the environment and to respond to those events.

For future work, ASMO's classical conditioning mechanism can be extended to further match the characteristics of human classical conditioning (with the aim to improve the mechanisms instead of imitating human classical conditioning). In addition, it can be extended to accommodate different types of learning, such as operant conditioning.

References

1. Bishop, C.: Pattern Recognition and Machine Learning. Information Science and Statistics. Springer (2006)
2. Chance, P.: Learning and behavior. Wadsworth (2013)
3. Furze, T.A.: The application of classical conditioning to the machine learning of a commonsense knowledge of visual events. Ph.d. dissertation, University of Leeds, United Kingdom (2013)
4. Novianto, R.: Flexible Attention-based Cognitive Architecture for Robots. Ph.d. dissertation, University of Technology Sydney, Sydney, Australia (2014)
5. Novianto, R., Johnston, B., Williams, M.A.: Attention in the ASMO cognitive architecture. In: Samsonovich, A.V., Jóhannsdóttir, K.R., Chella, A., Goertzel, B. (eds.) Proceedings of the First Annual Meeting of the BICA Society. Frontiers in Artificial Intelligence and Applications, vol. 221, pp. 98–105. IOS Press (November 2010)
6. Novianto, R., Johnston, B., Williams, M.-A.: Habituation and sensitisation learning in ASMO cognitive architecture. In: Herrmann, G., Pearson, M.J., Lenz, A., Bremner, P., Spiers, A., Leonards, U. (eds.) ICSR 2013. LNCS, vol. 8239, pp. 249–259. Springer, Heidelberg (2013)
7. Novianto, R., Williams, M.A.: The role of attention in robot self-awareness. In: The 18th IEEE International Symposium on Robot and Human Interactive Communication, RO-MAN 2009, pp. 1047–1053 (2009)
8. Novianto, R., Williams, M.A.: Innate and learned emotion network. In: Samsonovich, A.V., Jóhannsdóttir, K.R. (eds.) Proceedings of the Second Annual Meeting of the BICA Society. Frontiers in Artificial Intelligence and Applications, vol. 233, pp. 263–268. IOS Press (November 2011)
9. Pavlov, I., Anrep, G.: Conditioned Reflexes. Dover Publications (2003)
10. Powell, R.A., Honey, P.L., Symbaluk, D.G.: Introduction to learning and behavior, 4th edn. Cengage Learning (2013)
11. Russell, S.J., Norvig, P.: Artificial intelligence: a modern approach. Prentice Hall series in artificial intelligence. Prentice Hall (2010)
12. Sutton, R.S., Barto, A.G.: Toward a modern theory of adaptive networks: expectation and prediction. Psychological Review 88(2), 135–170 (1981)
13. Sutton, R.S., Barto, A.G.: Time-derivative models of pavlovian reinforcement. In: Gabriel, M., Moore, J. (eds.) Learning and Computational Neuroscience: Foundations of Adaptive Networks, pp. 497–537. The MIT Press, Cambridge (1990)
14. Wagner, A.R.: Sop: A model of automatic memory processing in animal behavior. In: Spear, N.E., Miller, R.R. (eds.) Information Processing in Animals: Memory Mechanisms, pp. 5–47. ch. 1. Lawrence Erlbaum Associates, Hillsdale (1981)
15. Williams, M.-A.: Robot social intelligence. In: Ge, S.S., Khatib, O., Cabibihan, J.-J., Simmons, R., Williams, M.-A. (eds.) ICSR 2012. LNCS, vol. 7621, pp. 45–55. Springer, Heidelberg (2012)

Social Robots in Postural Education: A New Approach to Address Body Consciousness in ASD Children

Giuseppe Palestra[1,2], Ilaria Bortone[3,4], Dario Cazzato[2], Francesco Adamo[2], Alberto Argentiero[4], Nadia Agnello[4], and Cosimo Distante[2]

[1] Department of Computer Science, University of Bari, Bari, Italy
[2] National Council of Research, National Institute of Optics (INO), Lecce, Italy
[3] Scuola Superiore Sant'Anna, Institute of Communication, Information and Perception Technologies (TeCIP), Pisa, Italy
[4] Euro Mediterranean Scientific Biomedical Institute (ISBEM), Brindisi, Italy
{giuseppepalestra,ilariabortone}@gmail.com
http://www.saracenrobot.com,
http://www.kisshealth.it

Abstract. Autism Spectrum Disorders (ASD) represent one of the most prevalent developmental disorders among children with different level of impairments in social relationships, communication and imagination. In addition, impaired movement is also observed in individuals with ASD and recent studies consider this factor as a limitation for fully engagement in the social environment. In the present work, we propose a new approach to promote postural education in autistic children with the involvement of a humanoid social robot and the therapist in a triadic interaction environment to better understand their motor development and body consciousness.

Keywords: Autism, children, social robot, postural education, human-robot interaction, game.

1 Introduction

Autism Spectrum Disorders (ASD) are now one of the most prevalent developmental disorders among children. Autistic children exhibit a wide variety of behaviors and developmental levels, from repetitive, ritualistic and stereotyped behaviors (RRBs) of posture and part of the body (e.g. hand flapping, rocking, swaying, ...), to limit in imagination, communication, to self-hetero-aggressive intent, lack of flexibility or hypersensitivity to anything that involves changes in the surrounding. However, difficulty with engagement, attention, and appropriate behavior in the classroom are common and interfere with students' ability to participate in the educational mainstream [1].

In addition to the main three core symptoms, impaired movement is commonly observed in individuals with ASD. In fact, it has been found that ASD individuals display atypical movement patterns during locomotion, reaching and

M. Beetz et al. (Eds.): ICSR 2014, LNAI 8755, pp. 290–299, 2014.
© Springer International Publishing Switzerland 2014

aiming [2]. Research suggests that the postural system in individuals with ASD is immature and may never reach adult levels, which can be a limiting factor on the execution of other motor skills, such as coordinated hand/head movements and inhibition of reflexes, and may constrain the ability to develop mobility and manipulatory skills. The origin of this lack in motor coordination in ASD individuals seems to be related to the relationship between mirror neuron system (MNS) activation, responsible of motor coordination and social skills, and the development of autism.

1.1 Motivation

The human-robot social interaction has become a popular research field in recent years. There are many robotic platforms that serve the goal of developing human-robot social interaction and there are numerous researches indicating robots can be used as a therapy medium to assist children with special needs [3]. In particular, early intervention is critical for the children inflicted with autism in order for them to lead productive lives with a higher degree of independence in their future years.

Several activities play an important role in child development. According to the International Classification of Functioning and Disabilities Version for Children and Youth (ICFCY), the World Health Organization remarks that the game play is one of the most important standpoints for a child in his/her life [4]. In fact, playing contributes the development of children by advancing their social skills, as well as their communication skills and also sensory and motor skills. Through the game play, children recognize their social environment and establish the necessary relationships.

On the other hand, education is considered the most effective therapeutic strategy. More specifically, computers in the education and therapy of people with autism has demonstrated to be beneficial for the development of self-awareness and self-esteem. In recent years, software-based systems include highly structured virtual environments [5] have been used by therapists and teachers as tools in order to teach social and other life skills (e.g. recognizing emotions, crossing the road, learning where and how to sit down in a populated cafeteria). However, interactions with an interactive physical robot can contribute important realtime, multimodal, and embodied aspects which are characteristic of facetoface social interaction among humans [6]. We believe that sharing the space with a robot is a motivational resource that aids the child to effectively engage in the task as opposed to a virtual character in a screen [7].

In this scenario, SARACEN (Social Assistive Robots for Autistic Children EducatioN) and KISSHealth (Knowledge Intensive Social Services for Health) propose a new approach to promote postural education in autistic children with social robots combining expertise in Biomechanics with expertise in Computer Science and its applications. SARACEN project proposes innovative methods for early diagnosis of ASD and therapy support for autistic children with socially assistive robots. On the other side, KISSHealth Project proposes an integrated approach to face up the problems of postural abnormalities and, as first step, it promotes the development of a postural consciousness in order to prevent and moderate postural diseases.

2   Social Robotics in Postural Education

The lack of social interaction is one of the most debilitating deficits associated with autism spectrum disorder [8]. As previously illustrated, children with autism often have trouble communicating and interacting with other individuals, their interests and activities may be limited. For these reasons, therapists are key figures in the education of children with ASD and they can be seen as parts of a larger team composed by parents, teachers and professionals. Emerging applications include social robots as tools to teach skills to children with autism, to play with them and to elicit desired behaviors from them [9], [10]. In the present work, the robot acts as a mediator for educational purposes, being able to both teach the correct postural behaviors and evaluate the learning process in a triadic interaction environment. Individual repeated freeform interactions are being used as education-therapy model, that allow the child to interact with the robot with no interaction from the therapist, unless necessary, and in presence of his/her parents for a comforting presence (Fig. 1).

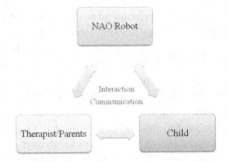

Fig. 1. Child, robot and therapist/parents interaction

The design and functionality of the robot have a significant influence on its effectiveness in therapy. Several research studies have been conducted to extract the requirements in relation to the end-user group (children with autism), categorizing robot according to appearance, functionality, safety requirements, autonomy, modularity and adaptability [11]. Considering these remarks, the NAO, a small humanoid made by Aldebaran Robotics (0.57 cm tall and 4 kg weighty), has been used. It has a total of 25 Degrees of Freedom (DoF), 11 DoF for the lower limbs, that include legs and pelvis, and 14 DoF for the upper limbs that include trunk, arms and head [12]. Furthermore, the manufacturer of NAO offers several software tools to use with the robot, as Choregraphe, an application that permit to create robot behaviors, test them on a simulated robot or directly on a real one.

The present paper represents a step forward in the application of social robots in the education of children with ASD focusing the attention to the aspects related to motor development as an essential condition to improve also in interaction and social development.

2.1 The Postural Education Application

The concept of education refers to the process of helping children in development and growth, giving them skills that they will use throughout life. For children with special needs, education and therapy are integrated and tailored to the needs of the individual child in order to achieve the child's physical, mental and social development [13]. In this sense, the play scenarios cover all the important domains of childrens development (intellectual, sensory, communication, motor, social and emotional [14]) and they can be used to provide different experiences and possibilities for developing aspects in all the developmental domains.

In this paper, we focused our attention to motor development, that includes all aspects of controlling the body, its muscles, and its movements. To fully engage in social interaction, an individual requires a full repertoire of movement behaviors for use in communication and for understanding the communicative nature of others' movements [2]. In this sense, posture can be considered the first step in improving childs abilities because of its importance for both gross and fine motor functions.

2.2 System Architecture

In the play scenario, we developed three levels of complexity including a teaching phase, a reinforcement phase and a game phase. During the teaching phase, the child was introduced with the robot and he/she familiarized with the body parts. The robot is provided with several lessons on human body composition, in relation to its principal components, as head, trunk, arms and legs that the robot shows with its own robotic body (in Fig. 2 some examples), and on human body posture when sitting, standing, bending and handling. As the robot is quite small and due to its kinematic limitations, some postures were unnatural so to support the process of knowledge, an inhouse cartoon is simultaneously projected on a monitor, that highlights figures and words the robot is talking about. Then, the child has reinforced and improved his/her performance with body posture by repeating them with the robot. The robot was configured to wait for the child to show the correct body part using the cards he/she has on the table. Finally, the child and the robot played an interactive game using the information he/she has acquired in the learning process.

The autonomous system is implemented through different software modules, that correspond to the designed levels:

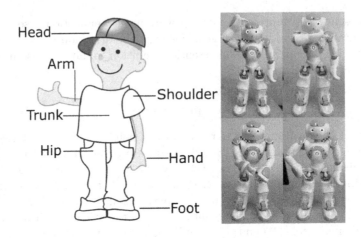

Fig. 2. The robot shows the human body parts: head, shoulder, hand, hip

1. Presentation module: the robot shows up and explains the purpose of the lessons and asks if the child is ready to begin;
2. Lesson module: the robot explains some fundamentals of human body, correct and wrong postures; the robot uses verbal and gestural communication and it shows the correct and incorrect postures;
3. Random choice module: the robot randomly chooses a body posture from postural lectures;
4. Ask module: the robot asks the child to pick a figure;
5. Recognize module: computer vision algorithms recognize the figures selected from the child;
6. Compare module: algorithms compare the figure picks from the third module with the child's one. If correct, the robot communicates it to the child; otherwise, it communicates the inaccuracy of the selection;
7. Ask module: the robot ask the child if he/she wants to play again go to the next lesson. If the child wants to play again, the system restarts with random module, otherwise the system call the lesson module.

The serious game, developed through the modules 3, 4, 5, 6 and 7, is necessary to reinforce the acquired information in an enjoyable way and to engage the children attention.

2.3   Experimental Methods

Participants. The target group in our experiments is the autistic children but we tested the designed approach with adults as a first step. It was performed with ten volunteers (5 female, 5 male). All participants were graduate students and the ages were distributed in the range of 25–33 with the mean $\mu = 28{,}6$ years and standard deviation $\sigma = 2{,}17$. None of the participants had any specific postural knowledge prior to the experiments. An ongoing experiment involves a group of twenty typically developing children (1113 years old) that have followed the "Postural Education at School" Project (the PoSE Project) implemented by the KISSHealth Project in the "MaterdonaMoro" Middle School of Mesagne. The preliminary results of this experiment will provide a guideline for more suitable experiment design in interaction between humanoid robot and children.

Material. The setup of the room used in the experiment is depicted in Fig. 3. The interaction takes place in the experiment room and the operators remain in the same room to deploy the overall system and operate the robot when required because the robot behaviors are completely autonomous. The therapist is close to the child and it acts as a support/encouragement while the main learning is done between the robot, TV screen and the child.

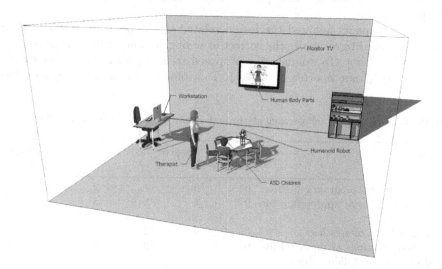

Fig. 3. An overview of the proposed scenario promoting a triadic interaction: the child, the therapist and the robot are involved in the session to both learn the required skills and generalize them to the people around him/her

In preliminary tests, the setup was prepared in the KISS-Health Laboratory located in the "San CamilloDe Lellis" Hospital of Mesagne. The scenario consisted of a 46" TV screen on which a video about the essential behaviors for a correct body posture, previously recorded, was projected; the NAO robot; the participant and the research team. The NAO robot was placed almost 0,5 m away from the participants on the table to avoid physical interaction.

Evaluation. The interaction was qualitatively assessed through selfreported measures based on the survey presented to the participants after the session. Each participants filled out a questionnaire which ranks the three phases of the educational program on a 15 Likert Scale. The possible answers were: strongly disagree, disagree, neutral, agree and strongly agree. It has been argued that using classic questionnaire with children does not always provide reliable feedback [7]; thus alternative tools will be used to explore the childrenrobot interaction.

3   Preliminary Results

It is important to emphasize that the preliminary experiments were performed in order to perfect the framework of proposed game before performing these tests with ASD children. Informed consents were collected for all the participants involved in the study. The proposed scenario was tested with ten graduate students without any prior knowledge of postural education. In this study, twenty cards of correct/wrong postures were proposed by the humanoid robot in order to test the participants.

The participants followed the lesson related to the body posture and then they were asked to recognize the correct or wrong card according to the request of the NAO robot. To evaluate the proposed scenario, a survey was presented to the participants reported in Table 1. To avoid social desirable responses the

Table 1. Participant's selfassessment through classic questionnaire of 15 Likert Scale

Questions	strongly disagree	disagree	neutral	agree	strongly agree
The educational program was clear.	0	0	1	2	7
The contents were appropriate for children.	0	1	3	5	1
The game was interesting.	0	0	2	7	1
The body motions of the NAO fitted what it was talking about.	0	0	0	4	6
The body postures of the NAO were clear to see.	0	1	3	6	0

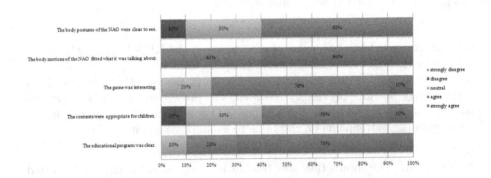

Fig. 4. Stacked Frequencies of 5categoryLiker Scaleitems

questionnaire were all anonymous. To ease the interpretation of the results, we display the distribution of observations in a bar chart measuring the frequency of the categories selected by the participants and illustrated in Fig. 4. The outcomes show how thirty nine of the fifty possible items are in agree with the statements in the survey, nine assume a neutral attitude under investigation and only two are in disagree with our assertions. More detailed experimental data, relative to the sample of typically developing children, are still ongoing and they will be presented in subsequent works.

4 Discussion

The purpose of conducting child-robot interaction sessions is to enable the children to overcome their deficiencies and gain a better understanding of the world in relation to their social skills, emotional awareness, and their communication with the environment and people around them. In this proposed scenario, the attention is focused on improving the children's movement development and body consciousness. To achieve these objectives, educationtherapy sessions are composed of activities that can result into positive behaviors from children with autism. Here we consider a triadic interaction, which is one that involves a child, a robot, and another companion, in this case the therapist.

Some authors [15] have highlighted the anecdotal results of introducing robots into experiments or therapeutic sessions with ASD individuals, but overall preliminary results suggested that the game-based approach for the children to learn, understand and correctly guess the emotions shown by NAO has been a success. Hence, a humanoid with moderate likelihood to actual human does have potential to teach children with autism about head and body postures that are associated with certain feelings or emotions. Consequently, a robot in human shape is a salient mediator to teach emotions to the children and this can easier be transferred from child-robot to human-human interaction in actual social scenarios [10].

These considerations suggest us that the proposed approach might be helpful with children with ASD in terms of:

- Sensory Problems. Many children with autism have difficulty with multiple sources of sensory input. The envisioned method can isolate specific stimuli and allows subjects to control how much they will experience.
- Lack of generalization. Difficulty generalizing behaviors learned in a single setting to similar appropriate situations limited treatment efforts in autism. Postural aspects involve a lot of daily life tasks and this can help subjects to recognize the correct situations.
- Visual thought patterns. Visual support for early learning has been effective for teaching young children with a range of special needs [16]. The proposed method seems to be an appropriate modality for people with autism and should give them an excellent opportunity for learning new concepts and behaviors.

– Responsiveness with computer technology. Although computers have not been adapted by special education programs as quickly as some would like, there is increasing evidence that they represent an effective new approach to education and learning for children with developmental disabilities [17].

Although a number of research issues need to be solved, we believe that the multidisciplinary approach here promoted develops a new experimental setting that can integrate interactions between children with ASD and robots, with the aim of analyzing and improving childrens behaviors. To our knowledge, no systematic empirical research exists addressing the question of how posture is perceived by autistic children. Our research builds on a larger body of theoretical and empirical work concerning socially interactive robots and their applications in therapy and education for autistic children.

5   Perspectives and Ongoing Work

Future developments will investigate the reliability and efficacy of the proposed method, drawing a comprehensive set of therapeutic and educational objectives in a closed loop with both therapists and teachers. An experimental study involving subjects of children with autism and typically developing children is underway. Current modules presented in this study will be improved based on inputs from medical experts. In the future, instead of randomly choosing the body posture lecture with equal probabilities, it will be chosen with some probability based on the child's downfalls, i.e. if their neck is the main issue, then it is more likely lectures will be on the neck (e.g. neck 50%, back 25%, hip 25%). Furthermore, future researches will also assess the possibilities to extend the solution here adopted to a school-based environment.

Acknowledgments. This work is partially supported by Italian Ministry for Education, University and Research (MIUR) and European Union in the framework of Smart Cities and Communities and Social Innovation in the framework of 2007-2013 National Operational Program for Research and Competitiveness under Grants PON04a3_00201 and PON04a3_00097.

References

1. Schilling, D.L., Schwartz, I.S.: Alternative seating for young children with autism spectrum disorder: Effects on classroom behavior. Journal of Autism and Developmental Disorders 34(4), 423–432 (2004)
2. Srinivasan, S.M., Pescatello, L.S., Bhat, A.N.: Current Perspectives on Physical Activity and Exercise Recommendations for Children and Adolescents With Autism Spectrum Disorders. Physical Therapy (2014)
3. Akalin, N., Uluer, P., Kose, H., Ince, G.: Humanoid robots communication with participants using sign language: An interaction based sign language game. In: 2013 IEEE Workshop on Advanced Robotics and its Social Impacts (ARSO), November 7-9, pp. 181–186 (2013)

4. World Health Organization (Ed.): International Classification of Functioning, Disability, and Health: Children & Youth Version: ICFCY. World Health Organization (2007)

5. Robins, B., Dautenhahn, K., Te Boekhorst, R., Billard, A.: Robotic assistants in therapy and education of children with autism: can a small humanoid robot help encourage social interaction skills? Universal Access in the Information Society 4(2), 105–120 (2005)

6. Dautenhahn, K., Werry, I.: Towards interactive robots in autism therapy: Background, motivation and challenges. Pragmatics & Cognition 12(1), 1–35 (2004)

7. Ros, R., Baroni, I., Demiris, Y.: Adaptive humanrobot interaction in sensorimotor task instruction: From human to robot dance tutors. Robotics and Autonomous Systems 62(6), 707–720 (2014)

8. Knott, F., Dunlop, A.W., Mackay, T.: Living with ASD How do children and their parents assess their difficulties with social interaction and understanding? Autism 10(6), 609–617 (2006)

9. Pioggia, G., Igliozzi, R., Sica, M.L., Ferro, M., Muratori, F., Ahluwalia, A., De Rossi, D.: Exploring emotional and imitational android-based interactions in autistic spectrum disorders. Journal of CyberTherapy & Rehabilitation 1(1), 49–61 (2008)

10. Boucenna, S., Narzisi, A., Tilmont, E., Muratori, F., Pioggia, G., Cohen, D., Chetouani, M.: Interactive technologies for autistic children: A review. In: Cognitive Computation. Springer (2014),
http://www.isir.upmc.fr/files/2014ACLI3112.pdf

11. Cabibihan, J.J., Javed, H., Ang Jr, M., Aljunied, S.M.: Why robots? A survey on the roles and benefits of social robots in the therapy of children with autism. International Journal of Social Robotics 5(4), 593–618 (2013)

12. Aldebaran–Robotics company, http://www.aldebaran-robotics.com

13. Ferrari, E., Robins, B., Dautenhahn, K.: Therapeutic and educational objectives in robot assisted play for children with autism. In: 18th IEEE International Symposium on Robot and Human Interactive Communication, RO-MAN 2009, pp. 108–114 (September 2009)

14. Tilstone, C., Layton, L.: Child development and teaching pupils with special educational needs. Routledge Falmer (2004)

15. Ricks, D.J., Colton, M.B.: Trends and considerations in robot-assisted autism therapy. In: 2010 IEEE International Conference on Robotics and Automation (ICRA), pp. 4354–4359 (2010)

16. Tarbox, J., Dixon, D.R., Sturmey, P., Matson, J.L.: Handbook of Early Intervention for Autism Spectrum Disorders: Research, Policy, and Practice. Springer (2014)

17. Smith, V., Sung, A.: Computer Interventions for ASD. In: Comprehensive Guide to Autism, pp. 2173–2189. Springer, New York (2014)

A Multi-modal Utility to Assist Powered Mobility Device Navigation Tasks

James Poon[1] and Jaime Valls Miro[2]

[1] University of Technology Sydney, Australia
james.poon@student.uts.edu.au
[2] University of Technology Sydney, Australia
jaime.vallsmiro@uts.edu.au

Abstract. This paper presents the development of a shared control system for power mobility device users of varying capability in order to reduce carer oversight in navigation. Weighting of a user's joystick input against a short-tem trajectory prediction and obstacle avoidance algorithm is conducted by taking into consideration proximity to obstacles and smoothness of user driving, resulting in capable users rewarded greater levels of manual control for undertaking maneuvres that can be considered more challenging. An additional optional comparison with a Vector Field Histogram applied to leader-tracking provides further activities, such as completely autonomous following and a task for the user to follow a leading entity. Indoor tests carried out on university campus demonstrate the viability of this work, with future trials at a care home for the disabled intended to show the system functioning in one of its intended settings.

Keywords: Shared control, co-autonomy, wheelchair.

1 Introduction

Powered mobility devices (PMD) including wheelchairs and scooters are quite widely used by aged and disabled people. The use of these aids becomes quite frequent after the age of 65, with nearly 4.5 million powered wheelchair users in the United States alone [1]. Combined with predictions indicating the global population of people aged over 60 is set to double between 2000 and 2050 [2], there is a drive to improve the methods through which services towards aged and disabled care is delivered. In order to fulfil this need, shared autonomy has become a well-explored research area for assisting PMD users in everyday mobility tasks [3–5]. The goal of shared autonomy is to mitigate the detrimental impact of poor vision or cognitive and physical deficiencies on PMD proficiency through environment-sensing and decision-making rather than relying purely on potentially dangerous user commands. Additionally, novel methods of communicating intent to the PMD including force-feedback modulation of a controller's available range of motion [6] as well as more advanced approaches such as gaze-tracking [7] and EEG headsets, although the latter may still present some difficulties to cognitively impaired users in cluttered living spaces [8]. These new

M. Beetz et al. (Eds.): ICSR 2014, LNAI 8755, pp. 300–309, 2014.

tools and platforms towards enabling disabled individuals to safely undertake everyday mobility tasks independently have the simultaneous effects of reducing the effort on their part, while also helping to mitigate the burden on carers by decreasing the level of supervision required for mobility oversight.

The outcomes of this work were primarily designed around the needs of Greystanes Disability Services (GDS); a care home supporting people with disabilities and complex health needs, with an aim of freeing up staff-hours better utilised in other aspects of patient care. As a carer is required to work with individual patients in everyday mobility activities, shared autonomy could provide a substantially less supervision-intensive alternative where one carer can oversee multiple PMDs simultaneously. Throughout the care home, a shared control mechanism for everyday mobility tasks will also be beneficial to prevent collisions with people, other PMDs or structure while encouraging independent driving on behalf of the user. Additionally, a regular activity is taking multiple wheelchair-bound patients on outdoor excursions through the surrounding bushland environment; currently a rather staff-intensive activity as one staff member is required per wheelchair for the duration of the exercise. It is hence desirable to have a system capable of encouraging patients capable of driving to follow the wheelchair ahead of them or the staff member leading the convoy. The remainder of this paper is structured as follows: Section 2 outlines the test platform and software configuration, Sections 3-4 outline the modes of operation developed, Sections 5-6 document and discuss experimental results, and Section 7 closes with concluding remarks.

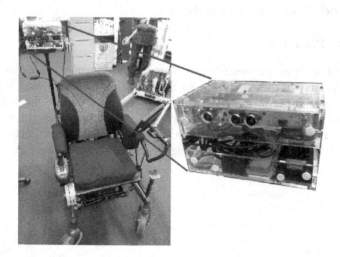

Fig. 1. Instrumented wheelchair with mounted sensor package

2   Experiment Setup

Three distinct modes of operation have been developed:

- Shared control driving with forward projection
- Fully autonomous leader following, disregarding user input
- A semi-autonomous leader-following 'exercise' with sliding-scale autonomy

At present each mode is a standalone item, designed to meet separate healthcare mobility needs. The first encourages disabled users to drive independently, and the second allows carers to move PMD users in cases where driving is prohibitive. The third mode is designed as an engaging activity for users partially capable of driving to safely follow a leader from one location to another; useful in places such as within GDS where groups of people are often moved between areas, for example between a recreational area and a sauna room where several users may wish to drive themselves, but punctuality may be of some priority to carers.

Development and experimentation was conducted on the Centre of Autonomous Systems instrumented wheelchair platform (Fig 1), fitted with drive motors and wheel encoders. An additional sensor module was added, containing a MS Kinect RGB-D camera, Hokuyo laser scanner and an Xsens inertial measurement unit. All sensors and the motor controller interface with the on-board Fit-PC, running Ubuntu 10.04 and utilizing the ROS (Robot Operating System - www.ros.org) middleware. Regardless of the mode of operation selected, at all times a collision prevention safeguard layer sits above the platform driver. Additionally, odometry pose information is combined with laser scanner data to hold environmental information beyond the scanner's field of view.

3   Shared Control Navigation

3.1   Local Planning

As users of PMDs may suffer from involuntary movements such as jerkiness or tremoring [9] it becomes necessary to filter out inputs which may not be indicative of desired platform behaviour. Noise reduction can be done through a multitude of signal processing methods such as weighted average [10] or Kalman filters [11]. There also exist advanced mechanisms for screening out of involuntary yet seemingly fluent input actions through learning frameworks [12], however for

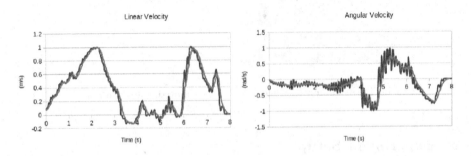

Fig. 2. Weighted average filter (blue to red)

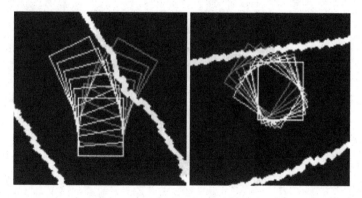

Fig. 3. Trajectory modulation of platform footprint pose from input (grey) to safe (white) when driving and turning in a corridor, indicated by the thicker white lines

$n_{frames} = f_{proj} \times \Delta t_{proj}$
project footprint poses based on joystick inputs
while *platform collision predicted* **do**
    increment angular velocity away from direction of collision
    if $joystick_{angular\%} \geq joystick_{linear\%}$ **then**
        if $joystick_{linear} \geq 0$ **then**
          | decrement linear velocity
        else
          | increment linear velocity
        end
    end
    re-project footprint poses until collision predicted
    velocities are acceptable if platform is collision-free for n_{frames}
end

Algorithm 1: Forward-projection pseudocode

the scope of this work filtering was carried out through more conventional noise mitigation approaches under the assumption of tremor suppression. Figure 2 demonstrates the output of a weighted moving-average filter showing significant noise reduction at the cost of a slight time delay. These filtered input velocities are then applied to the PMD's footprint for a brief forward-projection (Algorithm 1), as shown in Figure 3. This allows for responsive local control without prior map-building, enabling functionality in environments often subject to frequent change and/or have many other moving entities, such as a shopping area.

4 Leader Following

A blob-tracking algorithm was developed to be used with the laser scanner readings. After seeding with an initial pose, a cluster of closely spaced points can be tracked through sequential scans. 'Merging' with other objects in the environment such as walls is mitigated by a maximum search radius heuristic. Figure 4

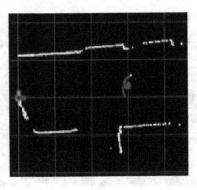

Fig. 4. Tracking (red contour with corresponding dot) using laser scanner

shows a sample frame from the tracking with the leader highlighted in red. These poses are directly fed as inputs into a Vector Field Histogram [13] (VFH), chosen for its suitability in planning to local goals. A VFH determines safe control speeds based on a polar obstacle grid of the platform's immediate surroundings. In the case of autonomous following, the resultant velocity commands are fed directly to the safeguard layer. For the following exercise, the user weight η (Section 4.1) is applied between the VFH output v, w_{vfh} and the user's filtered input v, w_{filter} (Eqns 1-2). Full autonomy is enabled if the user releases the joystick, in order to avoid interruptions to the platform's movement which the leader may not be able to notice immediately.

$$v = \begin{cases} v_{vfh} & \text{if } v_{filter} = w_{filter} = 0 \\ \eta \times v_{filter} + (1 - \eta) \times v_{vfh} & \text{otherwise} \end{cases} \tag{1}$$

$$\omega = \begin{cases} \omega_{vfh} & \text{if } v_{filter} = w_{filter} = 0 \\ \eta \times \omega_{filter} + (1 - \eta) \times \omega_{vfh} & \text{otherwise} \end{cases} \tag{2}$$

4.1 Weighting

$$\eta_{sa} = 1 - e^{-\alpha \times d_{min}} \tag{3}$$

$$\eta_{sm} = e^{-\beta \times \delta_{axis}} \tag{4}$$

$$\eta_{ob} = e^{-\gamma \times |\omega_{track} - \omega_{filter}|} \tag{5}$$

$$\eta = min(\eta_{sa}, \eta_{sm}, \eta_{ob}) \tag{6}$$

$$\eta = \begin{cases} 0 & \text{if } \eta \le \eta_0 \\ 1 & \text{if } \eta \ge \eta_1 \end{cases} \tag{7}$$

The user weight is primarily determined using two or three proficiency metrics [14, 15] of safety (Eqn 3), smoothness (Eqn 4) and 'obedience' (Eqn 5). d_{min} represents the closest obstacle to the platform footprint and δ_{axis} represents a percentage change in joystick axis position per second. Obedience is determined by the difference between the angular velocities of the VFH and the filtered joystick input, only taken into account when leader tracking is active. The gains α, β and γ were roughly determined from desired values of η at specific values; for instance β was chosen to be 0.0035 from a desired η_{sm} of 0.5 at 200% axis movement per second. A higher weighting gives the user a greater level of permitted deviation from what is considered a safe local trajectory to permit closer proximity to hazards, and a lower weighting tends to limit the user to a more conservative style of driving. Weighting (Eqns 6-7) is then used to blend filtered input velocities with the outputs from the local planner (Eqns 8-9). In our experiments, cutoff values for η_0 and η_1 were 0.25 and 0.75 respectively.

$$v = \eta \times v_{filter} + (1 - \eta) \times v_{planner} \tag{8}$$
$$\omega = \eta \times \omega_{filter} + (1 - \eta) \times \omega_{planner} \tag{9}$$

5   Results

Preliminary tests were conducted on the University of Technology, Sydney campus. As navigating areas such as doorways and narrow hallways presents particular difficulty [16] to the average PMD user, experimentation focused on suitably constrained spaces such as those that may be found in a 'normal' interior space not necessarily configured to accomodate PMDs. Figure 5 shows the trajectory difference in trajectories between the joystick input (red) and the output of the shared control algorithm (green) when roughly aimed at a doorway with $\sim$5 cm clearance on either side. Figure 6 shows the trajectories from the VFH (green) against the leader's path from laser scanner tracking (red) through another doorway with $\sim$10 cm clearance twice, with a tight on-the-spot turn in

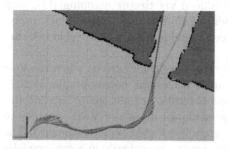 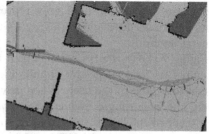

Fig. 5. Assisted driving through narrow doorway (green) compared to raw input (red)

Fig. 6. Autonomous following (green) through doorway via VFH behind leader entity (red)

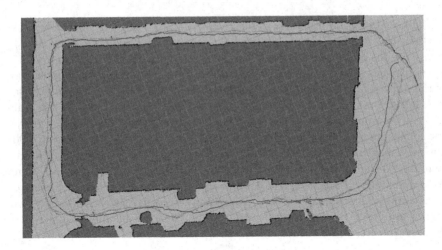

Fig. 7. Leader (blue) following exercise in counter-clockwise indoor loop

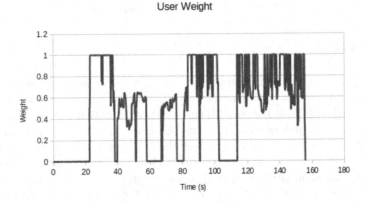

Fig. 8. User weighting results for following exercise

between. Trajectories and maps were produced via Hector mapping [17]; map regions are as follows: light grey represents known vacant space, black represents known surfaces and dark grey represents unknown space; the grid cells are 1 metre square.

Figure 7 shows the trajectory taken by the wheelchair (green) when the user attempts to follow a leader (blue). Figure 8 shows the corresponding control weight allocated to the user. When driving is erratic or tight spaces are encountered, the weighting is reduced to allow mediation from the VFH in order to stay truer to the leader trajectory or to safely bypass the potential hazard. The test user also allowed the VFH to fully take over momentarily in a few areas by releasing the joystick, resulting in the lengthier 0% weighting periods visible.

6   Discussion

Despite a major drawback of these results being that the manual driving components were conducted by an able-bodied user attempting to emulate a PMD user unable to smoothly or confidently maneuvre through tight spaces, the early results appear to positively address the project's needs. The modes of operation evaluated in these experiments all feature distinct value in healthcare, from enabling patients to safely move themselves or a single staff to convoy several PMDs, to a mobility exercise providing engagement to users while guiding them to their destination. The benefits of the latter extend beyond its intrinsic value by additionally providing a potential means of PMD training or basic proficiency assessment. Providing users accustomed to staff supervision with the opportunity to use intelligent PMDs unsupervised may likely require some training and initial acclimation, however the benefits to the patients' self-esteem from maintained independence and to staff by reducing time spent guiding PMD users would be considerable.

Testing these solutions at Greystanes Disability Services with disabled patients is presently of high priority to assess the system's performance in real test cases, as well as to obtain feedback and suggestions from the system's intended users when comparing the system's outputs to raw user driving. The development of a simple GUI to handle the execution of relevant softwares for each mode would be highly beneficial to staff and some PMD users, and to unify these solutions into a single utility. Other future planned developments include a more sophisticated noise mitigation algorithm for involuntary movements that may not be simple tremors, and an improved planner for more complex local trajectories beyond basic forward projection. Different sensors may be required to detect hazards on the ground or objects below the 2D laser scanner's height such as furniture items or household utilities. Obstacles on the ground may not be a significant concern indoors, however as outdoor use is intended a smooth traversable surface cannot always be ensured. Outdoor applications must also take into account obstacles identified by PMD users to be difficult to pass [18] such as groups of people. Downward-pitched stereo cameras or an IR depth imager could be used to provide height-maps and additional stability/traversability information [19, 20] for planning algorithms; a study left for future development and investigation.

7   Conclusions

This paper describes several solutions to assist caregiver staff by reducing the need to monitor PMD users in everyday mobility tasks related to patient care. Each approach has been demonstrated to be capable of providing navigational support across several different scenarios including safeguarded driving, autonomous following and a hybrid driving exercise. The reduced requirement for manual oversight would provide staff with more time towards less rudimentary tasks, while additionally enhancing the self-esteem and personal independence of disabled PMD users. Despite the shortcomings of the experiments conducted we believe the outcomes are relevant in the context of this work, and hope to follow these developments with a larger trial at a care home involving several disabled users under adequate supervision from qualified care staff.

Acknowledgements. The authors would like to thank Greystanes Disability Services - http://www.greystanes.org.au) for their collaboration with this research project.

References

1. Kaye, H.S., Kang, T., LaPlante, M.P.: Mobility Device Use in the United States. Disability Statistics Report 14. Disability Statistics Center, University of California, San Francisco (2000)
2. Congressional Budget Office (C.B.O.): Global population aging in the 21st century and its economic implications (2005), http://www.cbo.gov /sites/default/files/cbofiles/ftpdocs/69xx/doc6952/12-12-global.pdf (accessed: May 14, 2014)
3. Lankenau, A., Meyer, O., Krieg-Bruckner, B.: Safety in robotics: the Bremen Autonomous Wheelchair. In: 1998 5th International Workshop on Advanced Motion Control, pp. 524–529 (1998)
4. Pires, G., Honorio, N., Lopes, C., Nunes, U., Almeida, A.T.: Autonomous wheelchair for disabled people. In: Proceedings of the IEEE International Symposium on Industrial Electronics, vol. 3, pp. 797–801 (1997)
5. Oishi, M., Cheng, A., Bibalan, P., Mitchell, I.: Building a smart wheelchair on a flexible software platform. In: RESNA International Conference on Technology and Aging (2011)
6. Masone, C., Franchi, A., Bulthoff, H.H., Giordano, P.R.: Interactive planning of persistent trajectories for human-assisted navigation of mobile robots. In: 2012 IEEE/RSJ International Conference on Intelligent Robots and Systems (IROS), pp. 2641–2648 (2012)
7. Al-Haddad, A., Sudirman, R., Omar, C.: Gaze at Desired Destination, and Wheelchair Will Navigate towards It. New Technique to Guide Wheelchair Motion Based on EOG Signals. In: First International Conference on Informatics and Computational Intelligence (ICI), pp. 126–131 (2011)
8. Graham-Rowe, D.: Wheelchair makes the most of brain control (2010), http://www.technologyreview.com/news/420756/wheelchair-makes-the-most-of-brain-control/ (accessed: April 3, 2014)
9. van der Zwaag, B.J., Corbett, D., Jain, L.: Minimising tremor in a joystick controller using fuzzy logic. In: Third International Conference Knowledge-Based Intelligent Information Engineering Systems, pp. 5–8 (1999)
10. Lu, B.-Y., Liu, C.-Y., Luh, J., Shiao, C.-C., Chong, F.C.: Standard Deviation Based Weighted Average Function for Reducing Tremor Disturbance of the Joystick. In: 2007 9th International Conference on e-Health Networking, Application and Services, pp. 214–216 (2007)
11. Welch, G., Bishop, G.: An Introduction to the Kalman Filter (2006), http://www.cs.unc.edu/~welch/media/pdf/kalman_intro.pdf (accessed: April 22, 2014)

12. Demeester, E., Nuttin, M., Vanhooydonck, D., Van Brussel, H.: A model-based, probabilistic framework for plan recognition in shared wheelchair control: experiments and evaluation. In: Proceedings of the 2003 IEEE/RSJ International Conference on Intelligent Robots and Systems (IROS 2003), vol. 2, pp. 1456–1461 (2003)

13. Borenstein, J., Koren, Y.: The vector field histogram-fast obstacle avoidance for mobile robots. IEEE Transactions on Robotics and Automation 7(3), 278–288 (1991)

14. Li, Q., Chen, W., Wang, J.: Dynamic shared control for human-wheelchair cooperation. In: 2011 IEEE International Conference on Robotics and Automation (ICRA), pp. 4278–4283 (2011)

15. Urdiales, C., Peula, J.M., Fdez-Carmona, M., Barru, C., Prez, E.J., Snchez-Tato, I., Toro, J.C., Galluppi, F., Corts, U., Annichiaricco, R., Caltagirone, C., Sandoval, F.: A new multi-criteria optimization strategy for shared control in wheelchair assisted navigation. Autonomous Robots 30(2), 179–197 (2011)

16. Koontz, A.M., et al.: Design Features That Affect Maneuverability of Wheelchairs and Scooters. Archives of Physical Medicine and Rehabilitation 91(5), 759–764 (2010)

17. Kohlbrecher, S., Meyer, J., von Stryk, O., Klingauf Kolbrecher, U.: A Flexible and Scalable SLAM System with Full 3D Motion Estimation. In: Proc. IEEE International Symposium on Safety, Security and Rescue Robotics, SSRR (2011)

18. Torkia, C., et al.: Power wheelchair driving challenges in the community: a users perspective. Disabil Rehabil Assist Technol. (2014)

19. Murarka, A., Kuipers, B.: A stereo vision based mapping algorithm for detecting inclines, drop-offs, and obstacles for safe local navigation. In: IEEE/RSJ International Conference on Intelligent Robots and Systems (IROS 2009), pp. 1646–1653 (2009)

20. Norouzi, M., Miro, J.V., Dissanayake, G.: A statistical approach for uncertain stability analysis of mobile robots. In: 2013 IEEE International Conference on Robotics and Automation (ICRA), pp. 191–196 (2013)

Motion-Oriented Attention
for a Social Gaze Robot Behavior

Mihaela Sorostinean, François Ferland, Thi-Hai-Ha Dang, and Adriana Tapus

Computer Vision and Robotics Lab, Computer Science and System Engineering
Department, ENSTA-ParisTech, Palaiseau, France
{firstname.lastname}@ensta-paristech.fr

Abstract. Various studies have shown that human visual attention is
generally attracted by motion in the field of view. In order to embody
this kind of social behavior in a robot, its gaze should focus on key points
in its environment, such as objects or humans moving. In this paper, we
have developed a social natural attention system and we explore the
perception of people while interacting with a robot in three different
situations: one where the robot has a totally random gaze behavior, one
where its gaze is fixed on the person in the interaction, and one where
its gaze behavior adapts to the motion-based environmental context. We
conducted an online survey and an on-site experiment with the Meka
robot so as to evaluate people's perception towards these three types of
gaze. Our results show that motion-oriented gaze can help to make the
robot more engaging and more natural to people.

1 Introduction

During the last decade, a lot of research has focused on examining the effect
of a robot's gaze while interacting with a human, since it is known that gaze
plays an important role in human-human communication [1,2]. It has been found
that enriching a robot with a human-like gaze behavior helps it to be perceived
as more intelligent and social [3]. Furthermore, the attention that the robot is
showing to the person in the interaction has great influence in the way a person
understands the robot's messages [4].

A possible way to achieve more natural and human-like behavior in interaction tasks is to design a distributed robot attention system that allows the robot
to be distracted from the main interaction by external events. In a human-
human interaction setting, such events could include another person entering
the room or the noise of an object hitting the ground. While such events should
not disturb the interaction and communication process, it is entirely expected
that people would momentarily shift their attention through gaze to the external
disturbing factors. We believe that robots should behave similarly in order to
exhibit natural, human-like behavior. Previous works suggest that attention disturbance can encourage people to make the communication with a robot more
social, and adaptable attention enables the robot to be accepted as an intentional and proactive communication agent [5]. Furthermore, participants in an

M. Beetz et al. (Eds.): ICSR 2014, LNAI 8755, pp. 310–319, 2014.

online survey showing videos of a receptionist robot believed that it was more human-like when it exhibited random gazes in conjunction with person-directed gaze [6]. In [7], a multimodal, saliency-based bottom-up attention system has been used on the iCub humanoid robot to drive its exploratory behavior. The system guided the gaze of the robot to visually stimulating zones (e.g. moving targets, brightly-colored features) and localized sound sources, while inhibition and habituation mechanisms prevented the robot from having its attention fixed at a single target for a long period of time.

The purpose of this research is to discover how a robot embodied with a human-like visual attention system is perceived in a scenario of interaction with a person. The Meka robot plays the role of a receptionist, and interacts with people that come to ask some specific information regarding a meeting with a professor. During the interaction, the robot displays three types of gazing behavior: randomly gazing at people or objects in the background, a fixed gaze at the person in the interaction, and a motion-oriented gaze that focuses both on the person standing in front of the robot and on other possible targets in the background. Our interest is to find out in which of these three scenarios people would feel more comfortable to interact with the robot, and how much the gaze behavior of the robot affects the interaction. Impressions from people will be collected from both a third- and first-person perspective, first from an online video survey and then with an on-site experiment with the Meka robot.

2   Testbench Configuration

The experiments presented in this work have been conducted with the Meka humanoid robot (see Fig. 1). It has been designed to work in human-centered environments. The robot features compliant force control throughout its body, durable and strong hands, and an omnidirectional base with a prismatic lift. The head is a 7 Degrees-of-Freedom (DOF) robotic active vision head with high resolution FireWire cameras in each eye, integrated DSP controllers, and zero-backlash Harmonic Drive gearheads in the neck. Designed for a wide range of expressive postures, it is a platform particularly well suited for researchers interested in human-robot interaction and social robotics. Additionally, we use a separate webcam to augment the peripheral field of view of the robot. When the robot is focusing on a person, his/her body fills an important area of the Meka's eyes cameras field of view, hence rendering the tracking of the background motion difficult. We also use the microphone of this webcam for speech recognition purposes.

In our implementation of motion-oriented attention, which uses the Robot Operating System (ROS), the robot can alternate between two tracking sources for its gaze behavior:

- **OpenNI Tracking** To track the person interacting with the robot, we use an ASUS Xtion Live Pro, integrated into the torso of Meka, and the OpenNI2 ROS Package for full skeleton frames tracking. The head frame of the person,

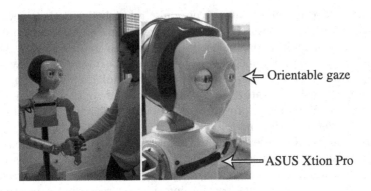

Fig. 1. The Meka humanoid robot

once it is detected, is transformed into the frame of the head of the robot. The origin of this transformed frame is continuously sent to Gaze Control.

- **Motion Tracking** Background motion detection is performed on the webcam camera. Based on [8], semi-dense point tracking is performed to detect and localize motion to the left of the robot. When motion is detected over an empirically-determined threshold quantity, a target is sent to Gaze Control. As the external webcam cannot provide depth information, this target is generated at a fixed, predetermined distance of 2 meters. While this might result in a slight disparity between the gaze target and the actual motion in the scene, this is generally imperceptible from the interacting person's point of view, since the webcam is relatively close to the robot.

Gaze Control generates smooth motion for 6 actuators of the head of the robot: neck pan and tilt, head tilt, and eyes pan (both eyes individually) and tilt. Motion Tracking has priority over OpenNI Tracking. As long as there is no motion in the field of view of the webcam, the robot is looking at the interacting person. However, as soon as motion is detected, the robot focuses on this new target for up to 3 seconds before returning control to OpenNI Tracking. Additionally, speech recognition as well as speech synthesis is performed through an Internet connection by the Google Chrome API.

3   Experimental Design

Figure 2 shows a schematic view of the experimental setup and a screenshot from the video used in the online survey. The Meka robot is positioned behind a tall and narrow desk. The prismatic lift of the robot sets it at a standing height of 1.75 m. The webcam serving the dual purpose of voice capture and motion detection is set on this desk. The person starts the experiment by standing approximately 2 meters from the desk, in front of the robot, for at most 10 seconds, to ensure proper detection by OpenNI. When ready, the experimenter signals the person the beginning of the experiment, and the person walks to the desk and greets the robot (by saying "Hi" or "Hello"). The Meka robot then asks the participant details about the professor's name and the purpose of his/her visit. Thereafter, it provides directions about how to reach the professor's office. The length of one interaction is approximately 1 minute. During this interaction, a person passes in the background for two round trips with the purpose of distracting the robot's attention for at least 4 times. A video (based on sequences from the online survey) showing the interaction with the robot in the three conditions is publicly available online[1].

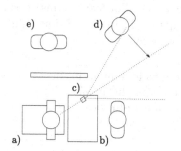 

Fig. 2. Experimental setup: (a) the Meka robot, (b) the interacting person, (c) the peripheral view webcam, (d) a person passing in the background, (e) the experimenter. A partition screen prevents the experimenter from accidentally entering the robot's field of view. The robot interacts with the person as someone else is about to enter the robot's field of view. The bottom-left part of the screenshot shows the point of view from the eyes of the Meka robot.

In order to investigate people's perception of the gaze of the robot, we chose to compare three types of behavior. People participating in the experiment were exposed to three corresponding experimental conditions:

- **C1-Random** Random Gaze Attention Robot. The robot looks randomly at the background while talking with the person. The gaze target changed every 6 seconds. This condition was used to highlight if any motion, regardless of frequency or targeting, was perceived as natural.
- **C2-Fixed** Fixed Gaze Attention Robot. The robot is embodied with fixed gaze attention, so its gaze is fixed on the face of the person standing in front of it.
- **C3-Distributed** Motion-oriented Gaze Robot. The robot has a distributed gaze behavior, looking both at the person in the interaction and at other moving targets in the background in a human-like manner.

The experiment was conducted in two identical phases with different sets of participants. For the first phase of the experiment, we created an online questionnaire in which we integrated three videos that illustrate the three gaze behaviors and a set of questions that allowed us to evaluate the people's perception

---

[1] http://perso.ensta-paristech.fr/~tapus/eng/media/videos/icsr2014.mp4

about the behavior of the robot and its attention system. The second phase
of the experiment was conducted in the laboratory with a face-to-face interaction between the robot and the participants. For both experimental stages the
participants filled-out a pre-experiment general questionnaire with 4 items (age,
gender, background, and robotics knowledge) that provided us with demographic
information. After each condition, the participants answered 9 questions (with
a total of 17 sub-items), all presented on a 7-point Likert scale. To avoid the
priming effect of knowing the content of the questionnaire after the first condition, the order in which the three conditions appeared was randomized for each
participant. An additional set of 3 questions (with a total of 5 sub-items) were
asked at the end of the whole experiment to compare all three conditions. The
following questions were relevant to the study in this paper:

- **Q1.** Was the robot's behavior human-like?
- **Q2.** Did its overall bodily behavior contributed to its human-likeness?
- **Q3.** Did its gaze movement contributed to its human-likeness?
- **Q4.** Did its head movement contributed to its human-likeness?
- **Q5.** Was the robot embodied with attention?
- **Q6.** Was the robot expressive?
- **Q7.** Did the robot appeared intelligent?
- **Q8.** Was the interaction engaging?
- **Q9.** Which condition was the most social?
- **Q10.** Which condition was the most natural?
- **Q11.** With the robot of which condition would you prefer interact with?

From the participants' responses to these questions, we wanted to test the
following hypotheses:

- **H1.** The motion-oriented gaze robot (C3-Distributed) is perceived as more
  natural and more human-like than the two other types of gaze (C1-Random
  and C2-Fixed).
- **H2.** The motion-oriented gaze robot (C3-Distributed) appears more attentive to its environment than the two other types of gaze (C1-Random and
  C2-Fixed).
- **H3.** The motion-oriented gaze robot (C3-Distributed) appears more expressive to the person in interaction than the two other type of gaze do (C1-Random and C2-Fixed).

4   Experimental Results

4.1   Online Survey

The online questionnaire was completed by 69 individuals (45 male, 24 female,
aged 19 to 74, average 30, 66.7% from Romania, 31.9% from Canada, 1.4%
from France). To analyze the data from the online questionnaire, a single factor
analysis of variance (ANOVA) test was conducted between the values rated on
a 7-points Likert scale for each two pairs of random, fixed, and motion-oriented
gaze behaviors. Figure 3 shows significant results from the questionnaire (results
from Q7 and Q8 were not significant).

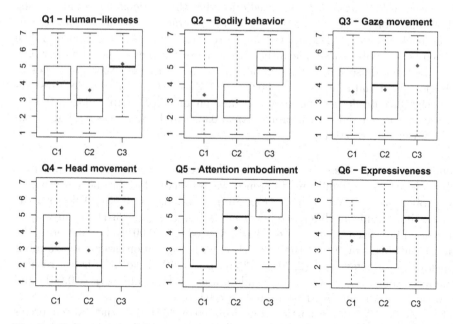

Fig. 3. Online survey Likert-scale results for each condition: C1-Random, C2-Fixed and C3-Distributed. The red dots represent the means.

Validation of hypothesis H1. The hypothesis that the motion-oriented gaze robot (C3-Distributed) would be perceived as more natural and human-like was strongly supported in the online questionnaire. The robot embodied with a motion-oriented gaze behavior was significantly more appreciated relatively to the random gaze and to the fixed gaze for its human-like behavior ($F(2, 204) = 20.18, p < 0.001$, Tukey's Honestly Significant Difference (HSD) comparison revealed significant difference between C1-Random and C3-Distributed ($p < 0.001$) and C2-Fixed and C3-Distributed ($p < 0.001$)). This was also the case for its overall bodily behavior ($F(2, 204) = 24.49, p < 0.001$, Tukey's HSD test results: significant difference between C1-Random and C3-Distributed ($p < 0.001$) and between C2-Fixed and C3-Distributed ($p < 0.001$)), its gaze movement ($F(2, 204) = 15.68, p < 0.001$, Tukey's HSD test results: significant difference between C1-Random and C3-Distributed ($p < 0.001$) and between C2-Fixed and C3-Distributed ($p < 0.001$)), and head movement ($F(2, 204) = 44.31, p < 0.001$, Tukey's HSD test results: significant difference between C1-Random and C3-Distributed ($p < 0.001$) and between C2-Fixed and C3-Distributed ($p < 0.001$)). In Q9, 50.73% of the participants rated the social-attentive robot with motion-oriented attention as more social than the robot embodied with fixed attention (36.23%) or the robot embodied with random gaze behavior (13.04%, $\chi_2 = 35.46$, $p < 0.001$). Moreover, in Q10, 82.6% of participants thought that the social-attentive robot was more natural than the other two robot gaze behaviors (random-7.25% and fixed-10.1%, C1-Random and C2-Fixed, respectively, $\chi_2 = 14.22$, $p < 0.001$).

Validation of hypothesis H2. Our prediction that the participants would rate the robot that looked both at the person in the interaction and at the movement in the background (C3-Distributed) as more attentive was also supported. The participants attributed significantly higher scores for the attention with which the robot in this scenario was embodied, in regard to the other two types of behavior, fixed and random ($F(2, 204) = 32.03, p < 0.001$, Tukey's HSD test results: significant difference between C1-Random and C3-Distributed ($p < 0.001$), between C2-Fixed and C3-Distributed ($p = 0.001$) and also between C1-Random and C2-Fixed ($p < 0.001$)).

Validation of hypothesis H3. Regarding the expressiveness between the three robot behaviors, a significant difference was observed for the motion-oriented gaze robot, as ANOVA returns ($F(2, 204) = 15.36, p < 0.001$, Tukey's HSD test results: significant difference between C1-Random and C3-Distributed ($p < 0.001$) and between C2-Fixed and C3-Distributed ($p < 0.001$)). Furthermore, the participants also noted that the interaction with the motion-oriented gaze robot appeared more engaging ($F(2, 204) = 33.66, p < 0.001$, Tukey's HSD test results: significant difference between C1-Random and C3-Distributed ($p < 0.001$) and between C2-Fixed and C3-Distributed ($p < 0.001$)).

Finally, 66.67% of the participants declared in Q11 that they would prefer to interact with the distributed motion-oriented gaze robot (C3-Distributed), compared to 24.6% and 8.7% for C2-Fixed and C1-Random, respectively ($\chi^2 = 72.74, p < 0.001$).

4.2   Robot Real-World Experiments

For the second phase, 21 participants (17 male, 4 female, aged 20 to 57, average 28.3) have been recruited from the university. 12 participants were from France, 2 from Romania, and one from each of these countries: Algeria, Argentina, Canada, China, Italy, Morocco, and Tunisia. Participants received a written copy of the script for the interaction and were asked to follow it as much as possible. They were allowed to refer to this script while interacting with the robot, but were told in advance that the dialogue was not evaluated in this experiment, and that they should focus on the overall behavior of the robot instead of the quality of the speech recognition or the answers received from it, as speech recognition and the dialogue are not central parts of this work. In case of any speech recognition errors, the operator could manually select the answers (in a manner that could not be perceived by the participant) so that the interaction would not be slowed down. As with the online survey result, we used a one-way within groups ANOVA test to analyze the significance of the participants' responses on a 7-points Likert scale. Figure 4 presents significant results from the questionnaire (results for Q2 to Q4 were not significant).

Validation of hypothesis H1. A significant difference was discovered in the human-like perception of the robot between the first and the third condition (ANOVA

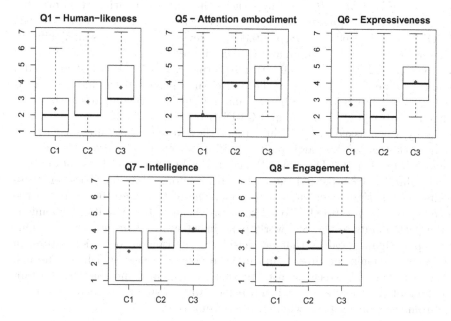

Fig. 4. Live trials Likert-scale results for each condition: C1-Random, C2-Fixed and C3-Distributed. The red dots represent the means.

test: $F(2,60) = 3.19, p = 0.04$, Tukey's HSD comparison: $p = 0.04$ between C1-Random and C3-Distributed). In Q9, while 52.38% of the participants appreciated the robot with motion-oriented gaze (C3-Distributed) as natural, compared to 23.80% for fixed gaze (C2-Fixed) and 19.01% for random gaze (C1-Random), these results were not significant ($\chi^2 = 4.26$, $p = 0.11$). However, the robot with motion-oriented gaze was significantly perceived in Q10 as more social than the robot embodied with the other two behaviors (52.38% -C3-Distributed, 33.30% -C2-Fixed, and 9.52% -C1-Random, $\chi^2 = 6.07$, $p = 0.04$, one of the participants perceived no natural or social behavior in any of the robots).

Validation of hypothesis H2. The ANOVA test showed a significant difference among the three gaze behaviors in the attention level that participants perceived in each robot ($F(2,60) = 10.12, p < 0.001$). Tukey's HSD comparisons indicated a significant difference between the robot embodied with random gaze behavior (C1-Random) and the robot embodied with motion-oriented gaze behavior (C3-Distributed, $p < 0.001$) and between the fixed and random behavior (C1-Random and C2-Fixed, $p = 0.003$), but no significant difference was observed between the fixed and the motion-oriented gaze behavior (C2-Fixed and C3-Distributed). Moreover, the robot in C3-Distributed was also considered more intelligent with respect to the one in C1-Random (ANOVA test: $F(2,60) = 3.43, p = 0.038$, Tukey's HSD comparison: $p = 0.029$ between C1-Random and C3-Distributed).

Validation of hypothesis H3. Our hypothesis that the robot with motion-oriented gaze appears more expressive was supported by the real-world experiment as well (ANOVA test: $F(2, 60) = 5.98, p = 0.004$, Tukey's HSD comparison: significant difference between C1-Random and C3-Distributed ($p = 0.02$) and between C1-Random and C2-Fixed ($p = 0.005$)). Furthermore, the ANOVA test revealed a significant difference on the way the participants perceived the interactions as engaging ($F(2, 60) = 5.19, p = 0.008$), and a Tukey's HSD comparison showed a difference between C1-Random and C3-Distributed ($p = 0.006$).

Finally, when asked in Q11 which robot they would rather interact with, an equal percentage of the participants (42.85%) selected the robot with motion-oriented attention (C3-Distributed) and the robot with fixed attention (C2-Fixed) while 14.3% preferred the robot with random attention. However, these results are not significant enough to prefer one of the conditions between the three ($\chi_2 = 3.41, p < 0.181$). To get some insight into these results, the opinions of the participants to the real-world experiment were taken into account. One of the participants told us that he considered the random behavior natural in a robot receptionist because it made the robot look busy, which is a normal behavior for a real life receptionist. Another participant found that the constant switching of the robot's gaze when a person was passing by tended to distract him from the conversation, which was also annoying.

5   Discussion and Conclusion

In this paper, we described an experiment evaluating embodied, socially attentive motion-oriented gaze control for a humanoid robot interacting as a receptionist. We have shown that a behavior distributing attention between the participant in the interaction and the motion in the background contributes in making a robot seem more natural, human-like, and attentive to various people as supported by an online survey and an on-site experiment. From this experiment we noticed that the percentage of people that considered a robot embodied with attentive behavior as more social in respect to the other two behaviors is approximately the same for both the online survey and the direct interaction with the robot. This leads to the conclusion that a receptionist robot should have a motion-oriented attention behavior in order to be perceived as social during an interaction with a person.

Nevertheless, there were differences between the online and the real-world experiments regarding how natural the robot behavior was perceived in each of the three conditions. It can be seen in the results above that in a direct interaction with the robot the percentage of people that preferred the fixed or the random behavior for the robot are considerably higher in comparison with the online survey. This is mainly due to the influence of the social physical embodied presence of the robot to the participants' perception. Furthermore, while the majority of participants in the online survey considered the robot with motion-oriented gaze as the one they would rather interact with in a receptionist-visitor scenario, that is not the case for the on-site experiment, where an equal percentage of people chose the fixed and the motion-oriented behavior.

From the results of this work, we believe that there are several ways of making social gaze for robots that interact with humans. Instead of expecting to have one unique social gaze for the robot in all kind of situations, an adaptive behavior would be more appropriate. In future work, we plan on studying automatically adaptive behaviors in applications such as tour guide robots and assistive care.

Acknowledgements. We would like to thank the ERASMUS program that funded the internship of Mihaela Sorostinean, Fabio Pardo for allowing us to use his code for speech rendering, and Sean Andrist for participating in the first phase of the experiment.

References

1. Exline, R.V., Gray, D., Schuette, D.: Visual behavior in a dyad as affected by interview content and sex of respondent. In: Tomkins, S., Izzard, C. (eds.) Affect, Cognition and Personality, Springer, New York (1965)
2. Kendon: Some functions of gaze direction in social interaction. Acta Psychologica 26(1), 1–47 (1967)
3. Koelemans, D.: The effect of virtual eye behaviour during human-robot interaction. In: Proceedings of the 18th Twente Student Conference on IT (2013)
4. Mutlu, B., Forlizzi, J., Hodgins, J.: A storytelling robot: Modeling and evaluation of human-like gaze behavior. In: Proceedings of the 6th IEEE-RAS International Conference on Humanoid Robots, pp. 518–523. IEEE (2006)
5. Muhl, C., Nagai, Y.: Does disturbance discourage people from communicating with a robot? In: The 16th IEEE International Symposium on Robot and Human interactive Communication, RO-MAN 2007, pp. 1137–1142. IEEE (2007)
6. Holthaus, P., Pitsch, K., Wachsmuth, S.: How can i help? International Journal of Social Robotics 3(4), 383–393 (2011)
7. Ruesch, J., Lopes, M., Bernardino, A., Hornstein, J., Santos-Victor, J., Pfeifer, R.: Multimodal saliency-based bottom-up attention a framework for the humanoid robot icub. In: Proceedings of the International Conference on Robotics and Automation, pp. 962–967. IEEE (2008)
8. Garrigues, M., Manzanera, A., Bernard, T.M.: Video extruder: a semi-dense point tracker for extracting beams of trajectories in real time. Journal of Real-Time Image Processing, 1–14 (2014)

Integrating Reinforcement Learning and Declarative Programming to Learn Causal Laws in Dynamic Domains

Mohan Sridharan[1] and Sarah Rainge[2]

[1] Department of Electrical and Computer Engineering, The University of Auckland, NZ
[2] Department of Computer Science, Texas Tech University, USA
m.sridharan@auckland.ac.nz, sarah.rainge@ttu.edu

Abstract. Robots deployed to assist and collaborate with humans in complex domains need the ability to represent and reason with incomplete domain knowledge, and to learn from minimal feedback obtained from non-expert human participants. This paper presents an architecture that combines the complementary strengths of Reinforcement Learning (RL) and declarative programming to support such commonsense reasoning and incremental learning of the rules governing the domain dynamics. Answer Set Prolog (ASP), a declarative language, is used to represent domain knowledge. The robot's current beliefs, obtained by inference in the ASP program, are used to formulate the task of learning previously unknown domain rules as an RL problem. The learned rules are, in turn, encoded in the ASP program and used to plan action sequences for subsequent tasks. The architecture is illustrated and evaluated in the context of a simulated robot that plans action sequences to arrange tabletop objects in desired configurations.

1 Introduction

Robots deployed in assistive roles in complex domains such as healthcare and disaster rescue face some fundamental learning and representation challenges. For instance, it is difficult to equip a robot assisting caregivers in an elder care home with accurate (and complete) domain knowledge; some of the rules governing the domain dynamics (e.g., "pain medication cannot be stacked in the top shelf") may be unknown to the robot or may change over time. Furthermore, the robot has to reason with qualitative and quantitative descriptions of knowledge, and the human participants may not have the time and expertise to provide elaborate and accurate feedback.

As a step towards addressing these challenges, this paper presents an architecture that combines the complementary strengths of Reinforcement Learning (RL) and declarative programming to support commonsense reasoning and incremental discovery of (previously unknown) rules that govern the domain dynamics. Specifically, we use Answer Set Prolog (ASP), a declarative language, to represent domain knowledge in the form of objects, relations between objects, and any known rules governing the domain dynamics. Inference with this knowledge is used to obtain the components of an RL formulation of the task of incrementally discovering unknown domain rules. We illustrate this architecture in the context of a simulated robot that plans action sequences to arrange tabletop objects in desired configurations.

M. Beetz et al. (Eds.): ICSR 2014, LNAI 8755, pp. 320–329, 2014.
© Springer International Publishing Switzerland 2014

The remainder of the paper is organized as follows. Section 2 motivates the proposed architecture by briefly reviewing related work. Section 3 describes the proposed architecture and its individual components. The experimental setup and results in a simulated domain are described in Section 4, along with directions for future research, and Section 5 presents the conclusions.

2 Related Work

We motivate our architecture by reviewing a representative set of related work on: (a) learning from non-expert human feedback and environmental interactions; and (b) knowledge representation for such learning.

Reinforcement learning provides an elegant mathematical formulation for agents to learn from repeated interactions with the environment, selecting actions that maximize a numerical reward signal [15]. For an agent engaged in a sequential decision making task, and occasionally receiving reinforcement signals from a human trainer, it is challenging to learn the best possible action policy. Several RL-based algorithms have been developed to address this interactive shaping problem [10], e.g., the use of RL and animal training insights for clicker training to support interactive learning of synthetic characters [4], and for action and behavior learning on a four-legged robot [9]. Other researchers have used RL-based frameworks for interactive shaping [10,16]. For instance, the TAMER framework allows an agent to receive feedback about specific tasks from a human trainer fully aware of the agent's state and action capabilities [11]. These algorithms, however, do not consider human training in conjunction with feedback that agents can receive by interacting with the environment.

The feedback signals obtained from non-expert humans and the environment may differ in format and frequency; human feedback may also be a function of a set of (previous or future) states and actions. RL-based algorithms have been proposed to address this challenge. For instance, different linear functions have been considered for combining human reward and environmental reward signals in some benchmark simulated domains [12]. A bootstrap learning algorithm has also been developed to enable agents to incrementally and continuously estimate the relative importance of human feedback and environmental feedback in the Tetris domain and the multiagent Keepaway Soccer domain [1,14]. More recently, a policy shaping algorithm has been developed for including human feedback in interactive RL formulations of agent domains [8]. However, these algorithms do require complete knowledge of the domain and the rules that govern domain dynamics.

Declarative languages provide appealing knowledge representation and commonsense reasoning capabilities that have been used for simulated and physical robots deployed in assistive roles in complex domains [6,18]. Algorithms have been developed to generalize from a limited number of samples by using knowledge representation in RL frameworks, e.g., relational RL incorporates a relational learner in a traditional RL algorithm [5]. However, the agent still needs to be provided accurate and elaborate knowledge about domain objects and rules. The architecture described in this paper is a step towards enabling robots to incrementally discover the rules governing the domain dynamics by integrating the commonsense reasoning capabilities of declarative programming with the incremental learning capabilities of RL.

3   Problem Formulation

This section describes our architecture for incrementally learning the rules governing the domain dynamics. As an illustrative example used throughout this paper, we consider scenes with a simulated robot in a tabletop domain with blocks characterized by three properties (color, shape, and size). The robot's objective is to plan an action sequence to achieve the desired arrangement (i.e., configuration) of blocks in collaboration with human participants (if available). The perception and actuation challenges are abstracted away to focus on the representation and learning challenges.

3.1   Architecture Overview

Figure 1 shows the proposed architecture. The robot is initially provided some domain knowledge in the form of objects (and their properties), relations between the objects, and some rules governing the domain dynamics. This domain knowledge is encoded in the ASP knowledge base (KB). Incremental learning of the (previously unknown) rules is formulated as an RL problem, and the current beliefs encoded in KB are used to define the components of the RL formulation that supports the use of high-level feedback (e.g., positive or negative reinforcement) that can be provided even by non-expert humans. This formulation and the action policy computed by RL are used to discover previously unknown rules governing the domain dynamics. These rules are encoded in the ASP KB and used for planning action sequences for subsequent tasks. Although the architecture's components are described below for the tabletop domain, it is applicable to other human-robot collaboration domains.

3.2   ASP Knowledge Base

ASP is a declarative language that can represent recursive definitions, defaults, causal relations, special forms of self-reference, and other language constructs that occur frequently in non-mathematical domains, and are difficult to express in classical logic formalisms [3]. ASP is based on the stable model semantics of logic programs; it builds on the research in non-monotonic logics and disjunctive databases [7].

The syntax, semantics and representation of the transition diagram of our illustrative domain are described in an *action language* AL [7]. Action languages are formal models of parts of natural language used for describing transition diagrams. AL has a sorted signature containing three *sorts*: *statics*, *fluents* and *actions*. Statics are domain properties whose truth values cannot be changed by actions, while fluents are properties whose truth values are changed by actions. Actions are defined as a set of elementary actions that can be executed in parallel. A domain property p or its negation $\neg p$ is a domain literal. AL allows three types of statements:

a **causes** l_{in} **if** $p_0, \ldots, p_m$	(Causal law)
l **if** $p_0, \ldots, p_m$	(State constraint)
impossible $a_0, \ldots, a_k$ **if** $p_0, \ldots, p_m$	(Executability condition)

where a is an action, l is a literal, l_{in} is an inertial fluent literal, and $p_0, \ldots, p_m$ are domain literals. The causal law states that action a causes inertial fluent literal l_{in} if the literals $p_0, \ldots, p_m$ hold true. A collection of statements of AL forms a system description.

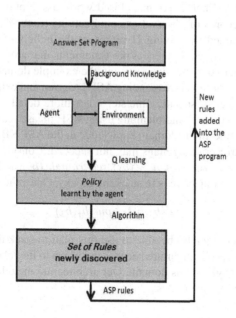

Fig. 1. Proposed Architecture: the closed loop of knowledge representation, reasoning and reinforcement learning enables discovery of new rules and their use in subsequent tasks

The domain representation consists of a system description $\mathscr{D}$ and history $\mathscr{H}$. $\mathscr{D}$ consists of a sorted signature and axioms used to describe the transition diagram τ; $\mathscr{H}$ stores the history of actions executed and observations received. The sorted signature is a tuple that defines the names of objects, functions, and predicates available for use in the domain. The sorts of the tabletop domain include: *block, location, color, shape,* and *size.* The domain's fluent: $on(block, location)$, defined in terms of the arguments, states that a specific block is at a specific location; this is an inertial fluent that obeys the laws of inertia. There are some defined fluents for block properties: $blockColor(block, color)$, $blockShape(block, color)$ and $blockSize(block, color)$. The action $put(block, location)$ puts a block at a specific location (*table* or another block). The dynamics are defined in terms of causal laws such as:

$$put(b_1, loc_1) \textbf{ causes } on(b_1, loc_1)$$

state constraints such as:

$$\neg on(b_1, loc_1) \textbf{ if } on(b_1, loc_2), \ loc_1 \neq loc_2$$

and executability constraints such as:

$$\textbf{impossible } put(b_1, loc_1) \textbf{ if } on(b_2, loc_1), \ b_2 \neq b_1$$

The domain representation $(\mathscr{D}, \mathscr{H})$ is translated into an ASP program Π, i.e., a collection of statements describing domain objects and relations between them. Π consists of the causal laws of $\mathcal{D}$, inertia axioms, closed world assumption for defined fluents, reality checks, and records of observations amd actions from $\mathcal{H}$. The ground literals in an *answer set* obtained by solving Π represent the beliefs of an agent associated with program Π. Program consequences are statements that are true in all such belief sets. Tasks such as planning and diagnostics in the example domain can be reduced to computing answer sets of the corresponding ASP program and extracting the sequence of actions to be executed at specific time steps. For more details about the translation (from AL to an ASP program) and planning using ASP, please see [7].

It is difficult to encode all the domain knowledge in the ASP KB. However, if the KB is incomplete, the corresponding plans may not succeed. Consider the scene in which three blocks of the same size: *red square* (b_1), *red triangle* (b_2), and *blue rectangle* (b_3), are on the table, and the objective is to stack the blocks. One valid plan is:

$$put(b_3, b_2), \quad put(b_1, b_3)$$

The robot should (theoretically) be able to use this plan to stack the blocks. However, execution of this plan results in failure because unknown to the robot, a block cannot be placed on top of a triangle in this domain. Our architecture includes an RL component to incrementally discover such rules.

3.3 RL Formulation and Rule Learning

A reinforcement learning problem is represented by the tuple $\langle S, A, T, R \rangle$, whose entries correspond to: a set of states, a set of actions, an unknown state transition function ($T : S \times A \times S' \rightarrow [0,1]$) and an unknown real-valued reward function ($R : S \times A \times S' \rightarrow \Re$). The objective is to find a policy $\pi^* : S \rightarrow A$ that maximizes the cumulative expected reward over a planning horizon. For the tabletop domain:

- States are the different configurations of blocks on the table. Constraints in the ASP KB eliminate impossible states. The desired configuration is the goal state.
- Actions move a block between locations, resulting in a state transition. Constraints in the ASP KB eliminate impossible actions and state transitions.
- The state transition and reward function are unknown to the robot; they are designed in the simulator to mimic the robot's interaction with the real world.
- The reward function provides a large utility (i.e., positive value) for achieving the desired configuration of blocks; a large negative utility is provided to actions that do not produce the desired effect.

For interactions in the real world, rewards will be assigned based on the robot's observations of action outcomes and the feedback provided by humans (when available).

Algorithm 1 summarizes the steps to discover new rules; it is based on the observation that actions that have much lower utility than other actions did not provide desired outcomes, and thus should not (or cannot) be performed. The algorithm takes as input the domain knowledge encoded in the ASP KB. To generate additional samples, new scenes may be created by randomly changing the property values of blocks in the scene. This optional step may be omitted to limit exploration and/or for scenes with a small number of blocks. For each scene, the robot generates components of the RL

Algorithm 1. Algorithm to discover domain rules.

Input: Domain knowledge in ASP KB; N=1.
Output: ASP KB with newly discovered rules.

1 Generate scenes (and initial and goal states) by randomly changing the property values of blocks in the scene; N= no. of scenes. // `optional step`
2 **for** $i \in [1,N]$ **do**
3 | Determine components of RL formulation from ASP KB.
4 | Learn policy π_i using Q-learning [15,17].
5 | Create table Tb_i that stores the relative (numerical) utility of combinations of property values of the blocks.
6 | **for** *each state* $s \in S$ **do**
7 | | Identify the best action (a_{best}) and worst action (a_{worst}) by selecting the highest and lowest (Q)values of actions corresponding to s.
8 | | Identify property values of the blocks involved in the execution of a_{best} and a_{worst}.
9 | | Increment the utility of entries (i.e., rows) in Tb_i that correspond to the property values identified for a_{best}.
10 | | Decrement the utility of entries in Tb_i that correspond to the property values identified for a_{worst}.
11 | **end**
12 **end**
13 $Tb_{total} = \Sigma_i Tb_i$.
14 Convert rows in Tb_{total} with larger negative utilities than other rows to ASP rules.
15 Merge with existing ASP rules and generalize.
16 **return** *ASP KB with new rules.*

---

problem and learns a policy to achieve the desired configuration using Q-learning; this learning may be achieved using a combination of simulation and real-world trials. A table is created whose rows correspond to combinations of property values of blocks that can be involved in a *put* action. For each state in the set of states, the best action and the worst action are selected based on the computed policy, and the property values of blocks corresponding to these actions are identified. Entries in the table corresponding to the property values identified for action a_{best} (a_{worst}) have their relative utilities incremented (decremented). If multiple tables were created, they are summed up and the entries in the resultant table corresponding to large negative utilities are considered to represent actions that should not occur. The corresponding rules are encoded and merged with existing rules in the ASP KB, and used in subsequent planning tasks.

4   Experimental Setup and Results

In the tabletop domain, the robot's objective is to stack blocks with different properties in desired configurations (Section 3). We use the knowledge representation language SPARC to write the ASP programs [2]; it expands CR-Prolog that includes consistency restoring rules in ASP [7], and uses DLV [13] to obtain answer sets.

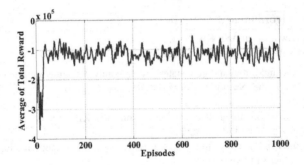

Fig. 2. Learning curve for three blocks with different shape (*square, rectangle, triangle*) but same color and size

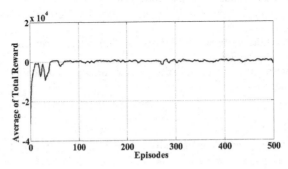

Fig. 3. Learning curve for three blocks with same shape (*triangle*), color, and size

4.1   Learning Curves

For Q-learning, we experimentally chose a learning rate $\alpha = 0.01$ and a discount factor $\gamma = 0.8$; these parameters influence the extent to which previously unseen regions of the state-action space are explored. The corresponding learning curves, convergence rates, and the average rewards are different based on the property values (and the number) of the blocks. Figure 2 and Figure 3 show learning curves obtained for two different scenes with blocks of the same size: (1) three blocks with different shapes but same color; (2) three triangles of the same color; the objective is to stack the blocks on top of each other. The curves are different because each initial and desired configuration of blocks determine the possible state transitions, actions and thus the rewards obtained during the learning phase. In all scenes, the policies are learned incrementally and efficiently.

4.2   Discovering New Rules

Once one or more policies are obtained, a table is constructed; each row of the table represents possible property values for two blocks involved in a *put* action. Table 1 shows an illustrative example of such a table, using a policy generated for the example in Section 3.2 with three blocks: *red square* (b_1), *red triangle* (b_2), and *blue rectangle* (b_3). The column of relative utilities ("Utility" in Table 1) is initialized to contain zeros.

The subsequent steps in Algorithm 1 are executed to update values in the table. For instance, for the best action in a specific state (a_{best}), the corresponding property values are identified and the utilities of the appropriate rows in the table are incremented.

Table 1. Table of relative utilities for a domain with three blocks of the same size (red square, red triangle and blue rectangle). The numbers in the last column represent the relative utility of placing a block of a specific shape (column 1) on another specific-shaped block (column 2) when they are of the same color or different color.

Shape1	Shape2	Color	Utility
Square	Square	Different	0
Square	Triangle	Different	0
Square	Rectangle	Different	5
Triangle	Square	Different	0
Triangle	Triangle	Different	0
Triangle	Rectangle	Different	10
Rectangle	Square	Different	-5
Rectangle	Triangle	Different	-20
Rectangle	Rectangle	Different	0
Square	Square	Same	0
Square	Triangle	Same	-10
Square	Rectangle	Same	0
Triangle	Square	Same	10
Triangle	Triangle	Same	0
Triangle	Rectangle	Same	0
Rectangle	Square	Same	0
Rectangle	Triangle	Same	0
Rectangle	Rectangle	Same	0

In Table 1, we observe that the largest negative utility corresponds to an action that would place a rectangle on top of a triangle of a different color. The next lowest utility corresponds to placing a square on top of a triangle with same color. The first observation can be translated into an ASP rule:

$$\neg occurs(put(b_1, b_2)) : -blockShape(b_1, rectangle), \quad blockShape(b_2, triangle).$$
$$blockColor(b_1, C_1), \quad blockColor(b_1, C_2), \quad C_1! = C_2.$$

Each such rule is merged with existing rules by matching common predicates and generalizing across different groundings of a predicate, and the ASP KB is revised. Over multiple experimental trials, the robot may determine, for instance, that *no block can be placed on a triangle*:

$$\neg occurs(put(b_1, b_2)) : -blockShape(b_2, triangle).$$

For the task of stacking the three blocks (in Section 3.2), the revised ASP program produces the new plan:

$$put(b_3, b_1), \quad put(b_2, b_3)$$

Unlike the previous attempt (in Section 3.2) before using RL to discover rules, executing this action sequence results in the robot successfully stacking the three blocks.

4.3 Discussion and Future Work

We have evaluated our architecture on scenes with different number of blocks that have different properties and property values. The experiments indicate that the robot is able to incrementally and efficiently identify rules governing the domain dynamics. As the robot acquires more (accurate) domain knowledge, the ability to use the corresponding plans to complete the assigned tasks increases (and approaches 100%). Although the architecture is demonstrated in a simplistic domain in this paper, it addresses key knowledge representation and learning challenges in robotics. The architecture is thus applicable to other domains in which robots collaborate with non-expert humans.

This architecture opens up multiple directions for future research. It is non-trivial to generalize from the discovered rules and revise existing rules in the KB. Although the domain considered in this paper simplifies this problem, it is an interesting topic for further investigation. Another direction for future research is to explore the architecture's extension to support formulations similar to relational reinforcement learning (RRL). However, unlike existing RRL formulations, the use of ASP will support commonsense reasoning while the robot uses abstractions across different states and actions to make learning computationally tractable. It may also be possible to use the diagnostics capabilities of ASP to focus on a specific subset of the state-action space (in the RL formulation) in response to the failure of specific steps in the plan. Finally, the current implementation abstracts away the perception and actuation challenges in robotics. For physical robots in real world application domains, we are investigating the architecture's extension to partially observable states and non-deterministic action outcomes, using probabilistic belief states in the RL formulation.

5 Conclusions

This paper described an architecture that integrates the complementary strengths of RL and declarative programming to support knowledge representation, commonsense reasoning, and incremental discovery of unknown rules governing the domain dynamics. The domain knowledge encoded in the ASP KB is used to formulate the incremental discovery of domain rules as an RL problem. The action policies obtained by RL are used to discover rules that are, in turn, encoded in the ASP KB and used to plan action sequences for subsequent tasks. This architecture is thus a significant step towards the long-term objective of designing robots that can collaborate with and assist non-expert humans in real world application domains.

Acknowledgments. The authors thank Michael Gelfond for his feedback on the architecture described in this paper. This research was supported in part by the U.S. Office of Naval Research (ONR) Science of Autonomy Award N00014-13-1-0766. Opinions and conclusions are those of the authors and do not necessarily reflect the views of the ONR.

References

1. Aerolla, M.: Incorporating Human and Environmental Feedback for Robust Performance in Agent Domains. Master's thesis, Department of Computer Science, Texas Tech University (May 2011)
2. Balai, E., Gelfond, M., Zhang, Y.: Towards Answer Set Programming with Sorts. In: Cabalar, P., Son, T.C. (eds.) LPNMR 2013. LNCS, vol. 8148, pp. 135–147. Springer, Heidelberg (2013)
3. Baral, C.: Knowledge Representation, Reasoning and Declarative Problem Solving. Cambridge University Press (2003)
4. Blumberg, B., Downie, M., Ivanov, Y., Berlin, M., Johnson, M.P., Tomlinson, B.: Integrated Learning for Interactive Synthetic Characters. In: International Conference on Computer Graphics and Interactive Techniques (SIGGRAPH), pp. 417–426 (2002)
5. Dzeroski, S., Raedt, L.D., Driessens, K.: Relational Reinforcement Learning. Machine Learning 43, 7–52 (2001)
6. Erdem, E., Aker, E., Patoglu, V.: Answer Set Programming for Collaborative Housekeeping Robotics: Representation, Reasoning, and Execution. Intelligent Service Robotics 5(4), 275–291 (2012)
7. Gelfond, M., Kahl, Y.: Knowledge Representation, Reasoning and the Design of Intelligent Agents. Cambridge University Press (2014)
8. Griffith, S., Subramanian, K., Scholz, J., Isbell, C., Thomaz, A.: Policy Shaping: Integrating Human Feedback with Reinforcement Learning. In: International Conference on Neural Information Processing Systems, Lake Tahoe, USA (2013)
9. Kaplan, F., Oudeyer, P.-Y., Kubinyi, E., Miklosi, A.: Robotic Clicker Training. Robotics and Autonomous Systems 38 (2002)
10. Knox, W.B., Fasel, I., Stone Design, P.: principles for creating human-shapable agents. In: AAAI Spring 2009 Symposium on Agents that Learn from Human Teachers (2009)
11. Knox, W.B., Stone, P.: Tamer: Training an Agent Manually via Evaluative Reinforcement. In: International Conference on Development and Learning, ICDL (2008)
12. Knox, W.B., Stone, P.: Combining Manual Feedback with Subsequent MDP Reward Signals for Reinforcement Learning. In: International Conference on Autonomous Agents and Multiagent Systems, AAMAS (2010)
13. Leone, N., Pfeifer, G., Faber, W., Eiter, T., Gottlob, G., Perri, S., Scarcello, F.: The DLV System for Knowledge Representation and Reasoning. ACM Transactions on Computational Logic 7(3), 499–562 (2006)
14. Sridharan, M.: Augmented Reinforcement Learning for Interaction with Non-Expert Humans in Agent Domains. In: International Conference on Machine Learning Applications, ICMLA (December 2011)
15. Sutton, R.S., Barto, A.G.: Reinforcement Learning: An Introduction. MIT Press (1998)
16. Thomaz, A., Breazeal, C.: Reinforcement Learning with Human Teachers: Evidence of Feedback and Guidance with Implications for Learning Performance. In: National Conference on Artificial Intelligence, AAAI (2006)
17. Watkins, C., Dayan, P.: Q-learning. Machine Learning 8, 279–292 (1992)
18. Zhang, S., Sridharan, M., Gelfond, M., Wyatt, J.: Integrating Probabilistic Graphical Models and Declarative Programming for Knowledge Representation and Reasoning in Robotics. In: Planning and Robotics (PlanRob) Workshop at ICAPS, Portsmouth, USA (2014)

Robot Pressure:
The Impact of Robot Eye Gaze
and Lifelike Bodily Movements
upon Decision-Making and Trust

Christopher Stanton and Catherine J. Stevens

MARCS Institute,
University of Western Sydney
c.stanton@uws.edu.au, kj.stevens@uws.edu.au
http://marcs.uws.edu.au

Abstract. Between people, eye gaze and other forms of nonverbal communication can influence trust. We hypothesised similar effects would occur during human-robot interaction, predicting a humanoid robot's eye gaze and lifelike bodily movements (eye tracking movements and simulated "breathing") would increase participants' likelihood of seeking and trusting the robot's opinion in a cooperative visual tracking task. However, we instead found significant interactions between robot gaze and task difficulty, indicating that robot gaze had a positive impact upon trust for difficult decisions and a negative impact for easier decisions. Furthermore, a significant effect of robot gaze was found on task performance, with gaze improving participants' performance on easy trials but hindering performance on difficult trials. Participants also responded significantly faster when the robot looked at them. Results suggest that robot gaze exerts "pressure" upon participants, causing audience effects similar to social facilitation and inhibition. Lifelike bodily movements had no significant effect upon participant behaviour.

Keywords: human-robot interaction, nonverbal communication, eye gaze, trust, compliance, persuasion.

1 Introduction

In coming years, it is expected that social robots will become increasingly common, assisting and collaborating with people in a wide variety of environments such as public spaces, the home, office, school, and health care. For such human-robot collaborations to be successful, social robots must be capable of fostering the trust and confidence of people they interact with. Between people, nonverbal communication plays a significant role in establishing rapport and influencing others. For example, doctors who sit with uncrossed legs with arms symmetrically side-by-side are rated more highly by patients [1], mirroring another's posture can increase rapport within groups [2], hand shaking has been shown to

M. Beetz et al. (Eds.): ICSR 2014, LNAI 8755, pp. 330–339, 2014.

increase compliance when requesting money [3], and eye gaze has been shown to increase likability, request compliance, and perceptions of truthfulness [4]. Thus, it is important to investigate whether nonverbal communication can have similar effects in interactions between robots and people.

Trust and rapport is also affected by a person's appearance. Initial judgments of a political candidate's facial appearance can predict the outcomes of political elections [5], while positive characteristics such as intelligence, competence, leadership, and trustworthiness are attributed to attractive persons [6]. Perhaps most importantly, in the context of human-robot interaction (HRI), is that people are most likely to cooperative with and trust others who are physically similar to themselves [7,8], thus providing clues for the physical design of humanoid robots. While some attention has been paid to humanoid robot form and appearance, especially with regards to androids (e.g. the uncanny valley [9]), less attention has been devoted to investigating the impact of robots imitating "human-like" movements during HRI, such as shifting postures, blinking or breathing. In this study we investigate the influence of robot eye gaze and two different "lifelike" bodily movements upon participants' willingness to trust and interact with the robot during a cooperative visual task.

2   Nonverbal Communication, Trust and HRI

A large body of research has discovered how particular forms of human to human nonverbal communication can influence trust, perceptions of truthfulness, and rapport [10]. For example, leaning forward, using eye gaze, nodding, and smiling can all help build rapport [11]. Even the nature of a smile can provide an indication of whether a person is telling the truth [12].

Gaze, in particular, is a powerful nonverbal cue, with every culture having strict but unstated rules governing eye contact [13]. Gazing at the eyes of another can signal willingness to interact [14]. When people first meet, gaze enhances attraction and liking [4]. In court rooms, witnesses are viewed as more credible when they employ eye gaze [15]. People who avert gaze are more likely to be perceived as lying [16]. However, liars actually increase eye contact [17], a cunning ploy playing on the widespread belief that liars avert eye gaze [18]. Gaze can also impact the likelihood of people complying with a request. People on the street are more likely to take a leaflet offered by a person who looks them in the eye [19], hitchhikers have more success in finding a ride when they gaze at drivers [20], and eye gaze can increase the amount of money people are willing to donate to charity [21].

HRI research concerning nonverbal communication has generally replicated the findings of human-human interaction research. For example, an android mirroring the posture of its human interaction partner increased the partner's ratings of likability towards the robot [22]. Between people, students who receive eye gaze have better recollections of a story told to them by their teacher [23], and a similar effect was found when people were told a story by a robot [24]. Gaze has been shown to increase the persuasiveness of a story-telling robot [25], and people are more likely to comply with a robot's suggestions when it uses nonverbal cues such as gaze and gesture [26]. Furthermore, it has been demonstrated that people respond to a humanoid robot's trust-relevant nonverbal signals (such as crossing the arms and leaning away) in the same manner as they respond to similar signals from people [27].

With regards to the nature of robotic movement and its ability to influence compliance, trust and perceptions of capability, this research question remains largely unexplored. A meta-analysis of trust in HRI found that although reliable and predictable task performance was the most important factor, robot anthropomorphism could also influence trust [28]. In a virtual environment where participants are represented by avatars with no movement, a lifelike avatar resulted in a poor social interaction as the degree of realism portrayed by the avatar raised participants' expectations about its' capabilities [29]. In a study using simulated robots in immersive virtual environments, where participants viewed smooth versus trembling motions of a robot performing a physical manipulation task, participants rated the smooth moving robot more trustworthy. However, in a second interactive experiment with the virtual robot, motion fluency had no impact upon trustworthiness [30].

2.1 Hypotheses

In the current study, participants complete repeated trials of a cooperative visual tracking task (the "shell game") with a humanoid robot, with trial difficulty ranging from easy to very hard. The robot acts as an assistant to the participant, with participants able to ask the robot for help, while on occasion the robot will volunteer an answer.

Hypothesis 1. As previous research indicates eye gaze can increase compliance and persuasion, and is also associated with truthfulness, we predict robot eye gaze will increase the likelihood of participants changing their answer to the robot's suggested answer.

Hypothesis 2. As eye gaze is a cue for indicating interest in another and willingness to interact, we predict robot eye gaze will increase the likelihood of participants asking the robot for assistance.

Hypothesis 3. While largely exploratory in nature, we hypothesise robot "lifelike" bodily movements will increase the likelihood of participants changing their answer to the robot's suggested answer due to these movements positively influencing participants' perceptions of the robot's capabilities.

Hypothesis 4. As task difficulty increases and participants become more unsure of the correct response, participants will be more likely to ask the robot for help and trust the robot's opinion.

3   Method

Experimental Design. A mixed design (2x2x2x4) was employed, with within-subjects variables Eye Gaze (2 levels, On/Off) and Task Difficulty (4 levels), and between-subject variables Breathing and Eye Tracking (both 2 levels, On/Off).

Independent Variables. Task Difficulty (four levels, ranging from easy to very hard) was manipulated to prevent ceiling and floor effects, and to aid in participant vigilance. Three robot behaviours were manipulated, described below.

Eye Gaze. When asking for the participant's answer, if Eye Gaze was On the robot would look directly at the participant (direct gaze). If Eye Gaze was Off, the robot would look at the monitor displaying the shell game (averted gaze). Eye Gaze On versus Off was randomised across 50% of trials.

Eye Tracking. During the cup shuffling process, if Eye Tracking was On the robot's head would move to create the appearance of tracking one of the moving cups. When Eye Tracking was Off, the robot's head would not move, and instead face the centre of the monitor displaying the shell game.

Breathing. When Breathing was On, the robot's body was never completely still, and instead it would rhythmically oscillate between two very similar poses to create the appearance of breathing. When Breathing was Off, the robot's body was still.

Stimuli. Participants played a graphical computerised version of the classic "shell game", in which an object is hidden under one of three cups, and those cups are quickly shuffled to create doubt and uncertainty as to the true location of the object (see Figure 1). Game trials were comprised of 4 levels of difficulty, ranging from easy to very difficult, with difficulty determined by the speed of cup movement (Slow, Medium, Fast, Very Fast). At total of 48 trials (12 trials of each level of difficulty) were presented to each participant, randomised for difficulty. No feedback was given to the participant regarding whether their answers were correct or incorrect after each trial, but a score update was displayed after every 12 trials for the purpose of keeping the participant interested in the game.

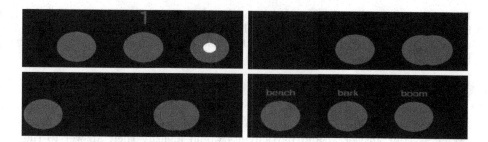

Fig. 1. Screen shots of the shell game stimuli. *Top left*: the game would initiate with a "3, 2, 1" countdown (countdown at time "1" is displayed), with the object of interest displayed as a white circle. *Top right & bottom left*: When the game begins the white circle disappears, and the cups are shuffled horizontally with overlap, occlusion and changes of direction creating doubt as to the object's true location. *Bottom right*: When the cups stop moving after 4 seconds words appear above each cup to identify the different cups.

Participants. A total of 59 first year psychology students, 51 female and 8 male, ranging in age from 18 to 49 years ($M=22.4$ years, $SD=7.1$ years), participated in the experiment in return for course credit.

Procedure. A cover story was used, with participants told the purpose of the experiment was to test the robot's vision and speech recognition systems, and that their participation would allow benchmarking of the robot's vision system against human performance. Participants were told to treat the robot as a team member, and they should aim to achieve the highest possible team score. In truth, the robot was controlled using a "Wizard of Oz" set up to which the participant was blind to. The robot, an Aldebaran Nao, sat on a chair-like box, with a computer mouse on either side of the robot, with the robot clicking a mouse button with its hand after each trial to create the illusion of logging the participants' answers. Participants sat facing the shell game display, with the robot situated to the left of the participant in a position that allowed the robot to move its head to either look at the shell game or the participant.

Fig. 2. Experimental setup. The Aldebaran Nao humanoid robot sits on a "chair" between the participant and game stimuli. In the picture displayed Eye Gaze is Off as the robot is looking at the shell game (rather than the participant) when asking for the participant's answer.

For each trial, the cup shuffling process took four seconds, after which a one syllable word appeared above each cup. The robot would ask the participant "What is your answer?", and participants would identify their answer to the robot using the word that appeared above the cup they believed to be hiding the object. Participants were informed they could ask the robot for help using key phrases such as "What do you think?" or "I don't know, please help me". Furthermore, on a total of 16 randomised trials per participant the robot was programmed to either help (8 trials) or deliberately hinder (8 trials). When helping the participant, if the participant had stated the correct answer the robot would say "I agree", while if the participant had given an incorrect answer the robot would say "Are you sure? I think it is <*correct answer*>. What is your final answer?". When hindering the participant, the robot would say "Are you sure? I think it is <*incorrect answer*>. What is your final answer?".

Dependent Variables. The following data were recorded: the frequency with which each participant asked the robot for help; the frequency with which each participant changed their answer to the robot's answer when it differed to their own; task accuracy (i.e. did the participant choose the correct answer); and the time taken by each participant to provide each answer.

4   Results

A total of 2829 trials were conducted (59 participants, 48 trials per participant, and 3 trials were discarded due to technical problems). Each participant's response means were calculated and mixed repeated measures analyses of variance (ANOVA) were conducted with Breathing and Eye Tracking as between-subjects factors and Eye Gaze and task Difficulty as within-subject factors.

Trusting the Robot's Opinion. As expected (**H4**), a main effect of task Difficulty was found, $F(3,324)=5.4, p=.001$, with participants more likely to change their answer to the robot's as cup movement speed increased. On the easiest difficulty level participants accepted the robot's advice on 16.4% of trials ($SD=.335$) versus 32.3% of the hardest trials ($SD=.418$). There was a significant interaction between Eye Gaze and Difficulty, $F(3,324)=2.827$, $p=0.039$, with participants more likely to trust the robot's opinion when it gazed at them on the hardest trials, but less likely to trust the robot on all easier difficulties (see Figure 3). There were no significant effects related to Eye Tracking or Breathing. The hypothesis (**H1**) that participants would trust the robot more when the robot gazed at them was not supported, nor was the hypothesis (**H3**) lifelike bodily movements would increase trust towards the robot.

Asking for the Robot's Opinion. As hypothesised (**H4**), a main effect of Difficulty was found, $F(3,162)=16.535$, $p=.000$, with participants asking for Nao's opinion more often as the speed of cup movement increased. On Easy trials participants asked for help on 9.9% of trials ($SD=.203$) compared to 25.0% of Very Hard trials ($SD=.286$). There was a significant interaction between Eye Gaze and Difficulty, $F(3,162)=5.424$, $p.=.001$, with participants more likely to ask the robot for help with Eye Gaze for Fast trials, but less likely for Medium trials. To further understand this interaction between Eye Gaze and Difficulty, a second ANOVA was conducted in which task Difficulty was determined not by speed of cup movement, but by grouping trials into quartiles using accuracy means. Using this new measure of task difficulty there was a significant main effect of Eye Gaze $F(1,54)=4.826$, $p=0.032$, with participants asking for help more often when Eye Gaze was On as opposed to Off (see Figure 4). Thus, there is some, but not unequivocal, support for the hypothesis (**H2**) participants would be more likely to ask the robot for help when the robot looks at them.

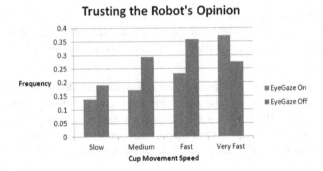

Fig. 3. Results for Eye Gaze and Trusting the Robot's Opinion. A significant interaction between Eye Gaze and Task Difficulty (cup movement speed) was found, with participants less likely to trust the robot when Eye Gaze is used on Slow, Medium and Fast trials, but more likely to trust the robot on Very Fast trials.

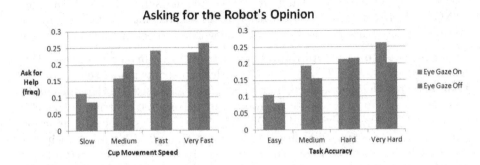

Fig. 4. Results for Eye Gaze and Asking for the Robot's Opinion. *Left*: difficulty is determined by cup movement speed, with a significant interaction between Eye Gaze and Cup Speed. *Right*: difficulty is determined by dividing the 48 trials into quartiles using each trial's accuracy mean. A significant main effect of Eye Gaze was found.

Task Performance. Two measures of task accuracy were used: 1) participants' initial answers, excluding answers which were changed in response to robot advice; 2) participants' final answers. For initial answers, there was a significant interaction between Difficulty and Eye Gaze, $F(3,162)=39.348, p=.000$, with participants more likely to choose the correct answer on easier trials when the robot looked at them, but less likely to choose the correct answer on harder trials when the robot looked at them. For participants' final answers, the same significant interaction between Eye Gaze and Difficulty was obtained, $F(3,162)=28.487$, $p=.000$. Results are shown in Figure 5.

Response Time. A main effect of Eye Gaze was found, $F(1,54)=24.73$, $p=.000$, with participants on average 0.6 seconds quicker to answer when Eye Gaze is On ($M=6.79, SD=4.05$) as opposed to Off ($M=7.39$, $SD=4.48$). A significant interaction was found between Difficulty and Eye Gaze, $F(3,175)=4.012$, $p=.008$, with the effect of eye gaze upon trial duration increasing as task difficulty

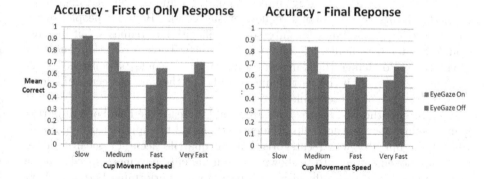

Fig. 5. A significant interaction between Eye Gaze and task Difficulty (cup movement speed) upon participants' accuracy was found. Eye Gaze has little effect on the easiest trials, assists performance on Medium trials, and hinders performance on the more difficult Fast and Very Fast trials.

increases. For example, the difference between Eye Gaze On and Off for Easy trials is just 0.2 seconds, but for Very Hard trials participants are on average 0.74 seconds quicker to respond when Eye Gaze is On.

5   Discussion

Eye Gaze had two unpredicted but powerful effects upon participant decision-making and behaviour. Firstly, robot gaze impacted participant performance, with direct gaze improving participant performance on easier trials, but hindering it on more difficult trials. We postulate this was caused by robot gaze creating "pressure" and anxiety in participants, generating audience effects similar to social facilitation and inhibition - a well researched effect in which people, when in the presence of others as compared to alone, perform better at easy tasks but worse at difficult tasks [31]. While social facilitation is usually studied as an effect of mere presence (as opposed to eye gaze), there is evidence that direct gaze versus averted gaze can induce social facilitation effects [32]. Furthermore, social facilitation arising from robot presence has been observed [33]. The notion of "robot pressure" is supported by response times, with participants markedly quicker to respond to the robot when the robot gazed at them. Interestingly, robot gaze occurs after the trial has completed but before the participant has provided their answer to the robot, demonstrating robot gaze is causing participants to doubt and rethink their initial response on difficult trials. A practical implication of these findings is that when people are performing difficult tasks or making difficult decisions, it may be best for robots to look the other way.

We hypothesised robot eye gaze would increase the likelihood of participants trusting the robot's opinion. Instead, a significant interaction was found between eye gaze and task difficulty, with participants more likely to comply with the robot's suggested answer when it gazed at them on the hardest trials, but conversely on easier trials direct gaze reduced trust. This suggests robot gaze can have either a positive or negative impact upon trust and compliance, depending upon the nature of the robot's request or suggestion. Between people, direct gaze can reduce compliance for unreasonable, illegitimate requests, but increase compliance for reasonable, legitimate requests [34]. Thus, a robot's request for a participant to change their answer on an easy trial could be construed as illegitimate, especially if the participant is confident they are correct, while for a difficult trial the opposite would be true.

We also hypothesised that robot gaze would increase the likelihood of participants asking the robot for help. Evidence was found to support this hypothesis when task difficulty was recategorised using quartile accuracy means, rather than cup movement speed. As shown in Figure 5, cup movement speed is not a perfect indicator of task difficulty, with participants performing better on Very Fast trials as opposed to Fast trials, highlighting an area for improvement when developing future shell game stimuli.

No support was found for the hypothesis that a humanoid robot's lifelike bodily movements of "breathing" and "eye tracking" would make participants more likely to trust the robot's judgments in a visual tracking task. During debriefing many participants reported they failed to notice the robot's eye tracking behaviour in their peripheral vision as they were focused on the shell game, perhaps explaining the absence of effects. While many participants reported noticing the robot's breathing motion, it had no impact on their behaviour.

Lastly, as task difficulty increased, participants were more likely to ask the robot for help and more likely to trust the robot's suggested answer, demonstrating people are willing to accept a robot's advice when making difficult decisions.

References

1. Harrigan, J.A., Oxman, T.E., Rosenthal, R.: Rapport expressed through nonverbal behavior. Journal of Nonverbal Behavior 9(2), 95–110 (1985)
2. Lafrance, M., Broadbent, M.: Group Rapport: Posture Sharing as a Nonverbal Indicator. Journal of Group & Organization Management 1(3), 328–333 (1976)
3. Guéguen, N.: Handshaking and compliance with a request: a door-to-door setting. Social Behavior & Personality 41(10), 1585–1588 (2013)
4. Kleinke, C.L.: Gaze and eye contact: a research review. Psyc. Bull. 100, 78–100 (1986)
5. Olivola, C.Y., Todorov, A.: Elected in 100 milliseconds: Appearance-based trait inferences and voting. Journal of Nonverbal Behavior 34(2), 83–110 (2010)
6. Feingold, A.: Good-looking people are not what we think. Psychological Bulletin 111(2), 304–341 (1992)
7. DeBruine, L.M.: Facial resemblance enhances trust. Proceedings of the Royal Society B: Biological Sciences 269, 1307–1312 (2002)
8. Farmer, H., McKay, R., Tsakiris, M.: Trust in Me: Trustworthy Others are Seen as More Physically Similar to the Self. Psych. Sci. 25(1), 290–292 (2013)
9. Mori, M.: The Uncanny Valley. Trans. by MacDorman, K.F. and Kageki, N. IEEE Robotics & Automation Magazine 19(2), 98–100 (2012)
10. Feldman, R.S.: Fundamentals of nonverbal behavior. Cambridge Uni Press (1991)
11. Tickle-Degnen, L., Rosenthal, R.: The Nature of Rapport and Its Nonverbal Correlates. Psychological Inquiry 1(4), 285–293 (1990)

12. Ekman, P., Friesen, W.V., O'Sullivan, M.: Smiles when lying. Journal of Personality & Social Psychology 54, 414–420 (1988)
13. DeVito, J.A.: Human Communication: The Basic Course. Pearson (2011)
14. Goffman, E.: Behavior in public places: Notes on the social organization of gatherings. Free Press, New York (1963)
15. Hemsley, G.D., Doob, A.N.: The Effect of Looking Behavior on Perceptions of a Communicator's Credibility. J. of Applied Social Psyc. 8(2), 1559–1816 (1978)
16. Taylor, R., Hick, R.F.: Believed cues to deception: Judgements in self-generated serious and trivial situations. Legal & Criminological Psychology 12, 321–332 (2007)
17. Mann, S., Vrij, A., Leal, S., Granhag, P.A., Warmelink, L., Forrester, D.: Windows to the Soul? Deliberate Eye Contact as a Cue to Deceit. Journal of Nonverbal Behavior 36, 205–215 (2012)
18. Mann, S., Ewens, S., Shaw, D., Vrij, A., Leal, S., Hillman, J.: Lying Eyes: Why Liars Seek Deliberate Eye Contact. Psychi., Psycho. & Law 20(3), 452–461 (2013)
19. Kleinke, C.L., Singer, D.A.: Influence of Gaze on Compliance with Demanding and Conciliatory Requests in a Field Setting. Personality & Social Psych. Bull. 5, 386–390 (1979)
20. Snyder, M., Grether, J., Keller, K.: Staring and compliance: A field experiment on hitchhiking. Journal of Applied Social Psychology 4, 165–170 (1974)
21. Bull, R., Gibson-Robinson, E.: The influence of eye-gaze, style of dress, and locality on the amounts of money donated to a charity. Hum. Rel. 34, 895–905 (1981)
22. Shimada, M., Yamauchi, K., Minato, T., Ishiguro, H., Itakura, S.: Studying the influence of the chameleon effect on humans using an android. In: Proc. Intelligent Robots and Systems, pp. 767–772 (2008)
23. Otteson, J.P., Otteson, C.R.: Effects of teacher gaze on children's story recall. Perceptual and Motor Skills 50, 35–42 (1980)
24. Mutlu, B., Forlizzi, J., Hodgins, J.: A Storytelling Robot: Modeling and Evaluation of Human-like Gaze Behavior. In: Proc. IEEE Conf. Humanoids, pp. 518–523 (2006)
25. Ham, J., Bokhorst, R., Cabibihan, J.: The influence of gazing and gestures of a storytelling robot on its persuasive power. In: Proc. Intl. Conf. Soc. Robotics (2011)
26. Chidambaram, V., Chiang, Y.H., Mutlu, B.: Designing Persuasive Robots: How Robots Might Persuade People Using Vocal and Nonverbal Cues. In: Proc. ACM/IEEE Intl. Conf. on HRI, pp. 293–300 (2012)
27. DeSteno, D., Breazeal, C., Frank, R.H., Pizarro, D., Baumann, J., Dickens, L., Lee, J.J.: Detecting the trustworthiness of novel partners in economic exchange. Psychological Science 23, 1549–1556 (2012)
28. Hancock, P.A., et al.: A meta-analysis of factors affecting trust in human–robot interaction. Human Factors 53(5), 517–527 (2011)
29. Slater, M., Steed, A.: Meeting People Virtually: Experiments in Shared Virtual Environments. In: Schroeder, R. (ed.) The Social Life of Avatars, pp. 145–171 (2002)
30. van den Brule, R., et al.: Do Robot Performance and Behavioral Style affect Human Trust? A Multi-Method Approach. Int J. of Soc. Robotics (2014)
31. Zajonc, R.B.: Social Facilitation. Science 149, 269–274 (1965)
32. Markus, H.: The effect of mere presence on social facilitation: an unobtrusive test. Jnl. of Exp. Soc. Psyc. 14(4), 389–397 (1978)
33. Riether, N., Hegel, F., Wrede, B., Horstmann, G.: Social facilitation with social robots? In: Proc. Intl. Conf. on HRI 2012, pp. 41–48 (2012)
34. Kleinke, C.L.: Interaction between gaze and legitimacy of request on compliance in a field setting. Jnl. of Nonverbal Behav. 5(1), 3–12 (1980)

A Novel Culture-Dependent Gesture Selection System for a Humanoid Robot Performing Greeting Interaction

Gabriele Trovato[1], Martin Do[2], Masuko Kuramochi[3], Massimiliano Zecca[4],
Ömer. Terlemez[2], Tamim Asfour[2], and Atsuo Takanishi[5,6]

[1] Faculty of Science and Engineering, Waseda University, #41-304, 17 Kikui-cho,
Shinjuku-ku, Tokyo 162-0044, Japan
[2] Institute of Anthropomatics, Karlsruhe Institute of Technology, Karlsruhe, Germany
[3] Kanda Institute of Foreign Languages, Tokyo, Japan
[4] School of Electronic, Electrical and Systems Engineering, Loughborough University, UK
[5] Department of Modern Mechanical Engineering, Waseda University
[6] Humanoid Robotics Institute (HRI), Waseda University
contact@takanishi.mech.waseda.ac.jp

Abstract. In human-robot interaction, it is important for the robots to adapt to our ways of communication. As humans, rules of non-verbal communication, including greetings, change depending on our culture. Social robots should adapt to these specific differences in order to communicate effectively, as a correct way of approaching often results into better acceptance of the robot. In this study, a novel greeting gesture selection system is presented and an experiment is run using the robot ARMAR-IIIb. The robot performs greeting gestures appropriate to Japanese culture; after interacting with German participants, the selection should become appropriate to German culture. Results show that the mapping of gesture selection evolves successfully.

Keywords: Social Robotics, Culture, Gestures, Greetings, HRI.

1 Introduction

The relationship between acceptance of robots into human societies and its background culture is an idea that has been debated since long time. According to the traditional view in literature, anxiety towards robots is more common in Western countries. As a matter of fact, differences between East and West in cognition, due to differing social structures, philosophies, and educational systems, trace back to ancient Greece and China [1]. Stereotypes are not always true, as there are positive examples of robotic heroes in Western science fiction too; however, technology acceptance, for instance, depends also on the country that is the producer, since the culture of that country may bias some aspects of the product. As a consequence, localisation of products may be done [2]. In our previous research [3], a comparative study carried out with Egyptian and Japanese participants, culture-dependent acceptance and discomfort were found relating to greeting gestures of a humanoid robot. As the importance of culture-specific customisation for acceptance of robots was confirmed, the need of a system of greeting selection for robots was highlighted.

M. Beetz et al. (Eds.): ICSR 2014, LNAI 8755, pp. 340–349, 2014.
© Springer International Publishing Switzerland 2014

1.1   Greeting Interaction and Related Works

Greeting is the basic way of initiating and closing an interaction. Hoffman-Hicks [4] states that greetings function primarily as formulaic exchanges which serve to acknowledge another person's presence. We desire that robots are able to greet, same as humans. Moreover, greetings are a form of interaction where cultural differences are evident. Depending on cultural background, there can be different rules of engagement in human-human interaction, gap in recognition of facial expressions and gestures, chances of misunderstanding and difficulty in communication. It is then necessary to understand from sociology studies which factors influence greetings between humans. A unified model for greeting does not exist; therefore the study has to be done through a survey of different sources from different countries.

Intimacy and Politeness are two important keywords in sociology, and both influence the choice of a greeting gesture [5, 6]. Intimacy is apparently influenced by Physical Distance, Eye contact [7], Gender [8], Location [5] and Culture [9]. Brown and Levinson [10] were the first to think of a formula for calculating Politeness. Even though they did not define numerically any coefficient, they represented Politeness as a function of Power Relationship, Social Distance and a cultural factor.

Some humanoid robots can perform programmed greetings. Among others is ARMAR-III [11], which met the German Prime Minister Angela Merkel. ASIMO [12] is capable of performing a wider range of greetings: handshake, wave both hands, and bow, and can recognise such gestures among others. MAHRU [13] is another example of humanoid robot which can greet through a simple bow.

While greeting gestures have been programmed, to the best of our knowledge, so far only a few greeting interaction experiments with robots have been conducted to test the impression on humans. Experiments done in [14], which focused on timing, rather than on culture; and experiments featuring the social robot ApriPoco, in which data from biological signals of subjects looking at Japanese, Chinese, and French greetings were compared [15]. Our intention in this experiment is instead to make the robot choose the right greeting, rather than assess human reaction. Another important difference is that our studies were done with a human sized humanoid social robot.

1.2   Objectives of this Paper

The main idea behind this study is a typical scenario in which a foreigner in a country visited for the first time (e.g. a Westerner in Japan) greets local people in an inappropriate way as long as he is unaware of the rules that define the greeting choice. For example, he might want to shake hands or hug, and will receive a bow instead. However, in a limited number of interactions, the foreigner can understand the rules and correct his behaviour. In the current experiment, we want a robot to be able to do the same: be trained with sociology data related to one country, and evolve its behaviour engaging with people of another country in a small number of interactions. For the implementation of the gestures and the interaction experiment, we used the humanoid robot ARMAR-IIIb [11].

As the experiment is carried out in Germany, the interactions are done with German participants, while preliminary training is done with Japanese data, which is culturally extremely different. Participants' feedback is also collected, but it is not the main goal of this research, because as previously stated, culture-dependent acceptance and discomfort were already found in the previous Egyptian-Japanese study [3]. The point of interest in this paper is about the evolution of behaviour itself, from Japanese to German.

The rest of the paper is organized as follows: in Section 2 we describe the system of greeting selection; in Section 3 the hardware implementation; in Section 4 we describe the experiment and show and discuss the results; in Section 5 we conclude the paper and outline future works.

2   Greeting Selection

2.1   Model of Greetings

It is necessary to identify the main factors that influence the choice of gesture in human greetings. In Section 1.1 the main factors are reported, and following a process of simplification, the resulting factors are summarised in Figure 1. Simplification consisted in dropping some factors, such as physical distance and eye contact, assuming that they are always fixed or guaranteed. The remaining factors (Culture, Gender, Location, Power relationship, Social distance) are therefore relative to only social aspects of interaction. They are listed on the left of Figure 1 with their possible values. These factors are the features of the problem of mapping an input containing this social context information into a greeting gesture selection on the right of Figure 1. The possible values of the features are categorical data, as they can assume only 2 or 3 values, and are given as input to the mapping problem. Culture is a special case as it can be considered a discriminant for switching among different mappings between the other factors and the outputs.

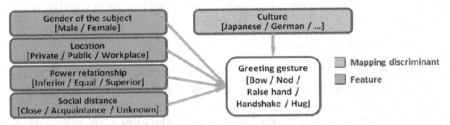

Fig. 1. The model of greeting selection synthesised in four features on the left, one mapping discriminant and the output on the right. Each block has some possible categorical values.

The outputs can also assume only a limited set of categorical values, the classes of a mapping problem. Greeting gestures list (Bow, Nod, Handshake, Raise hand, Hug) has been defined from relevant sources [5], [16, 17]. Originally, the set contained six gesture types, including kissing, which was dropped, because it was not possible to implement on the robot ARMAR-IIIb, which does not have a mouth. Waving and raising a hand or two hands were also considered as broadly the same type of gesture.

2.2   Greeting Selection System

Figure 2 contains the overview of the greeting selection system. It takes context data (Gender, Location, and so on) as input and produces the appropriate robot posture (the configuration for the chosen gesture) for that input. The context is the set of features shown in Figure 1. Inside the mapping box there is an algorithm that will be described in Section 2.3. The gesture chosen from this mapping is turned into robot configuration through the Master Motor Map (in short MMM) [18], which will be described more in detail in Section 3. The output gets evaluated by the participants of the experiment through written questionnaires. These training data that we can get from experience are given as feedback to the mapping, which is originally trained   only with data extracted from sociological studies. This model is generic: it is potentially implementable on any robot, with the exception of the robot-specific MMM mapper.

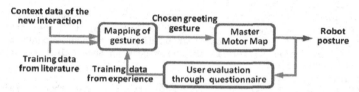

Fig. 2. Overview of the greeting selection system. Green arrow: input; red arrow: output.

2.3   Mapping Algorithm

Mappings can be trained with data taken from literature of sociology studies. We used data from [5], [16, 17] among others. Training data should be classified through some machine learning method or formula; nevertheless, data taken from these studies feature some properties that may limit the possible choice of classifying methods. In particular, their incompleteness: the focus of sociology papers is set only on specific aspects (such as gender-related studies, which do not provide any information regarding Power relationship) and the resulting data, put into a table, has some missing parts. Missing data makes inconvenient to use techniques, among others, like Principal Component Analysis or Neural Networks. Another constraint in the choice of the method is that context variables are categorical values: they cannot be assigned values like 0, 1, and 2, because applying a mapping method that assumes that $0 < 1 < 2$ would falsify the results.

Considering all these limitations, we decided to use conditional probabilities: in particular Naive Bayes formula, to map the data. The Naive Bayes classifier applies the Bayes theorem with the assumption that the presence or absence of each feature is unrelated to other features. This is appropriate to the features of the present problem. Moreover, Naive Bayes only requires a small amount of training data to estimate the parameters necessary for classification. The generic formula of posterior probability is shown in Equation 1 for the class variable C_j and the features x_k from the set X. Our modified version of the classifier takes also into account the possibility of missing data, assigning less weight to them.

$$p(C_j \mid X) \propto p(C_j) \prod_k p(x_k \mid C_j) \qquad (1)$$

The algorithm also includes rewards or penalties depending on the feedback collected from the experience (namely, participants' questionnaires). This was done because the algorithm has to learn quickly: as this is a real world problem rather than an abstract one, the desired amount of iterations necessary for a complete adaptation from the initial mapping to another one should be comparable to the number of interactions human need to understand behaviour rules. The process should not require hundreds or thousands of steps. The whole concept of the algorithm is shown the following pseudo-code:

```
begin
D ← training_data;   //the dataset (a table containing weights w)is built
for (each participant) begin
    f* ← new_input_context_data;     //a vector containing the current context
    if (∃ f* in D) then begin      //is f* already contained in the dataset?
        P_f*_g* ← w_f*;        //classification directly through weights w for f*
        g* ← argmax(w_f*);    //g* is the greeting with the maximum weight
    end;
    else begin
        P_f*_g* ← Naive_Bayes(f*);     //probability calculated through Naive Bayes
        g* ← argmax(P_f*_g*);    //g* is the greeting with the maximum likelihood
    end;
    bContinueUpdating ← calculate_stopping_conditions();
    if (bContinueUpdating == True) then begin
        eval_g* ← questionnaire_data; //from the participant, on a scale from 1 to 5
        P_f*_g* ← P_f*_g* * (1 + r*l);   //positive/negative reward * learning factor
        if (eval_g* ≤ 3) then begin      //if evaluation was negative
            g** ← suggested_data //g** is the suggested greeting type appropriate for f*
            P_f*_g** ← P_f*_g** * (1 + l);   //its vector gets a positive reward
            f** ← suggested_data //f** is the suggested context where g* is appropriate
            P_f**_g* ← P_f**_g* * (1 + l);   //its vector gets a positive reward
        end;
        update_dataset(D);
    end;
end; end;
```

The stopping conditions consist in calculating: - whether all possible values of all features have been explored; - whether the moving average of the latest 10 state transitions has decreased below a certain threshold, arbitrarily defined as 2 divided by the number of total states. This means that if mapping has already stabilised, no additional learning algorithm will be performed.

3   Implementation on ARMAR-IIIb

The implementation of the set of gestures on the robot was done in a way that it is not strictly hardwired to the specific hardware. Rather than defining manually the patterns of the gestures, the Master Motor Map [18] was used as intermediate passage.

The MMM is a reference 3D kinematic model that provides a unified representation of various human motion capture systems, action recognition systems, and so on. This representation can be subsequently converted to other representations, such as action recognisers, 3D visualisation or implementation on different robots. In this framework, the MMM is the interface for the transfer of motion knowledge between different embodiments.

The kinematic model of MMM is expanded with statistic/anthropomorphic data, such as: segment properties (e.g. length, mass and so on) defined as a function of global parameters (e.g. body height, weight). The body model of MMM is based on Winter's biomechanical model [18]. It contains some joints, such as the clavicula, which are usually not implemented in robots. A conversion module is necessary to perform a transformation between this kinematic model and ARMAR-IIIb kinematic model. The converter used [18] is based on non-linear optimization to maximise the similarity between the demonstrated human movement and the imitation by the robot.

The simplest and ideal way to reproduce a movement from given joint angles would consist in a one-to-one mapping. However, due to the differences in the kinematic structures of a human and the robot, one-to-one mapping can hardly show acceptable results in terms of humanlike appearance of the movement. In this converter, this problem is addressed by applying a post-processing procedure in joint angle space. The joint angles, given in the MMM format, are optimised concerning the tool centre point position and the kinematic structure of the robot through a non-linear algorithm. A feasible solution is estimated by using the joint configuration of the model on the robot, which serves as an initial solution for a further optimisation step.

We programmed the postures directly on the MMM model (Figure 3, left), and processed them by the converter. As the human model contains many joints, like pelvis, and clavicula, which are not present in the robot configuration, the conversion was not trivial.

The results we obtained with this algorithm needed some retouch, due to some part of the body (e.g. the neck) not implemented in the algorithm.

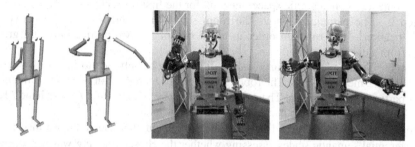

Fig. 3. MMM model (left) and implementation on ARMAR (right) of Raise hand and Hug

4   Description of the Experiment

4.1   Participants

The experiment was performed in Germany. Participants were 18 German people of different age, gender, workplace, knowledge of the robot, in order to ensure that the mapping could be trained with various combinations of context.

Not all combinations of feature values were possible to use in the experiment. For example, there cannot be a profile with both [Location: Workplace] and [Social distance: Unknown]. Moreover, the [Location: Private] case was left out, because it is impossible to simulate the interaction in a private context (such as one's home: the experiment took place in the laboratory).

Some participants repeated the experiment more than once. In this way, we could collect more data, just manipulating the value of one feature: e.g. for the Social distance feature: a participant who meets the robot for the first time can repeat the experiment later on, and will be considered "Acquaintance" instead of "Unknown".

The demographics of the 18 participants are as follows: M: 10; F: 8; average age: 31.33; age S.D.: 13.16. However, the number of interactions, taking repetitions into account was 30. M: 18; F: 12; average age: 29.43; age S.D.: 12.46. The number of participants was determined by the stopping condition of the algorithm.

4.2   Experimental Setup

The objective of the experiment was to adapt ARMAR-IIIb greeting behaviour from Japanese to German culture. Therefore, the algorithm working for ARMAR was trained with only Japanese sociology data and a mapping M0J was built. After interacting with German people, the resulting mapping M1 was expected to synthesise the rules of greeting interaction in Germany. A mapping M0G made from German sociology data was built but used only for verification.

The experiment protocol was as follows:

Step 1:   The mapping is trained with Japanese data.

Step 2:   Contextual data about the encounter is given as input to the algorithm and the robot is prepared. In the meantime, the participant is instructed about what to do: enter the room, turn left and greet the robot naturally considering the current context (e.g. in a public space, meeting for the first time, etc.).

Step 3:   The participant enters the room shown in Figure 4. A curtain covers the location of the robot in order to avoid one of the two parties initiating greeting from a distant location.

Step 4:   Turning left, the participant will find him/herself face to face with the robot, about 2 meters distant. The robot greeting is triggered by an operator as the human participant approaches. The possible choices are: [Bow / Nod / Raise hand / Handshake / Hug]. The two parties have greeted each other

Step 5:   The robot is turned off, and the participant fills questionnaire made of differential semantic scales assessing whether the chosen greeting was appropriate in the actual context. Further details are provided in the algorithm in Section 2.3.

Step 6: The mapping is updated using subject's feedback. The new mapping will be used in the next interaction.

Steps 2-6 to be repeated for each participant until stopping conditions are satisfied.

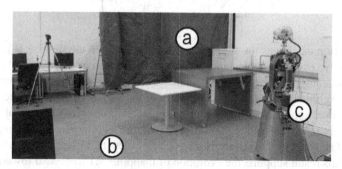

Fig. 4. Setup of the room of the experiment. The participant turns left after the curtain that covers the entrance (a) and finds him/herself in (b) face to face with the robot (c).

4.3   Results

The experiment was carried out through 30 interactions, when the moving average of state changes decreased below the threshold, and all greeting gestures had the chance to be selected at least once. Any behaviour mismatching with German participants' expectations did not influence their reactions, as they stuck with their own way of greeting, e.g. they would just respond raising a hand or nodding to a bow.

In Tables 1 and 2 it is possible to see the evolution of the mapping of gestures. The counter T, defined as the current number of learning iterations, corresponds to the steps 2 to 6 of the experimental protocol.

Table 1. M0J: MAPPING FOR T=0

		public male	public female	workp. male	workp. female
close	inf.				
close	equal				
close	sup.				
acquain.	inf.				
acquain.	equal				
acquain.	sup.				
unknown	inf.				
unknown	equal				
unknown	sup.				

Top row: Location; second row: Gender. Left column: Social distance; second column: Power relationship.

Table 2. M1: MAPPING FOR T=30

		public male	public female	workp. male	workp. female
close	inf.				
close	equal				
close	sup.				
acquain.	inf.				
acquain.	equal				
acquain.	sup.				
unknown	inf.				
unknown	equal				
unknown	sup.				

Yellow/vertical lines: bow; grey: nod; blue/diagonal lines: handshake; green/horizontal lines: raise hand; red/grid: hug

This new mapping was verified through an objective function V described in equation 2, which compares two different mappings M1 and M2.

$$V = \sum_f \sum_j \left(w_j^{(M1_f)} - w_j^{(M2_f)} \right)^2 \tag{2}$$

The function calculates the sum of the variance between the weights w in the same cell f (namely, every possible input value) in two different mappings M1 and M2. Each variance in the weights is calculated not only comparing the greeting with maximum likelihood, but considering the sum of the variances for each greeting j.

The function applied to M0J (Japanese initial mapping) and M1 (final mapping) gives 0.636 as result. Instead, comparing M1 with M0G (German initial mapping) we get 0.324. The t-test of the variances for each f proves the difference to be significant ($p < .05$). This result supports the evolution of mapping M1 from M0J towards M0G.

4.4   Discussion

It can be noticed from the evolution of mapping that after the interactions, the amount of states in which bowing is preferred has greatly decreased, while handshake is much more spread. Hug, not present in the Japanese mapping, appears after some participant expressed their feedback indicating that hugging would be appropriate.

Another observation is related to patterns present in the mappings: judging from the patterns in the rows in Table 1, it is clearly visible that a strict categorisation is present in the Japanese mapping in regards to Social distance, whereas the same pattern is not present in the German mapping. This fact seems to go in accordance with the more hierarchical view of the society in Japan. Both resulting German and Japanese mappings may not be 100% accurate compared to reality, but they are a simplification that is consistent respectively with German participants' feedback and Japanese sociology literature. After the end of learning phase, the robot can now potentially use two different mappings with human partners of different nationality.

5   Conclusion

In human-robot interaction, it is important for a robot to greet using gestures that are appropriate to specific human cultures in order to improve acceptance and reduce discomfort. For this reason, a system for greeting selection was made. From sociology studies, relevant context features were selected and an algorithm was created to update the mapping that selects the best gesture for each context. Gestures were implemented on the humanoid robot ARMAR-IIIb through the Master Motor Map framework and an experiment was performed with German participants. Through their feedback, ARMAR-IIIb could successfully learn a new mapping (German) of greeting selection given a defined context, starting from a Japanese mapping. This work is a step towards culture-related robots customisation and introduces a model of greetings that can be used with other robots. Ideally, robots will be able in the future to switch between different modes depending on the human cultural background. Future work can carry on towards different directions. Implementation on other robot platforms, and even non-human-like embodiments could be considered. Humanoid robots could be varying in shape, size and capabilities: using lights, playing sounds and so on. Different channels of communication could lead to different strategies of greeting.

Acknowledgements. This study was conducted as part of the Research Institute for Science and Engineering, Waseda University, and as part of the humanoid project at the Humanoid Robotics Institute, Waseda University. The experiment was carried out in Karlsruhe Institute of Technology, thanks to InterACT, the Waseda/KIT exchange network. We thank all staff and students involved for the support received.

References

1. Nisbett, R.E.: The geography of thought: how Asians and Westerners think differently–and why. Free Press, New York (2004)
2. Rogers, E.M.: Diffusion of Innovations, 5th edn. Free Press (2003)
3. Trovato, G., Zecca, M., Sessa, S., Jamone, L., Ham, J., Hashimoto, K., Takanishi, A.: Cross-cultural study on human-robot greeting interaction: acceptance and discomfort by Egyptians and Japanese. Paladyn. Int. Journal of Behavioral Robotics 4, 83–93 (2013)
4. Hoffman-Hicks, S.: The longitudinal development of French foreign language pragmatic competence. Doctoral Dissertation, Indiana University (1999)
5. Friedman, H.S., Riggio, R.E., Di Matteo, M.R.: A classification of nonverbal greetings for use in studying face-to-face interaction. JSAS Catalog of Selected Documents in Psychology 11, 31–32 (1981)
6. Hickey, L., Stewart, M.: Politeness in Europe. Multilingual Matters (2005)
7. Scherer, S.E., Schiff, M.R.: Perceived intimacy, physical distance and eye contact. Percept. Mot. Skills 36, 835–841 (1973)
8. Riggio, R.E., Friedman, H.S., DiMatteo, M.R.: Nonverbal Greetings: Effects of the Situation and Personality. Personality and Social Psychology Bulletin 7, 682–689 (1981)
9. Marshall, T.C.: Cultural differences in intimacy: The influence of gender-role ideology and individualism-collectivism. J. of Social and Personal Relat. 25, 143–168 (2008)
10. Brown, P., Levinson, S.C.: Politeness: Some Universals in Language Usage. Cambridge University Press (1987)
11. Asfour, T., Regenstein, K., Azad, P., et al.: ARMAR-III: An Integrated Humanoid Platform for Sensory-Motor Control. Humnoids 2006, 169–175 (2006).
12. Sakagami, Y., Watanabe, R., Aoyama, C., et al.: The intelligent ASIMO: system overview and integration. In: IROS 2002, vol. 3, pp. 2478–2483 (2002)
13. Bum-Jae, Y.: Network-Based Humanoids 'MAHRU' As Ubiquitous Robotic Companion. In: Presented at the 17th IFAC World Congress (2008)
14. Yamamoto, M., Watanabe, T.: Time delay effects of utterance to communicative actions on greeting interaction by using a voice-driven embodied interaction system. In: Int. Symp. on Computational Intelligence in Robotics and Automation, vol. 1, pp. 217–222 (2003)
15. Suzuki, S., Fujimoto, Y., Yamaguchi, T.: Can differences of nationalities be induced and measured by robot gesture communication? In: HSI 2011, pp. 357–362 (2011)
16. Greenbaum, D.P.E., Rosenfeld, H.M.: Varieties of touching in greetings: Sequential structure and sex-related differences. J. Nonverbal Behav. 5, 13–25 (1980)
17. Sugito, S.: Aisatsu no kotoba to miburi (あいさつの言葉と身振り). Bunkachou (1981)
18. Do, M., Azad, P., Asfour, T., Dillmann, R.: Imitation of human motion on a humanoid robot using non-linear optimization. Humanoids 2008, 545–552 (2008)

Socially Impaired Robots: Human Social Disorders and Robots' Socio-Emotional Intelligence

Jonathan Vitale, Mary-Anne Williams, and Benjamin Johnston

Innovation and Enterprise Research Lab, QCIS
University of Technology Sydney, Ultimo, NSW 2007, Australia
Jonathan.Vitale@student.uts.edu.au,
Mary-Anne@TheMagicLab.org,
Benjamin.Johnston@uts.edu.au

Abstract. Social robots need intelligence in order to safely coexist and interact with humans. Robots without functional abilities in understanding others and unable to empathise might be a societal risk and they may lead to a society of socially impaired robots. In this work we provide a survey of three relevant human social disorders, namely autism, psychopathy and schizophrenia, as a means to gain a better understanding of social robots' future capability requirements. We provide evidence supporting the idea that social robots will require a combination of emotional intelligence and social intelligence, namely *socio-emotional intelligence*. We argue that a robot with a simple socio-emotional process requires a simulation-driven model of intelligence. Finally, we provide some critical guidelines for designing future socio-emotional robots.

Keywords: social robots, socio-emotional intelligence, empathy, theory of mind, simulation theory, autism, psychopathy, schizophrenia.

1 Introduction

Social robots are embodied intelligent agents designed to coexist and interact with humans or with other social robots [9]. In order to avoid risks for the society their behaviour must to be safe and conform to social norms. Social intelligence, defined as the ability to make sense of others' actions and react appropriately to them [23], plays a crucial role in regulating acceptable interactions between people. Thus, social robots will require a form of social intelligence too [34].

In psychology studies, together with social intelligence, we find a subtly different form of intelligence, namely emotional intelligence. *Emotional intelligence* is defined as the ability to perceive, manage, and reason about emotions, within oneself and others [14]. Emotional intelligence is generally considered part of social intelligence [29]. However, in the robotic literature social intelligence is commonly introduced without considering this important need of emotional states elicitation and understanding.

M. Beetz et al. (Eds.): ICSR 2014, LNAI 8755, pp. 350–359, 2014.
© Springer International Publishing Switzerland 2014

In our everyday society there are people with an abnormal social behaviour. This population exhibits brain disorders involving deficits in social intelligence. It seems plausible that future social robots with high level cognitive capabilities, but lacking in social intelligence skills will develop similar social deficits [34]. In this work we want to provide a survey of human social disorder concerning deficiencies in social intelligence, so to gather significant information that can be used to trace a design guideline for future social robots. This is necessary in order to avoid the possibility of developing a society of socially impaired robots.

How can people achieve social intelligence? If we turn to psychology, philosophy, or the cognitive sciences in general, Theory of Mind (ToM) is the most shared and common strategy for gaining social intelligence abilities. ToM is defined as the ability to attribute mental states to oneself and others [28]. It provides mechanisms for comprehending/explaining everyday social situations, for predicting and anticipating others' behaviour, and even for manipulating other individuals.

Two main models of ToM are provided in the literature: Theory-Theory (TT) and Simulation Theory (ST). In the TT account the mind-reader deploys a naïve psychological theory to infer mental states in others from their behaviour, the environment, and/or their other mental states [18]. On the other hand, according to ST, the mind-reader select a mental state to attribute to others after reproducing or enacting within himself the very state in question, or a relevantly similar one [18]. In this way the mind-readers do not need theories; instead, they use their own body as a model of others.

Both ST and TT provide valid theories of how people can master the ability of making sense of others' social behaviour. However, from a social perspective there is a remarkable difference: ST requires an embodiment and the use of phenomenological mechanisms in order to "put the mind-reader in other's shoes", however, TT does not. ST contributes in *resonating* and so *empathising* with others, thus modulating socially acceptable behaviours, whereas TT works in a more mechanical and 'cold' vision. More specifically, a simulation-driven approach provides phenomenological bases for the development of *social reputation*, since it allows us to think about what others think of us and feel the corresponding positive or negative sentiment (i.e. they see me as a *good* person vs. *bad* person). The feeling elicited from the social reputation process via simulation, in turn, might become an incentive for individuals to conform to social norms [21].

Individuals with disorders related to social deficits show difficulties in empathising with and mentalising about others [15]. As suggested by Dautenhahn: "the better we understand human psychology and human internal dynamics, the more we can hope to explain embodiment and empathic understanding on a scientific basis" [8, p. 22]. Following this advice, and motivated by the previously exposed problematic scenario, in Section 2 we propose a brief survey of the most documented human social disorders and related deficits functional to social intelligence, namely autism, psychopathy and schizophrenia. In Section 3 we discuss this evidence, we define *socio-emotional intelligence* and then relate it to necessary sub processes. Finally, in Section 4 we provide our conclusions and a research agenda that will lead to the development of socio-emotional robots. To our knowledge this is the first study that tries to investigate human social disorders as a means to provide design principles for future social robot development.

2   A Survey of Human Social Disorders

2.1   Autism Spectrum Disorders

Autism Spectrum Disorders (ASD) individuals are characterized with deficits in three main areas: (i) communication, (ii) social interaction, and (iii) restrictive and repetitive behaviours and interests [1].

ToM is one of the main social skills in which ASD population shows deficiencies [15]. Compared with normally developed persons, ASD individuals are poorer at reasoning about what others think, know, or believe, recognizing emotional expressions and gestures, and making social attributions and judgements [2]. However, deficits in ToM ability are only the tip of the iceberg.

Indeed, ASD individuals show other social deficits functional to ToM, for example they are poor in understanding the emotional content of face expression, gestures and vocalizations and they fail in using these social signals as a way to express their own internal emotional state (e.g. arm around shoulder, hand over mouth, signalling embarrassment, ...) [5]. These deficits in emotion recognition/responding often lead to an impoverished facial affect. Thus, ASD individuals are perceived as unable to feel emotions. However, studies with electrodermal responses and self-report measures suggest that ASD individuals have appropriate emotional responsiveness to others [12]. Hence, ASD individuals seem to be able to normally experience such phenomenological internal states at least.

One of the earliest signs of ASD is a lack of sensitivity to social cues. For example, they exhibit poor eye contact, they have difficulties in joint attention (either using eyes, head pose or pointing gestures) and they show disinterest to other people [7,5,15]. Many studies investigated the gaze direction of ASD individuals using eye tracking systems. They found that this population look less at the eyes relatively to control participants (for a review see [31]). Birmingham and her research group suggest that perhaps the abnormality in ASD people lies in the likelihood that they will seek out and select social information from a scene; if such a population does not consider important social signals like others' gaze orientation, they will not be able for example to infer where others are looking. Due to this perceptual deficit, they might have less evidence to use during a mind-reading process [7].

Aligned to this perspective, Dawson et al. propose that social orienting deficits might cause ASD development and subsequent ToM deficiencies [10]. This hypothesis suggests that individuals developing with ASD fail to attend social stimuli from an early age. This lack of crucial social information during the normal development provokes later social cognitive deficits, such as facial expression processing and mind-reading.

Indeed, social orienting deficits can also explain their lacks in emotion recognition from social signals. Perhaps this population possesses embodied mechanisms to feel others' emotions, as well as mechanism for 'resonating' to others facial expressions and body movements (i.e. a functioning Mirror Neuron Systems, see [4,20] for a discussion). However, they lack of a social reward process and they cannot direct the attention on stimuli necessary for promoting mind-reading abilities [21]. Without focusing on important social signals, like the eyes, they might have severe limitations in ToM [4].

2.2 Psychopathy

The World Health Organization classifies *Psychopathy* as a form of antisocial (or dissocial) personality disorder [25]. Characteristics of such disorder are: (i) callous unconcern for the feelings of others; (ii) incapacity to maintain enduring relationships, though having no difficulty in establishing them; (iii) very low tolerance to frustration and a low threshold for discharge of aggression, including violence; (iv) incapacity to experience guilt or to profit from experience, particularly punishment; (v) marked proneness to blame others, or to offer plausible rationalizations, for the behaviour that has brought the patient into conflict with society [25].

In contradistinction to what is commonly believed, psychopathic individuals do not always present violent and criminal behaviour. Indeed, this population is mainly characterized by a lack of 'emotional empathy' [15]; they have a reduced ability to feel other people's emotional state, especially sadness and fear [11]. Psychopathic subjects have deficits in moral emotions such as remorse and guilt and they are usually indifferent to shaming and embarrassing situations [14].

Antisocial personalities usually exhibit a poor executive control, that is normally necessary for socially appropriate conduct [11]. This dysfunction might be due to non-responding violence inhibition mechanisms that are normally triggered during the feeling of others' distress in order to prevent the execution of antisocial behaviours [3]. Indeed, psychopathic individuals own a poor behavioural control, leading often to impulsivity. Furthermore, a study on startle reflex modulation of visual attention demonstrates that psychopathic individuals, compared to normal population, present an abnormal valence pattern [27]. The authors suggest that even if psychopaths express different subjective judgements to positive vs. aversive visual stimuli, they may find such stimuli equally inviting from an attention controlling perspective. This may be due to a dysfunction in attention reflex reactions when perceiving unpleasant content [27]. These evidences well support the existence of emotion regulation deficiencies in such population [14].

Deficits in emotion regulation seem to affect also face processing abilities. Psychopaths are impaired when processing fearful, sad and disgusted facial expressions, whereas it seems that they do not have impairments with happy facial expression, even if this should be due to the ease with which such expression is recognized [15]. Furthermore, this population has deficits in other emotional processing skills, such as failure to show normal response differentiation to emotional and neutral words, and abnormal reactions to emotional stimuli and events [14].

Surprisingly, this mental disorder does not involve abnormal levels of intelligence [14]. In fact, in contrast to other disorders like autism, psychopathic individuals successfully complete ToM tasks and currently there is no evidence of impairments in 'cognitive' (i.e. not emotional or empathic) ToM ability [15]. However, due to deficits in experiencing emotions, psychopath individuals cannot simulate them, and must rely exclusively on cognitive inputs in order to fulfil a mind-reading task [11].

2.3 Schizophrenia

Schizophrenia is a severe psychiatric disorder altering emotional, cognitive, and social functions [26]. In particular, significant impairment in social functioning is considered one diagnostic characteristic of schizophrenia [1]. Such impairment can have seriously impacts on social relationships [22]. Schizophrenic subjects suffer also from delusions and hallucinations, however in this survey we will consider only their deficits primarily related with social intelligence abilities (for a discussion on ToM and correlations with these symptoms, see [6]).

Similarly to autistic and psychopathic individuals, schizophrenic individuals lack general abilities in ToM and empathy [32,15]. In fact, some current models of schizophrenia suggest that this disorder can be understood as a deficit in representing others' mental states (i.e. cognitive ToM) and of 'resonating' to others' emotional states (i.e. empathy) [15]. However, even in this case deficits in these abilities are just the tip of the iceberg; indeed different cognitive sub processes seem to be affected in schizophrenia leading to a differentiation of such deficiencies respect to psychopathy and ASD. For example, whereas schizophrenic individuals seem to be able at least to understand the intended meaning of sincere interpersonal exchanges (differently from ASD population), they show deficiencies in insincere interactions, such as in understanding sarcastic conversations, that indeed lie more on emotional features such as prosody and intonation [32].

Schizophrenic individuals exhibit blunted feeling and they usually have inappropriate affective responses in social situations [15]. They show abnormalities of skin conductance response and they mostly respond with negative affect (e.g. depression) [15]. Furthermore, this population exhibits deficits in subjective experience of emotion [24]. Studies demonstrate that schizophrenic patients emotionally respond with fewer positive and negative facial expressions in response to emotional stimuli compared to normal population [13]. However, evidence from other studies support the idea that schizophrenic people can indeed feel emotional states, but they cannot sustain attention over the emotional stimuli and thus maintaining such emotional state during time.

Horan *et al.* used affective pictures as emotional stimuli in order to record the Event Related Potentials (ERP) of schizophrenic subjects [19]. The results show that schizophrenic individuals experience comparable amounts of similar emotions with respect to normal populations during the initial ERP components, but not in Late Positive Potentials (LPP). The authors suggest that this population may have functioning emotional response mechanisms, but a disruption in a later component associated with sustained attention processing of the observed emotional stimuli [19]. This inability to maintain the correct emotional response over time is correlated with deficits in executive control. In fact, without a sufficient elicitation of emotional processing over time it becomes critical to guide future behavioural choices [19].

The previous studies demonstrate attention deficits in schizophrenic populations. Indeed, these individuals perform poorly on nearly all tests of sensory and cognitive vigilance and some studies also demonstrate deficits in selective attention [26]. It has also been shown that there are abnormalities in eye movements during the scanning of emotional facial expressions. Similarly to ASD people, schizophrenic individuals look less at the eye region of the face [15,7]. Again, in a similar way as in ASD populations, schizophrenic patients show partial gaze avoidance specific to human faces, whereas they do not avoid gaze when they look to non-human faces [33]. However, a study by Sasson *et al.* provides evidence for a differentiation of emotional processing deficits in schizophrenics and ASD people [30]. Their results demonstrate that, whereas autism and schizophrenia share an impairment in fixating social stimuli (i.e. avoidance of eye region), the schizophrenic individuals show a delay in orienting the gaze to informative emotional stimuli (in this study faces). Thus it seems that, whereas autistic populations fail in the specificity of selective attention concerning emotional stimuli, schizophrenic population exhibited a generalized orienting delay [30].

Given this evidence it seems that schizophrenic individuals' inability to maintain sustained attention and their delay of selective attention over emotional stimuli are strictly correlated with deficiencies in social intelligence.

3   A Socio-Emotional Robotic Intelligence

In the introduction we suggested the need for social robots to possess social intelligence. We identified ToM as a crucial strategy to achieve such intelligence. However, we also mentioned that we need to prevent a society of 'socially impaired robots'. Under this perspective creating robots able to perfectly understand humans' intentions and react appropriately to them in a pure rational way is not enough. In fact, we have seen that psychopaths, that indeed are able to understand others' intentions, are a risk for the society as they can use ToM to manipulate or hurt people because unable to empathize or to feel sense of guilt.

We provided evidence demonstrating that the elicitation and regulation of emotions (i.e. emotional intelligence) are crucial skills needed to avoid, for example, psychopathic traits. Thus, given the importance of exhibiting both social and emotional intelligences, we provide a clearer and explicit definition of socio-emotional robot. We define a *socio-emotional robot*[1] as a robot able to direct attention over others' social behaviour, to make sense of it and to elicit correct

---

[1] In robotic literature the term socio-emotional robot was already widely used; however, to our knowledge, nobody provided an explicit definition of it.

emotional processes regulating and learning the expression of its behaviours in order to conform to society's culture, ethics, morality and common-sense.

Making sense of others' social behaviour can be achieved using ToM, whereas eliciting emotional processes for behaviour regulation and development is more related with empathy. Indeed, empathy may be a central feature of emotionally intelligent behaviour and it can be used to relate positively to others, thus increasing life satisfaction, reducing stress and motivating altruistic behaviour [29]. More specifically, a person can feel what the other person is feeling and so behave conformably to past experiences related with very similar feelings. Furthermore, a socio-emotional robot might learn correct behaviour through direct experience of its emotional states. Thus, having processes for emotion elicitation might facilitate the learning of society norms through a first-person experience.

3.1   The Need for a Simulative Mechanism

At the beginning of this paper we mentioned two possible approaches to master ToM: simulation-driven approaches and theory-driven approaches. Although both the approaches can be an acceptable explanation for mind-reading ability, when we look at empathy (and more in general at emotional intelligence) simulation-driven approaches play a crucial role in allowing this ability, as for example Goldman contends [17]. Further support comes also from neuroscience studies of Gallese [16]. Thus, it seems plausible that in order to exhibit emotional intelligence a simulative mechanism is needed. As our previous survey on human social disorders demonstrate the need of both social and emotional intelligence (i.e. socio-emotional intelligence) in order to avoid social disorders, we suggest the need of simulative mechanism in social robots. With this recommendation we are not saying that theories about the world are not necessary for a fully understanding of others, but rather that at some preliminary levels a simulation process is needed in developing empathy and promoting socio-emotional intelligence.

A simplified socio-emotional process can be described as: (a) detecting a social behaviour, (b) enacting a simulation process given such stimulus and allowing an *as-if* internal representation of it, (c) activate an appropriate viscero-emotional internal state (again through a simulation process), (d) use past experience and theories in order to give an interpretation to the perceived stimulus, (e) properly regulate the appraisal of the emotional state and the expression of an appropriate behaviour through (c), (d) and other theories about culture, ethics, morality and common-sense.

In most cases robotic studies on social intelligence make use of only two of such processes, namely (a) and (d). In fact, aligned with a pure information-based approach, researchers make use of datasets of social stimuli (face expressions, gestures, etc.) in order to create theories or models (d) to use for the interpretation of new social stimuli (a). Given the previous survey we can argue that this approach potentially leads to possible social disorders and thus justify the need of a simulation mechanism in socio-emotional robots.

We have seen that ASD individuals have a deficit in directing attention over social stimuli (a), thus leading to deficit in representing them internally (b). On the contrary they do not show deficits in activating appropriate visceral states (c), if properly stimulated. Deficits in (a) throughout their life lead to learn poor social-life theories (d) in conjunction with their visceral states (c). This might explain their inability in fulfilling a successful socio-emotional process, since they poorly direct attention over social stimuli, thus reducing evidence for a mind-reading process, and as they develop a poor learning of (d) given (c).

On the other hand, psychopathic subjects show normal capacities in perceiving and processing social stimuli (a,b), but they cannot elicit viscero-emotional states (c). This again might explain their ability in mind-reading people (d) using social evidences (a,b) but their deficits in regulating empathic and moral behaviours (e) because unable to empathise with others (c).

Finally, schizophrenic population suggests the importance of synchronizing the processes of socio-emotional intelligence. A delay or dysfunction in sustain and selective attention over social stimuli (a) might lead to non-synchronized or distorted internal and cognitive processes (b,c,d) over time. This in turn leads to deficiencies in mind-reading and emotion regulation (e).

4   Conclusions and Guidelines for Socio-Emotional Robots

In this work we motivated the need for robots to be able to coexist in human spaces avoiding risks and costs for the society. In order to understand better how to design such a kind of safe robots we proposed a brief survey of human social disorders. We proposed the need of emotional intelligence together with social intelligence, and in order to clarify better these necessary intelligences in robots we provided an explicit definition of socio-emotional robot. We suggested the need of socio-emotional intelligence in order to avoid socially impaired robots. We provided a simple model of a socio-emotional process and we used the evidences from the survey as a way to motivate the need of a simulation process in order to avoid social deficits.

Given the need of socio-emotional intelligence for development of future robots, we suggest an agenda of necessary future research. First, social roboticists will need to provide appropriate mechanisms of attention modulation over social stimuli (a). In order to fulfil this target we will need to understand the mechanisms underlying social rewarding of social stimuli. Second, some kinds of simulation models will be necessary in order to represent perceived stimuli and activate an appropriate internal response in the robot (b,c). The internal representation of the stimuli might require a mapping from the external multimodal representation to an internal unimodal one; this is necessary in order to integrate different modalities under a unique and more computationally tractable form of representation. This is similar to our capacity of mapping multimodal external stimuli to unimodal neural activations. Third, we will need learning mechanisms allowing the association between internal representation and appropriate mental attributions (d). A further process of decision making is then necessary in order to drive the robot's executive attention and regulate its behaviour (e).

We want to conclude mentioning some limitations of the current study. First, the proposed survey is limited to three social disorders. There are others syndromes and brain dysfunctions that worth a discussion and enrich our argument, but for the sake of simplicity we limited our work to the most investigated social dysfunctions. Second, studies on ASD individuals, psychopaths and schizophrenics are not so linear as proposed in this survey. There are many controversies and open questions, but in order to provide a readable manuscript and an argument easy to follow we proposed some limited studies about hypotheses commonly shared in the related literature. We are confident that studies like the one reported in this manuscript will allow a better understanding of the human brain. This in turn is an essential knowledge if we want to develop intelligent machines.

References

1. American Psychiatric Association, A.P.A.: Diagnostic and statistical manual of mental disorders: DSM-IV-TR®. American Psychiatric Pub. (2000)
2. Bachevalier, J., Loveland, K.A.: The orbitofrontal–amygdala circuit and self-regulation of social–emotional behavior in autism. Neuroscience & Biobehavioral Reviews 30(1), 97–117 (2006)
3. Blair, R.J.R.: A cognitive developmental approach to morality: Investigating the psychopath. Cognition 57(1), 1–29 (1995)
4. Bons, D., Van Den Broek, E., Scheepers, F., Herpers, P., Rommelse, N., Buitelaar, J.K.: Motor, emotional, and cognitive empathy in children and adolescents with autism spectrum disorder and conduct disorder. Journal of Abnormal Child Psychology 41(3), 425–443 (2013)
5. Brothers, L., et al.: The social brain: a project for integrating primate behavior and neurophysiology in a new domain, pp. 367–385. Foundations in Social Neuroscience (2002)
6. Brüne, M.: Theory of mind in schizophrenia: a review of the literature. Schizophrenia Bulletin 31(1), 21–42 (2005)
7. Burack, J.A., Enns, J.T., Fox, N.A.: Cognitive Neuroscience, Development, and Psychopathology. Oxford University Press (2012)
8. Dautenhahn, K.: I could be you: The phenomenological dimension of social understanding. Cybernetics & Systems 28(5), 417–453 (1997)
9. Dautenhahn, K., Billard, A.: Bringing up robots or–the psychology of socially intelligent robots: From theory to implementation. In: Proceedings of the Third Annual Conference on Autonomous Agents, pp. 366–367. ACM (1999)
10. Dawson, G., Meltzoff, A.N., Osterling, J., Rinaldi, J., Brown, E.: Children with autism fail to orient to naturally occurring social stimuli. Journal of Autism and Developmental Disorders 28(6), 479–485 (1998)
11. Decety, J., Moriguchi, Y.: The empathic brain and its dysfunction in psychiatric populations: implications for intervention across different clinical conditions. Bio. Psycho. Social Medicine 1(1), 22 (2007)
12. Dziobek, I., Rogers, K., Fleck, S., Bahnemann, M., Heekeren, H.R., Wolf, O.T., Convit, A.: Dissociation of cognitive and emotional empathy in adults with asperger syndrome using the multifaceted empathy test (met). Journal of Autism and Developmental Disorders 38(3), 464–473 (2008)

13. Earnst, K.S., Kring, A.M.: Emotional responding in deficit and non-deficit schizophrenia. Psychiatry Research 88(3), 191–207 (1999)
14. Ermer, E., Kahn, R.E., Salovey, P., Kiehl, K.A.: Emotional intelligence in incarcerated men with psychopathic traits. Journal of Personality and Social Psychology 103(1), 194 (2012)
15. Farrow, T.F., Woodruff, P.W.: Empathy in mental illness. Cambridge University Press, Cambridge (2007)
16. Gallese, V.: Embodied simulation: From neurons to phenomenal experience. Phenomenology and the Cognitive Sciences 4(1), 23–48 (2005)
17. Goldman, A.I.: In defense of the simulation theory. Mind & Language 7(1-2), 104–119 (1992)
18. Goldman, A.I., Sripada, C.S.: Simulationist models of face-based emotion recognition. Cognition 94(3), 193–213 (2005)
19. Horan, W.P., Wynn, J.K., Kring, A.M., Simons, R.F., Green, M.F.: Electrophysiological correlates of emotional responding in schizophrenia. Journal of Abnormal Psychology 119(1), 18 (2010)
20. Iacoboni, M., Dapretto, M.: The mirror neuron system and the consequences of its dysfunction. Nature Reviews Neuroscience 7(12), 942–951 (2006)
21. Izuma, K., Matsumoto, K., Camerer, C.F., Adolphs, R.: Insensitivity to social reputation in autism. Proceedings of the National Academy of Sciences 108(42), 17302–17307 (2011)
22. Kennedy, D.P., Adolphs, R.: The social brain in psychiatric and neurological disorders. Trends in Cognitive Sciences 16(11), 559–572 (2012)
23. Langton, S.R., Watt, R.J., Bruce, V.: Do the eyes have it? cues to the direction of social attention. Trends in Cognitive Sciences 4(2), 50–59 (2000)
24. Myin-Germeys, I., Delespaul, P.A., et al.: Schizophrenia patients are more emotionally active than is assumed based on their behavior. Schizophrenia Bulletin 26(4), 847–854 (2000)
25. Organization, W.H.: The ICD-10 classification of mental and behavioural disorders: clinical descriptions and diagnostic guidelines, vol. 1. World Health Organization (1992)
26. Parasuraman, R.: The attentive brain. MIT Press, Cambridge (1998)
27. Patrick, C.J., Bradley, M.M., Lang, P.J.: Emotion in the criminal psychopath: startle reflex modulation. Journal of Abnormal Psychology 102(1), 82 (1993)
28. Premack, D., Woodruff, G.: Does the chimpanzee have a theory of mind? Behavioral and Brain Sciences 1(04), 515–526 (1978)
29. Salovey, P., Mayer, J.D.: Emotional intelligence. Imagination, Cognition and Personality 9(3), 185–211 (1989)
30. Sasson, N., Tsuchiya, N., Hurley, R., Couture, S.M., Penn, D.L., Adolphs, R., Piven, J.: Orienting to social stimuli differentiates social cognitive impairment in autism and schizophrenia. Neuropsychologia 45(11), 2580–2588 (2007)
31. Sasson, N.J.: The development of face processing in autism. Journal of Autism and Developmental Disorders 36(3), 381–394 (2006)
32. Sparks, A., McDonald, S., Lino, B., O'Donnell, M., Green, M.J.: Social cognition, empathy and functional outcome in schizophrenia. Schizophrenia Research 122(1), 172–178 (2010)
33. Williams, E.: An analysis of gaze in schizophrenics. British Journal of Social and Clinical Psychology 13(1), 1–8 (1974)
34. Williams, M.-A.: Robot Social Intelligence. In: Ge, S.S., Khatib, O., Cabibihan, J.-J., Simmons, R., Williams, M.-A. (eds.) ICSR 2012. LNCS, vol. 7621, pp. 45–55. Springer, Heidelberg (2012)

Modelling of Pneumatic Air Muscles for Direct Rotary Actuation of Hand Rehabilitation Glove

Boran Wang, Kean C. Aw, Morteza Biglari-Abhari, and Andrew McDaid

University of Auckland,
Faculty of Engineering, New Zealand
{bwan055,k.aw,m.abhari,andrew.mcdaid}@aucklanduni.ac.nz

Abstract. Pneumatic muscle actuator (PMA) has the virtue of lightness, high power to weight ratio, small size, simple customization, easy fabrication, no stiction, inherent compliant behaviour and low cost. Hence, it has great potential as an actuator for hand rehabilitation device. Its stretchability and soft nature makes it easy to be mounted on curved surface and bend around corner. It is also able to produce very high forces over reasonable contraction. Here, we will present the use of PMA as a direct rotary actuator for hand rehabilitation glove. A model will be developed and verified experimentally for the PMA used in this configuration.

Keywords: pneumatic air muscle, hand rehabilitation, rotary actuator.

1   Introduction

Stroke is a common condition, which contributes substantially to disability around the world. Although the death rate per capita in New Zealand has declined steadily over the last 20 years, the total number of deaths from stroke increases as a result of increase in the total population [1]. Many reports suggest that the most effective way of relearning motor function is through carefully directed, well-focused, intensive and repetitive practice of the impaired hand [2-4]. With an aging population and limited hospital resources, the demand for robot-aided training program to replace the conventional labour-intensive rehabilitation technique increases. The idea of developing a robotic exoskeleton for hand rehabilitation has been widely recognized only over the last decade [5]. Existing grounded exoskeletons are sophisticated enough to facilitate a systematic post-stroke hand training program but they cannot be adapted to an easy-manipulated home-based training protocol because of their large volume, heavy weight or high cost [6-8]. They need to be robust but mechanical lightweight, resourceful control technology and compact power unit [9-11]. Without considering the human hand anatomy in depth, each single joint is actuated by one set of actuators, resulting in a device with high mechanical complexity. More recent designs take the natural coupling of distal interphalangeal (DIP) and proximal interphalangeal (PIP)

M. Beetz et al. (Eds.): ICSR 2014, LNAI 8755, pp. 360–369, 2014.

joints into consideration by activating PIP and metacarpal-phalangeal (MCP) joints only, hence reducing the weight and volume further. However, most of them are still using rigid components to build exoskeleton, which would possibly arouse heightened discomfort during training. This project aims to design a jointless rehabilitation glove completely using flexible actuators by treating user's hand itself as a skeletal structure. Targeted users are those with hypertonia and require long term self-motivated post stroke recovery training. As a proof of concept, this paper mainly focuses on the validation of using pneumatic air muscle (PMA) to generate rotary movements of user's affected unhealthy hand..

2   Construction of PMA

There are two primary components in fabricating an air muscle: a soft stretchable inner rubber tube and braided polyester mesh sleeve (Fig.1 (a)). The mesh sleeve is made in a crescent shape; as such it could bend around finger joint with its far side to human hand longer than the proximal side (Fig.1 (b)).

| (a) | (b) |

Fig. 1. Pictures of PMA (a) external braid with singed end, (b) fully constructed

3   Preliminary Testing

The optimum result in preliminary test have shown that a PMA of 60 mm long, 8 mm inner diameter and 25 mm outer diameter at 3.5 bar could lift up a test plastic finger with 50 g load by ~50° from a vertical flexed position.

4   Characterisation and Modelling in Bending Moment Application

Empirical modelling, which focuses on the concepts of observation and data fitting from real experiments was used to characterize the behaviour of the PMA. A mathematical model was established and validated both deductively based on its geometric structure and inductively through empirical findings. A finite element analysis will be performed in the future to optimize the dimension of PMA, but it is not within the scope of this paper.

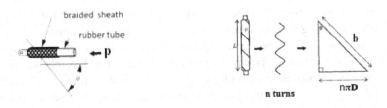

Fig. 2. Braid angle of PMA   **Fig. 3.** Geometry of PMA

There are mainly five factors that determine the behaviour of PMA: pressure, sleeve diameter, sleeve length, diameter of inner tube and braided angle (Fig. 2 and 3). The air pressure can determine the stiffness of PMA, while greater sheath diameter increases the force. Longer PMA yields larger working range of movement but were unable to lift heavy load at the finger. The experiment set-up in Fig.4 (a) shows the PMA was attached to artificial hand and regulated by solenoid valves. Fig.4 (b) shows a picture of the set-up. Experimentally, a finger with a PMA of 40 mm length and 10 mm diameter with a perpendicular distance of 1 cm (d_3) from the joint could sustain a weight of 80 g with a pressure of 3.5 bar. The change in stiffness due to the pressure and shape provides an upward force to lift the finger.

4.1   Force and Contraction

For a given amount of air flow (p), work input to inflate the PMA and work output to generate deformation is $dW_{in} = p \times dV$, and the work output produced in contracting the PMA is $dW_{out} = -F \times dx$, where x is contracted length of PMA; F denotes the surface tension in PMA; V is the PMA volume. By ignoring energy loss due to frictions, the input energy should be equal to output energy, i.e. $dW_{in} = dW_{out}$, which then gives

$$p \times dV = -F \times dx \qquad (1)$$

According to Fig. 2 and 3, the geometric relationship of L (PMA length), D (diameter of PMA at specific pressure), b (PMA's braid length) and n (the number of turns of the braid) can be derived as:

$$D^2 = \frac{b^2 - L^2}{(n\pi)^2} \qquad (2)$$

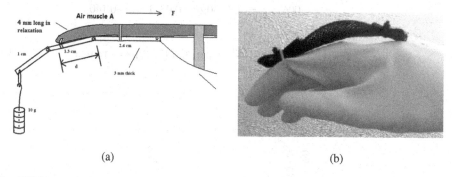

(a)   (b)

Fig. 4. PMA attached to artificial finger (a) a schematic of experiment setup, (b) a picture of the experiment set-up

By assuming the PMA as a perfect cylinder, its volume can be expressed as:

$$V = \frac{\pi(b^2 - L^2)}{4(n\pi)^2} L = \frac{b^2 L - L^3}{4\pi n^2} \tag{3}$$

In a bending moment application, the shape of air muscle must not be a perfect cylinder. Thus the force tangential to bending surface is derived as

$$F = -p\frac{dV}{dL} = -p\frac{(b^2 - 3L^2)}{4\pi n^2} \tag{4}$$

The tension force produced by PMA can be expressed as

$$F = -p\frac{\left(b^2 - 3\left(L(1-\varepsilon)\right)^2\right)}{4\pi n^2} \tag{5}$$

where L_m is the maximum length of PMA; ε is percentage of change in length due to contraction. Equation (5) can be rearranged as a polynomial expression, i.e.

$$F = c_0 \frac{pb^2}{4\pi n^2} + \frac{3pL}{4\pi n^2}(c_1 + c_2\varepsilon + c_3\varepsilon^2) \tag{6}$$

To validate the hypothesis, a series of experiments were run with different p, b, L, n and ε. The coefficients in Equation (6) were found with Matlab® data fitting tool. Experiments were carried out with very limited amount of pre-loading ($< 5\%$) of the PMA to reduce discomfort to the wearer and Fig. 5 shows the result. Coefficients are computed with a 95% confidence interval. With the absolute pressure varying from 1 to 3.5 bar, the coefficients remained reasonably consistent. The final coefficients with a goodness of fit (R^2) ranging from 0.8974 to 0.9662 are:-

$$c_0 = 0.00625, c_1 = -0.0130, c_2 = 0.0321, c_3 = -0.140$$

$$F = 0.00625 \frac{pb^2}{4\pi n^2} + \frac{3pL}{4\pi n^2}(-0.0130 + 0.0321\varepsilon - 0.140\varepsilon^2) \tag{7}$$

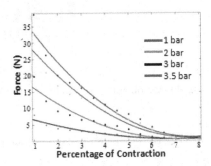

Fig. 5. Fitting of force against percentage contraction in PMA

As shown in Fig. 6 (a) the tension force along the PMA can be separated into two components: normal to the interacting surface ($F \times sin\alpha \times d_1$) and parallel to the surface ($F \times cos\alpha \times d_2$). Both components generate clockwise moment. However, when PMA is straightened, the bending moment decreases very quickly due to a drop in tension force. When α is ~ 50°, the resultant bending moment is ~ 0 and is unable to lift the finger up to a horizontal position. To increase bending moment, the distance between parallel component ($F \times cos\alpha \times d_2$) and pivot ($d_2$) was raised up by increasing d_3 as in Fig. 6 (b). The simulated active torque against bending angle (α) with increasing d_3 from 0 to 3 cm with 1 cm increment is shown in Fig.8. A greater perpendicular distance would generate more torque, but could also result in a higher profile. Typical value for hypertonia/spasticity is ~0.4 Nm to rotate the finger joint by 90° [12] and can be achieved with $d_3 = 3$ cm (Fig. 7). It is able to fully extend a finger with a load of 50 g to a horizontal position.

4.2   Influence of Pressure

The contraction force generated by PMA is positively proportional to the internal air pressure (p) but there will be a limit for p in a real application as the PMA will burst. Here, the use 3.5 bar with the inner tube diameter of 6 mm is found to be safe.

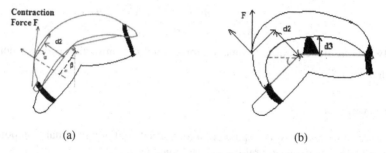

(a)   (b)

Fig. 6. A schematic diagram (a) for force analysis, (b) showing the increase in active torque by increasing d_3

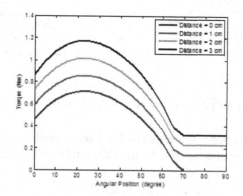

Fig. 7. Simulated result when d_3 has been changed

4.3   Influence of Braid Angle

Dynamic Analysis

To generate a model between braid angle and force, the relation between PMA's volume and braid angle was developed by assuming the pressurized PMA as a perfectly symmetrical cylinder with zero wall thickness. By assuming θ is the PMA's braided angle, the PMA's volume can be expressed as

$$V = \frac{1}{4}\pi D^2 L = \frac{b^3}{4\pi n^2}\sin^2\theta\cos\theta = \frac{b^3}{4\pi n^2}(\cos\theta - \cos^3\theta) \tag{8}$$

To find the maximum angle, both sides of equation were differentiated.

$$\frac{dV}{d\theta} = \frac{b^3}{4\pi n^2}\times(-\sin\theta + 3\cos^2\theta\times\sin\theta) \tag{9}$$

Therefore the tension force could be expressed as follows.

$$F = -p\frac{dV}{dL} = -p\frac{dV/d\theta}{dL/d\theta} = \frac{pb^2(3\cos^2\theta - 1)}{4\pi n^2} \tag{10}$$

By solving the equation, the maximum amount of contraction occurs when **F = 0** when $\theta = 54.7°$.

Static Analysis

The force in the PMA is separated into normal and longitudinal components by ignoring frictions between braids and braid thickness.

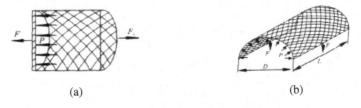

(a)   (b)

Fig. 8. Force analysis (a) in longitudinal direction, (b) in normal direction

In longitudinal direction, the contraction force and normal component of the tension force in the sleeve are balanced (Fig. 8 (a)). If the total number of threads is t, then

$$F + \frac{\pi D^2}{4}p = F_N = t \times f_N \tag{11}$$

In normal direction, force yielded by internal air pressure is balanced by the longitudinal component of tension force in sleeve (Fig. 8 (b)).

$$pDL = 2F_L = t \times f_L \tag{12}$$

Given $tan\theta = f_L/f_n$, an expression of tension force can be obtained as

$$F = \frac{\dfrac{PDL \times \cos\theta}{2} + \dfrac{\pi}{4}D^2 P\sin\theta}{\sin\theta} = \frac{pb^2(3\cos^2\theta - 1)}{4\pi n^2} \tag{13}$$

Equations (13) and (10) are similar, which means both dynamic and static methods generate the same result. The expression of force in terms of braided angle indicates as braid angle θ increases, the tension force generated by PMA will decrease, and is consistent with the experimental result. To verify the model, experiments were carried out by making the number of turns (n) constant, but changing the diameter and length of air muscle, so that the braid angle can be obtained by rearranging Equation (2).

The simulated plot (Fig. 9) of volume vs. braided angle indicates volume of air muscle keeps increasing until it reaches 54° when it begins to drop again. In practice, the PMA can burst when it reaches the maximum volume at a braided angle of approximately 50°. In Fig. 10, the experimental results match reasonably well with simulation plot. Systematic error between them could be due to internal frictions between rubber tubes and braid sheath, as well as different constraint conditions on producing the air muscle. The fluctuation in force response could be due to sharp angles in finger joints when the PMA was first actuated from its vertical flex position.

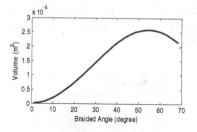

Fig. 9. Simulated result of volume vs. braided angle

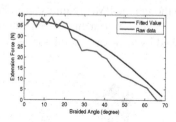

Fig. 10. Plot of force against braid angle

4.4   Influence of Braid Length

The experimental data matches well with the geometric model that there is a quadratic correlation between extension forces and braid length (Fig. 11). Mathematical modelling gives a goodness of fit (R^2) of 0.9606 to 0.9874. Experiments were carried out by setting pressure and braided angle constant, and measuring the forces generated by the PMA of different lengths. The coefficient (c_0) did not remain consistent as in Section 4.1 and this could be due to the random variation in the PMA. The final optimised PMA is 80 mm long and which is capable of lifting up a single finger digit from full flexion to a horizontal position.

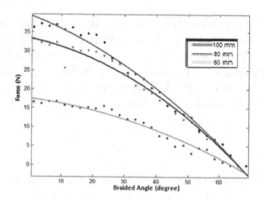

Fig. 11. Plot of contraction against sleeve length

Discussions and Future Work

The main limitation of a typical PMA is that it can only be actuated in a single direction. Since flexor hypertonia (the hand can only maintain a flexed position due to extensor weakness) is the most prevalent syndrome in stroke patients, the solution presented here could have a significant contribution to post-stroke hand rehabilitation. Even though the PMA requires an additional air compressor as its power source, the air pressure needed to actuate each finger was estimated to be 3.5 bar, which could be supplied by a compact, miniaturized air compressor.

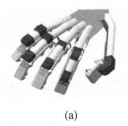 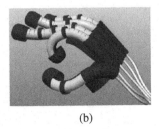

(a)   (b)

Fig. 12. Illustrations of (a) PMAs attached to testing rig, (b) a complete rehabilitation glove

As illustrated in Fig. 12. (a) and (b), the ultimate goal of this project is to develop and validate a multi-fingered model of a hand rehabilitation glove. Based on the model developed here, the dimension, material and shape of PMA can be optimized to minimize the weight of device and maximize the extension force. The proposed design has the potential to distinguish itself from other similar designs because of its highly complaint feature, low mechanical complexity and low profile.

Conclusions

This paper proposes a rotary movement of human finger joints for rehabilitation using flexible PMAs. The mathematical model was developed based on the dynamic and static analysis, verified and optimised through iterative tuning. Characterization of PMA considering the three most important factors; braided angle, pressure and braid length shows that a 80 mm long PMA, with one of its end fixed to the dorsal side of palm, is capable of fully rotating a finger joint with a pressure of 3.5 bar .

References

1. Mortality and Demographic Data (2009), http://www.health.govt.nz/publication/mortality-and-demographic-data-2006
2. Peter Langhorne, F.C.: Alex Pollock: Motor recovery after stroke: a systematic review. Lancet Neurol. 8, 741–754 (2009)
3. Krichevets, A.N., Sirotkina, E.B., Yevsevicheva, I.V., Zeldin, L.M.: Computer games as a means of movement rehabilitation. Disabil. Rehabil. 17(2), 100–105 (1995)

4. Biitefisch, C., Hummelheim, H., Denzler, P., Maurit, K.-H.: Repetitive training of isolated movements improves the outcome of motor rehabilitation of the centrally paretic hand. J. Neuro Sci. 130, 59–68 (1995)

5. Pilwon, H., Soo-jin, L., Kyehan, R., Jung, K.: Current Hand Exoskeleton Technologies for Rehabilitation and Assistive Engineering. Int. J. Precis. Eng. Man. 13(5), 807–824 (2012)

6. Dovat, L., Lambercy, O., Roger, G., Thomas, M., Ted, M., Teo Chee, L., Etienne, B.: HandCARE. IEEE Trans. Neural Syst. Rehabil. Eng. 16(6), 582–591 (2008)

7. Christopher, N.S., Rahsaan, J.H., Peter, S.L.: Development and pilot testing of HEXORR: Hand EXOskeleton Rehabilitation Robot. J. Neuroeng. Rehabil. 7(1), 1–16 (2010)

8. Satoshi Ueki, H.K., Satoshi, I., Yutaka, N., Motoyuki, A., Takaaki, A., Yasuhiko, I., Takeo, O., Tetsuya, M.: Development of a Hand-Assist Robot With Multi-Degrees-of-Freedom for Rehabilitation Therapy Satoshi. IEEE Intelligent Robots and Systems 2012 17(1), 136–146 (2012)

9. Devaraj, H.: Design and Development of a Portable Five-Fingered Hand Rehabilitation Exoskeleton, in Mechanicial Engineering Department, The University of Auckland I (2012)

10. John, B.M., Ralph, O.B., Dawn, B.M.: The SMART(R) Wrist-Hand Orthosis (WHO) for Quadriplegic Patients. J. Prosthet. Orthot. 5, 73–76 (1993)

11. Hommel, A., Wang, G.: Development and Control of a Hand Exoskeleton for Rehabilitation of Hand Injuries. In: IEEE Intelligent Robots and Systems 2005, pp. 3046–3051 (2005)

12. Popescu, D., Ivanescu, M., Popescu, D., Vladu, C., Vladu, I.: Force observer-based control for a rehabilitation hand exoskeleton system. In: 9th Asian Control Conference (2013)

Using a Gaze-Cueing Paradigm to Examine Social Cognitive Mechanisms of Individuals with Autism Observing Robot and Human Faces

Eva Wiese[1], Hermann J. Müller[2,3], and Agnieszka Wykowska[2,4]

[1] Department of Psychology, George Mason University, MSN 3F5 Fairfax, VA 22030, USA
[2] General and Experimental Psychology, Dept. of Psychology,
Ludwig-Maximilians- Universität, Leopoldstr. 13, 80802 Munich, Germany
[3] Birkbeck College, Department of Psychological Sciences, University of London,
Malet Street, London WC1E 7HX, UK
[4] Institute for Cognitive Systems, Technische Universität München,
Karlstr. 45/II, 80333 Munich, Germany
{wiese@psy.lmu.de,hmueller@lmu.de,agnieszka.wykowska@psy.lmu.de}

Abstract. This paper reports a study in which we investigated whether individuals with autism spectrum disorder (ASD) are more likely to follow gaze of a robot than of a human. By gaze following, we refer to one of the most fundamental mechanisms of social cognition, i.e., orienting attention to where others look. Individuals with ASD sometimes display reduced ability to follow gaze [1] or read out intentions from gaze direction [2]. However, as they are in general well responding to robots [3], we reasoned that they might be more likely to follow gaze of robots, relative to humans. We used a version of a gaze cueing paradigm [4, 5] and recruited 18 participants diagnosed with ASD. Participants were observing a human or a robot face and their task was to discriminate a target presented either at the side validly cued by the gaze of the human or robot; or at the opposite side. We observed typical validity effects: faster reaction times (RTs) to validly cued targets, relative to invalidly cued targets. However, and most importantly, the validity effect was larger and significant for the robot faces, as compared to the human faces, where the validity effect did not reach significance. This shows that individuals with ASD are more likely to follow gaze of robots, relative to humans, suggesting that the success of robots in involving individuals with ASD in interactions might be due to a very fundamental mechanism of social cognition. Our present results can also provide avenues for future training programs for individuals with ASD.

Keywords: Autism Spectrum Disorder, Human-Robot Interaction, Social Cognition, Social Interactions.

1 Introduction

Research in the area of social robotics and autism has greatly expanded in recent years. Robots have been shown to be effective in evoking social behavior in individuals with ASD (for review, see [3]). This has led many researchers to design social robots that could ultimately be used for training social skills in those who are impaired in this domain.

M. Beetz et al. (Eds.): ICSR 2014, LNAI 8755, pp. 370–379, 2014.
© Springer International Publishing Switzerland 2014

Robots that are designed for training social skills in individuals with ASD are typically tailored to match the needs of these particular populations. That is, they are designed to have simplified, not too overwhelming features; they are usually sufficiently human-like to be able to train *social* skills, but not too human-like to be intimidating for individuals with ASD; they offer sensory rewards for achievements (attractive sensory feedback related to behaviors that are being trained); they are designed to be safe in interaction (e.g., no sharp edges or jerky movements) and to offer control options to the interacting individual, which should enhance the ability to initiate an interaction (for list of design characteristics of robots for autism, see [3]).

Several case studies in which children with ASD responded well to an interaction with a humanoid robot have been described in the literature. For example, the robot Kaspar – designed at the University of Hertfordshire – has been shown to train children with ASD in emotion recognition, imitation games, turn-taking, and triadic interactions involving other humans [6]. Another robot, NAO (Aldebaran Robotics), was able to elicit eye contact in a child with ASD, and has also been reported to help in improving social interaction and communication skills [7]. Also Keepon – a robot simple in form and appearance developed by Hideki Kozima at the National Institute of Information and Communications Technology, Japan – has proved to evoke in children with autism social behaviors, interest, interpersonal communication [8-12] and even joint attention [13, 14].

These documented examples show that creating social robots for the purpose of training social skills in individuals with ASD is a promising avenue. To date, however, researchers have not unequivocally answered the question why robots are well accepted as social companions by individuals with ASD. The reported cases of social interactions between individuals with ASD and robots are, in most parts, qualitative data (video recordings, caregivers' reports or observation of an unconstrained interaction) and only a few studies have quantitatively investigated social interaction patterns [14-17]. Stanton and colleagues [15], for instance, found that children with ASD spoke more words with and were more engaged in interactions with social robots compared to simple toys that did not react to the children's behavior. As another example, Robins and colleagues [16] investigated whether the robot's appearance affects the patients' willingness to interact with them. It was found that children with ASD prefer robots with reduced physical features over very human-like robots. The studies provide evidence that children with ASD benefit from interacting with social robots resulting in improved social skills. However, the studies do not inform about the basic cognitive mechanisms that are triggered during interactions with social robots.

In order to answer the question of what cognitive mechanisms are actually at stake during interactions with robots – and what is the reason why the interactions with robots are more successful than those with other humans, one needs to conduct well-controlled experimental studies that are designed to examine selected cognitive mechanisms.

For example, the gaze-cueing paradigm [4, 5] is a well-established protocol to examine one of the most fundamental mechanisms of social cognition – gaze following. Gaze following occurs when one agent directs their gaze to a location; and another agent attends to that location (being *spatially* cued by the gaze direction of the first agent). Gaze following has been postulated to underlie important social cognitive processes such as mentalizing and joint attention [2]. Gaze following is an evolutionary adaptive mechanism [18], as attending to where others attend (as signaled by their gaze direction) informs about potentially relevant events in the environment (such as the appearance of a predator or prey). It also serves the purpose of establishing a common social context for joint action [19], among other types of interactions.

Individuals with ASD sometimes do not exhibit the typical pattern of results when reading out mental states from gaze behavior or in gaze cueing studies [1, 2]. A gaze cueing paradigm typically consists of a trial sequence in which first a face is presented centrally on a computer screen with gaze straight-ahead (in the direction of the observer). Subsequently, the gaze is shifted to a location and, then, a stimulus is presented either at a location in the direction to which the gaze is pointing (validly cued trials) or at a different location (invalidly cued trials). Participants are typically asked to detect, discriminate or localize the target stimulus. The logic behind this paradigm is that if participants follow the gaze of the observed agent on the screen, their focus of attention should be allocated to where the gazer gazes. Therefore, when the target stimulus appears at the attended location, its processing should be prioritized (due to attention having been already focused there), relative to when the target stimulus appears elsewhere. This has indeed been demonstrated by observing shorter reaction times [4, 5, 20-22] or lower error rates [22] to the target stimulus at the validly cued location, relative to invalidly cued locations. Moreover, brain responses (as measured by target-locked event-related potentials of the EEG signals) have been shown to be more enhanced for validly cued targets, relatively to invalidly cued targets [22, 23].

Interestingly, in our previous studies [20, 22], we have shown that gaze cueing effects were larger for human faces, as compared to robot faces when healthy adult participants were tested. We attributed this effect to humans adopting the so-called Intentional Stance [24] towards the observed human agent, but not towards the robot. Adopting the Intentional Stance is understood as "treating the object whose behavior you want to predict as a rational agent with beliefs and desires and other mental states exhibiting (…) intentionality" [24, p. 372]. In other words, adopting the Intentional Stance is simply attributing 'a mind' to the observed agent. In case of healthy adult participants, gaze following might make more sense when mind is attributed to the observed agent, relative to when the agent is treated only as a mechanistic device – because the gaze behavior of an agent with a mind might carry socially relevant content [18], while the gaze of a mechanistic device is devoid of such content. Accordingly, healthy adult participants follow the gaze of humans to a larger extent than that of mechanistic agents.

Aim of the Present Study

In the present study, we adopted a controlled paradigm targeted at a particular cognitive mechanism that can play a role in social interactions between individuals with ASD and robots. The aim was to test – using the gaze cueing paradigm involving human and robot agents – whether individuals with ASD would follow the gaze of a robot, even if they are reluctant to follow the gaze of humans [2]. The logic behind this was that since gaze following is one of the most fundamental mechanisms of social cognition, it might be affected by the general aptitude of individuals with ASD to interact with robots. If that were to be the case, this would cast light on the question of why robots are effective in eliciting social interactions in individuals with ASD.

2   Methods

2.1   Participants

18 patients (Mean age = 19.67, SD = 1.5, 3 women) diagnosed with ASD took part in this experiment as volunteers for monetary compensation. Participants were recruited at the St. Franziskus Berufsbildungswerk in Abensberg, where individuals with ASD are trained on a job to be integrated in a normal working environment. Participants received 8 Euros per hour as honorarium for participation. We decided to test social skills in adult participants diagnosed with autism, so that we would be able to conduct the study with the same paradigm (i.e., not modified for purposes of conducting the study with children) that has been designed earlier [20, 22] to test social attention in healthy adult participants.

2.2   Stimuli and Apparatus

Stimuli were photos 5.7° × 5.7° of visual angle in size. We decided to use static photographs in a very structured experimental setup in order to have properly controlled experimental conditions to examine the fundamental mechanisms of cognition. This is typically a first step to subsequent more realistic protocols, where experimental paradigms are introduced into real-life environments. First, however, it is necessary to answer the question of what are the exact cognitive mechanisms that are at stake and that are pinpointed by the controlled experimental setup.

In the human condition, the face of the same female individual was presented (source: Karolinska Directed Emotional Faces database [25]). In the robot condition, photos of an anthropomorphic robot (EDDIE, LSR, TU München) were presented, see Figure 1. EDDIE is a robot face developed to express emotions and thus has salient colourful facial features relevant for emotional expressiveness (big eyes, eyebrows, prominent lips) with additional animal-like characteristics (folding and extending cockatoo-like comb on top of its head as well as lizard-like ears on the sides) [26]. For our purposes, these features were not used: the robot had an invariable happy expression, with the comb and ears folded (almost not visible). The only "dynamic" part of the robot was the eyes that could move leftward or rightward. In fact, these were all static images, but the sequence of the images was rapidly presented producing an impression of a dynamic movement of the eyes.

Fig. 1. The human face (left) and the robot face (right) used in the present paradigm

Leftward or rightward gaze direction deviated by 0.2° from straight-ahead, in both the human and the robot condition. Stimuli were presented centrally on a white background, with eyes positioned on the central horizontal axis of the screen. Peripheral target letters were always presented at the same level as the eyes of the human or robot face. The target stimulus was a black capital letter (F or T), $0.2° \times 0.2°$ in size, which was presented at an eccentricity of 5.7° relative to the screen center (Fig. 2). Target positions (left or right) were determined pseudo-randomly.

Gaze direction was not predictive of the target position: gaze was directed either to the side on which the target appeared (valid trials, 50% trials) or to the other side (invalid trials, 50% of trials).

2.3   Procedure

Each experimental trial began with presentation of a fixation point (2 pixels) for 850 ms. The fixation display was followed a display with a face gazing straight-ahead (in the direction of the observer, 850 ms). The fixation dot remained visible (in-between the eyebrows of the face). The next event in the trial sequence consisted of a directional gaze shift to the left or the right. Subsequently, after 500 ms, the target letter was presented on either the left or the right side of the screen, with the face remaining present in the centre. Upon target presentation, participants responded as quickly and as accurately as possible to the identity of the target letter (F or T) using the 'd' or 'k' key on a standard keyboard, with response assignment counterbalanced across participants (d=F/k=T vs. d=T/k=F; the d/k letters were covered with F and T stickers). The target letter remained visible on the screen until a response was given or a time-out criterion (1200 ms) was reached. Figure 2 depicts an example trial sequence. The experiment consisted of 596 experimental trials preceded by 20 practice trials. All conditions were pseudo-randomly mixed.

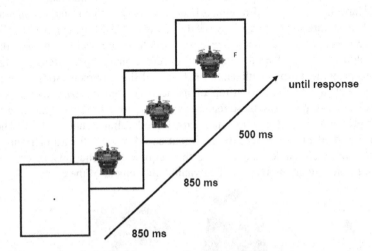

Fig. 2. An example trial sequence with validly cued condition. Proportions of stimuli relative to the screen are represented as they were in the experiment.

3   Results

Median reaction times (RTs) were computed for each participant. Individual median RTs were submitted to a 2×2 repeated-measures analysis of variance (ANOVA) with the factors *cue type* (human vs. robot face) and *validity* (valid vs. invalid). This ANOVA revealed a significant main effect of validity, $F (1, 17) = 10.848$, $p = .004$, $\eta_p^2 = .390$, with validly cued trials eliciting faster RTs (M = 541 ms, SEM = 24) than invalidly cued trials (M = 551 ms, SEM = 26). Importantly, however, this effect interacted significantly with the type of cue (human vs. robot face), $F (1, 17) = 5.104$, $p = .037$, $\eta_p^2 = .231$: while the human face did not elicit a gaze cueing effect ($M_{valid} = 543$ ms, SEM = 24 vs. $M_{invalid} = 547$ ms, SEM = 25, $t (17) = 1.203$, $p = .246$), the robot face did ($M_{valid} = 538$ ms, SEM = 23 vs. $M_{invalid} = 555$ ms, SEM = 27, $t (17) = 3.331$, $p = .004$), see Figure 3. The main effect of cue type was not significant, $F < .5$, $p > .55$.

An analogous analysis of error rates revealed no significant effects or interactions, all $Fs < 2$, all $ps > .22$. The mean error rate was 2%, with a standard deviation of .15.

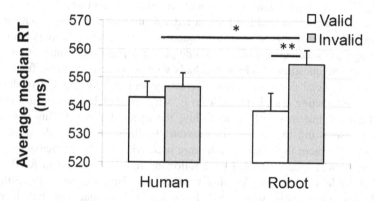

Fig. 3. Average median reaction times (RTs) required to discriminate the target (F vs. T) as a function of the validity of the gaze cue (valid vs. invalid) and the cue type (human vs. robot). Error bars represent standard errors of the mean adjusted to within-participants designs, according to [27]. *p < .05, **p < .005.

4   Discussion

The present findings reveal that individuals diagnosed with ASD are more likely to follow the eye movements of a robot face than those of a human face. Interestingly, this is in contrast with results obtained with healthy individuals [20], who – with the same paradigm – were more likely to attend to the direction of human gaze, relative to a robot gaze. This shows that while healthy participants are more ready to follow attention of human agents than of mechanistic devices, individuals with autism are more inclined to follow "eyes" of a non-human entity. This finding is of particular importance for social robotics, for several reasons. It demonstrates empirically that people with ASD can indeed process the eye movements displayed by a robot and shift their attentional focus to the gazed-at location. In doing so, individuals with ASD appear to react to robots in a similar way as healthy participants do to human partners: they share attention with them and attend to where others are attending. Thus, our data provide empirical evidence that robots have the capability of inducing attentional shifts in people with ASD and can thus be used to train people with ASD in the more general ability of gaze following. Sharing attention with others is also an important prerequisite for mentalizing and understanding others' actions – two social skills that are known to be impaired in ASD [2]. Since eye gaze directly informs about internal states, such as preferences or interests, and helps predicting what other people are going to do next [28], it seems that robots can be used to train individuals with ASD to understand others' intentions and predict others' actions – through gaze following.

While many robot systems have proved to be very successful in engaging individuals with ASD into an interaction [3], it is as yet little understood what the underlying cognitive mechanisms are. Our study reveals that it might be fundamental mechanisms (such as shared attention/gaze following) that are the basis for other higher-order social cognitive processes that are elicited in interactions with robots, but are not activated during interactions with humans. Therefore, the phenomenal experience of pleasantness and fun [3] that individuals with ASD seem to have when interacting with robots might be a consequence of more basic (and perhaps even implicit) cognitive mechanisms that come into play in human-robot interaction.

This raises the question of why individuals with ASD activate those fundamental mechanisms of social cognition when interacting with robots, but not to the same extent when interacting with humans. That is, why are they less likely to follow the eyes of humans, but more likely to follow the eyes of robots? This question is particularly interesting in the light of previous findings of Wiese, Wykowska and colleagues [20], where the same stimuli were used with healthy participants, but the opposite effect was found: stronger gaze following for the human than for the robot face. A possible explanation for this comes from Baron-Cohen [29, 30], who proposed that individuals with ASD have reduced mentalizing but increased systemizing skills, which makes them more interested in understanding the behavior of machines rather than of minds. Thus, it appears that the degree to which eye gaze is followed depends on how meaningful it is to the observer: Healthy controls make more sense of human-like eye movements and show stronger gaze following for human-like agents (presumably due to the behavior of human agents carrying socially informative content, [18]), while individuals with ASD make more sense of robot-like eye movements and show stronger gaze following for robot-like agents, presumably due to their aptitude for mechanistic systems and systemizing in general.

It might also be the case that both patterns of results are attributable to the same mechanism. That is, the differential cueing effects for human vs. robot faces (in both healthy participants and individuals with ASD) might be related to pre-activating certain representations of the observed stimulus: when a human face is observed, a whole representation of a human being might be activated; while a representation of a robot is activated when a robot face is seen. These representations include various characteristics. One of the characteristics of a human is that humans possess minds and their behavior is driven by mental states. In the case of healthy, typically developed people, this might produce a higher incentive to follow human gaze (relative to following gaze of a robot), because mental states and intentions carry socially informative meaning [18]. However, for individuals with ASD, the representation of a human might be associated with complex and probabilistic (hard to determine) behavior [31, 32]. A mechanistic device, by contrast, might be associated with a deterministic (and thus more predictable) behavior [30, 31]. Hence individuals with ASD may be more comfortable in the presence of systems with more predictable behavior, and thus be more ready to engage fundamental mechanisms of social cognition in interactions with them.

5   Concluding Remarks and Future Directions

There are two main conclusions that can be drawn from this research: First, social robots can be used to train people with ASD to follow eye gaze and understand that objects of interest are usually looked at before an action is performed with/on them. In doing so, one would hope that gaze following behavior shown with robots would generalize to human-human interactions and help people with ASD to develop basic mentalizing skills. Second, the present study casts light on the mechanisms that might be the reason for the success of robots in involving individuals with ASD into interactions with them [3]. We show that it might be the most fundamental mechanisms of social cognition that are elicited by robots, but that are not activated when individuals with ASD interact with other humans. As a consequence, interactions with robots are more efficient and smooth, and hence robots are successful in engaging individuals with ASD.

Acknowledgments. This work was supported by the German Research Foundation (Deutsche Forschungsgemeinschaft, DFG) EXC142 grant (DFG-Excellence Cluster 'Cognition for Technical Systems', sub-project #435: SocGaze awarded to HM) and a DFG grant awarded to AW (WY-122/1-1). We would also like to thank the St. Franziskus Berufsbildungswerk in Abensberg for enabling us to collect the data.

References

1. Ristic, J., et al.: Eyes are special but not for everyone: The case of autism. Cognit. Brain Res. 24, 715–718 (2005)
2. Baron-Cohen, S.: Mindblindness: an essay on autism and theory of mind. MIT Press/ Bradford Books, Boston (1995)
3. Cabibihan, J.J., Javed, H., Ang Jr., M., Aljunied, S.M.: Why Robots? A Survey on the Roles and Benefits of Social Robots for the Therapy of Children with Autism. Intl. J. of Social Robotics 5, 593–618 (2013)
4. Friesen, C.K., Kingstone, A.: The eyes have it! Reflexive orienting is triggered by nonpredictive gaze. Psychon. B. Rev. 5, 490–495 (1998)

5. Driver, J., Davis, G., Ricciardelli, P., Kidd, P., Maxwell, E., et al.: Gaze perception triggers reflexive visuospatial orienting. Vis. Cogn. 6, 509–540 (1999)
6. Robins, B., Dautenhahn, K., Dickerson, P.: From isolation to communication: a case study evaluation of robot assisted play for children with autism with a minimally expressive humanoid robot. In: Proc. 2nd Intern. Conf. Adv. Comp.–Human Inter., pp. 205–211. IEEE Press, New York (2009)
7. Shamsuddin, S., Yussof, H., Ismail, L.I., Mohamed, S., Hanapiah, F.A., Zahari, N.I.: Initial response in HRI-a case study on evaluation of child with autism spectrum disorders interacting with a humanoid robot Nao. Proc. Eng. 41, 1448–1455 (2012)
8. Kozima, H., Nakagawa, C., Yasuda, Y.: Interactive robots for communication-care: A case-study in autism therapy. In: IEEE International Workshop on Robot and Human Interactive Communication (ROMAN 2005), Nashville, TN, USA, pp. 341–346 (2005)
9. Kozima, H., Michalowski, M.P.: Keepon: A socially interactive robot for children. In: IEEE International Conference on Robotics and Automation (ICRA 2008),), The ICRA Robot Challenge Workshop, Pasadena, CA, USA (2008)
10. Kozima, H., Yasuda, Y., Nakagawa, C.: Social interaction facilitated by a minimally-designed robot: Findings from longitudinal therapeutic practices for autistic children. In: IEEE International Workshop on Robot and Human Interactive Communication, ROMAN 2007, Jeju, Korea (2007)
11. Kozima, H., Yasuda, Y., Nakagawa, C.: Robot in the loop of therapeutic care for children with autism. In: IEEE/RSJ International Conference on Intelligent Robots and Systems (IROS 2007), Workshop on Assistive Technologies: Rehabilitation and Assistive Robotics, San Diego, CA, USA (2007)
12. Kozima, H.: Robot-mediated communication for autism therapy. In: International Conference on Infant Studies, ICIS 2008, Vancouver, Canada (2008)
13. Kozima, H., Nakagawa, C., Yasuda, Y.: Children–robot interaction a pilot study in autism therapy. Prog. Brain Res. 164–385 (2007)
14. Zheng, Z., Bekele, E., Swanson, A., Crittendon, J.A., Warren, Z., Sarkar, N.: Impact of Robot-mediated Interaction System on Joint Attention Skills for Children with Autism. In: IEEE International Conference on Rehabilitation Robotics (2013)
15. Stanton, C.M., Kahn, P.H., Severson, R.L., Ruckert, J.H., Gill, B.T.: Robotic animals might aid in the social development of children with autism. In: Proc. HRI 2008, pp. 271–278. ACM Press (2008)
16. Robins, B., Dautenhahn, K., Dubowski, J.: Does appearance matter in the interaction of children with autism with a humanoid robot? Interaction Studies 7(3), 479–512 (2006)
17. Kim, E.S., Berkovits, L.D., Bernier, E.P., Leyzberg, D., Shic, F., Paul, R., Scassellati, B.: Social robots as embedded reinforcers of social behavior in children with autism. J. Autism Dev. Disord. 43, 1038–1049 (2013)
18. Tomasello, M.: Origins of Human Communication. MIT Press, Cambridge (2010)
19. Sebanz, N., Knoblich, G.: Prediction in joint action: what, when, and where. Topics in Cognitive Science 1, 353–367 (2009)
20. Wiese, E., Wykowska, A., Zwickel, J., Müller, H.I.: see what you mean: how attentional selection is shaped by ascribing intentions to others. PLoS ONE 7(9), e45391 (2012)
21. Wiese, E., Wykowska, A., Müller, H.: What we observe is biased by what other people tell us: Beliefs about the reliability of gaze behavior modulate attentional orienting to gaze cues. PLoS ONE 9(4), e94529 (2014)
22. Wykowska, A., Wiese, E., Prosser, A., Müller, H.: Beliefs about the minds of others influence how we process sensory information. PLoS ONE 9(4), e94339 (2014)

23. Schuller, A.M., Rossion, B.: Spatial attention triggered by eye gaze increases and speeds up early visual activity. Neuroreport 12, 2381–2386 (2001)

24. Dennett, D.C.: True believers: the intentional strategy and why it works. In: O'Connor, T., Robb, D. (eds.) Philosophy of Mind: Contemporary Readings, pp. 370–390. Routledge, London (2003)

25. Lundqvist, D., Flykt, A., Öhman, A.: The Karolinska Directed Emotional Faces (KDEF). Department of Neurosciences Karolinska Hospital, Stockholm (1998)

26. Kühnlenz, K., Sosnowski, S., Buss, M.: The Impact of Animal-like Features on Emotion Expression of Robot Head EDDIE. Advanced Robotics 24 (2010)

27. Cousineau, D.: Confidence intervals in within-subjects designs: A simpler solution to Loftus and Masson's method. Tutorial in Quantitative Methods for Psychology 1, 42–45 (2005)

28. Frith, C.D., Frith, U.: How we predict what other people are going to do. Brain Res. 1079, 36–46 (2006)

29. Baron-Cohen, S.: The extreme male brain theory of autism. Trends Cogn. Sci. 6, 248–254 (2002)

30. Baron-Cohen, S.: Autism occurs more often in families of physicists, engineers, and mathematicians. Autism 2, 296–301 (1998)

31. Watson, J.S.: Detection of self: The perfect algorithm. In: Parker, S.T., Mitchell, R.W., Boccia, M.L. (eds.) Self-awareness in Animals and Humans. Developmental Perspectives, pp. 131–148. Cambridge University Press, Cambridge (1994)

32. Schilbach, L., Timmermans, B., Reddy, V., Costall, A., Bente, G., et al.: Toward a second-person neuroscience. Behav. Brain Sci. 36, 393–462 (2013)

Adaptive Object Learning for Robot Carinet

Shengtao Xiao[1], Shuzhi Sam Ge[1,2] and Shuicheng Yan[1]

[1] Department of Electrical and Compture Engineering, National University of Singapore,
Singapore 119613
[2] Social Robotics Laboratory, Interactive Digital Media Institute,
National University of Singapore, Singapore 119613

Abstract. In this paper, an adaptive object learning method based on deep neural network is developed for a robot to learn features of moving objects, e.g., humans and vehicles, via observation. The proposed method provides a solution for the robot to learn unknown moving objects in a real-time scenario. A hybrid scheme of learning and identification is proposed to recognize the moving object by fusion of foreground segmentation and identification.

1 Introduction

It is believed that the robots will play a closer role in human daily life as they have been playing in the previous decades in the field of automation industry. Social robots are expected to play an even more important role in the fields of elderly care, home/hotel services, and health care in the near future as what Gates once envisioned [1]. Among all the tasks that social robots have competence to accomplish before being seamlessly integrated into human society, visual recognition is one of the most challenging ones to be solved. How can a robot visually recognize different objects even though some of them are not defined initially? To recognize unknown/undefined objects is an interesting topic which draws lots of attention from the research community [2, 3].

In all detection/recognition problems, pedestrian detection is one of the key technologies in automotive safety, human robot interaction and intelligent video surveillance. Much research work has been devoted in this area and the performances are sound. Various pedestrian datasets, e.g., the Inria dataset and the Caltech dataset, are used to train the features and the classifiers for detection. An arbitrary database is essential for learning representative features and precise classifiers. However, recognition of certain objects may be challenging as there may be no special dataset.

Deep learning has been extensively studied and applied in object classification and detection ever since the paper by Hinton [4], among which Convolutional Neural Network (CNN) [5] has been very popular. In [6], Zeiler introduced a method to effectively learn high level features with a newly proposed deconvolutional neural network. Additional switches introduced to the subsampling process break the traditional "output of the previous layer as input of the next" scheme. High level features learned with deconvolutional neural network yield impressive accuracy in classification tasks. The work [7] took traditional CNN as the basic architecture for feature detection and introduced a deformation to detect different parts of a pedestrian, such as upper body,

M. Beetz et al. (Eds.): ICSR 2014, LNAI 8755, pp. 380–389, 2014.

lower body, head, etc. Recent development in sparse CNN [8, 9] provided an unsupervised method for learning sparse convolutional features by combining the advantages of sparse coding and convolutional neural networks. Impressive performance has been obtained with this method in various tasks.

How to effectively and efficiently learn the human features and identify the person with whom the robot is interacting worths our research attention. Instead of handcrafting the human detector, we intend to build an intelligent learning scheme which automatically create dataset of an certain object and learn the classifier. In this paper, a human detector is to be learned with the proposed scheme without using external information/dataset. We aim to provide a universal learning algorithm for robots which can learn a variety of objects with the proposed way. In our scheme, the frame/robot is fixed in the observation stage, which ensures a static background for better object extraction for the demonstration. Database of unknown moving objects can also be constructed with moving robots by using of some techniques such as [10]. In this paper only focuses on those with static background.

A learning scheme which enables a social robot to gradually expand its knowledge pool and allows it to learn more objects in an adaptive manner has great potential in real world applications. In this paper, we propose such a scheme for a robot to self-learn unknown moving objects and use the learned neural network to recognize the object of the same class after training the network in a supervised manner with labels from humans. The proposed scheme allows the robot to associate the learned features with the physical meanings which can be understood by humans. We propose to use foreground segmentation to extract the moving object (human), followed by a supervised training of a convolutional neural network to learn the human classifier.

The remainder of the paper is organized as follows. Section 2 introduces some related works in foreground detection and convolutional neural networks. Section 3 describes the newly proposed learning system. In Section 4, experiments to verify the effectiveness of the proposed learning scheme are presented. The conclusion and future work are given in Section 5.

2 Related Work

2.1 Foreground Segmentation for Object Detection

Foreground segmentation/background segmentation has been widely studied in the literature in the context of video surveillance, optical motion capture and multimedia. The basic idea is to use background methods and do background subtraction to obtain the foreground information. Various approaches, such as Basic Background Modeling [11–13], Statistical Background Modeling including Single Gaussian (SG) [14], mixture of Gaussians (MOG) [15, 16] and Kernel Density Estimation [17], have been intensively investigated in the literature. Challenging factors mainly include dynamic background, sudden and gradual illumination changes, camera noises and moving background, which are also major problems to be solved in recent development of background segmentation. A texture based background segmentation methods was proposed in [18].

An adaptive background mixture model with online EM learning [16] is implemented in our system to subtract the background and retrieve the moving foreground for learning at later stages. The model is highly robust to illumination variations and shadows.

2.2 Convolutional Neural Network

Typical Structure: A typical Convolutional Neural Network [5] is shown in Fig. 1. Cx represents the convolutional layer, Sx denotes the subsampling layer and Fx the fully-connected layer. The first layer is the input layer which takes the training and testing samples as input data. A followed convolutional layer outputs the results of convolving k filters with the input layer, resulting in k feature maps. After convolution operation, the feature map will be shifted and distorted to the same amount as the input is shifted or distorted. Layer S2 is a subsampling layer which is crucial for obtaining the relative position of a certain feature of the samples. In this layer, a nonlinear mapping is implemented with a activation function. C3 may have input from several outputs from S2 (full connection is also possible). This results in higher level complex features, as complex features are usually formed by lower level features. F6 is a fully connected layer with C5 and its output will be fed into the output layer via a softmax function.

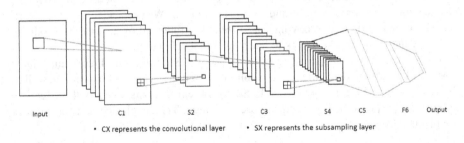

Input   C1   S2   C3   S4   C5   F6   Output

- CX represents the convolutional layer   - SX represents the subsampling layer

Fig. 1. Structure of a typical CNN

Convolution Operation: For convolutional neural network, 2-D convolution operation is used. For the 2-D convolution operation,

$$o[m, n] = f[m, n] * g[m, n] = \sum_{u=-\infty}^{\infty} \sum_{v=-\infty}^{\infty} f[u, v]g[u - m, v - n] \qquad (1)$$

where * denotes the convolution operation. Then a feature map is obtained by convolving the input image/feature with a linear filter, plus a bias and then followed by a non-linear mapping. This operation can be expressed by Eq. 2,

$$h_{i,j}^k = \varphi((W^k * x)_{i,j} + b_k) \qquad (2)$$

where the k-th feature map of a given layer is denoted as h^k. W^k and b_k are the k-th linear filter and bias term. $\varphi(\cdot)$ is the activation function, e.g. tanh, sigmoid, ReLU [19], etc. In this paper, rectified linear units (ReLU) is used due to its advantages in performing classification tasks [20]. ReLU suffers much less from the widely existing vanishing gradient problem. It was also reported in [20] that ReLU can substantially accelerate the training process as compared with the cases when other activation functions are used. The activation function of a ReLU is given in Eq. 3.

$$\varphi(u) = \max(0, u) \tag{3}$$

Dropout: The dropout technique [21] is adopted to prevent the overfitting problem. In the dropout network, as shown in Fig. 2, each hidden unit may be randomly omitted from the network with a user defined probability (usually, P=0.5). The output of the dropout network can be expressed as

$$h_{i,j}^k = \text{compare}(V_x, P)\varphi((W^k * x)_{i,j} + b_k) \tag{4}$$

where P is the user defined dropout probability and V_x is a randomly generated real positive number ranging from 0 to 1. If $V_x \geq P$, compare$(V_x, P) = 1$. Else when $V_x < P$, compare$(V_x, P) = 0$. When a neuron is selected to be "dropped out", it will not contribute to the forward pass and does not participate in the back-propagation. Dropout keeps sampling the entire network and uses the sampled network to do forward prediction and backward propagation. Therefore, this effectively prevents the co-adaptation of neurons as the structure varies every time and a neuron can hardly rely too much on a particular set of neurons. This forces the neurons to learn more robust features.

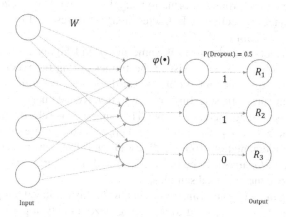

Fig. 2. Dropout neural network

3   Adaptive Object Learning System for Robot

The learning system has two steps. The first step is to generate a dataset and the second step is to train a CNN-based classifier. There are existing pedestrian datasets Algorithm

1 is therefore proposed. For demonstration, the object to be learned is the human. Other objects like moving vehicles, animals, etc. can also learned with similar algorithms. These are moving objects which can be easily subtracted and from the background [14–16] and formed as a sample dataset.

Algorithm 1. Object Learning System

Video Recording

Foreground Extraction

Extract foreground images as positive samples and background samples as negative samples.

Automatically Resize and crop images average scale, in this case 28×76.

Supervised Training for Classifier Learning

Feed the obtained dataset into the a user configured Convolutional Neural Network and train the classifier.

3.1 Removing Highly Occluded Foreground

A robot, Carinet (details of the robot will be provided in Section 4), is ordered to record some videos using its camera. To save memory, the camera will only be triggered when moving objects exist in the frame, and the robot will only save the frames when there are moving persons. There are 7 persons in the workspace and their movements are not pre-assigned. After sufficient frames are obtained, the robot will then pre-process the videos to extract foreground information, e.g. Fig. 4, with the foreground detection algorithm mentioned in Section 3. Extracted images are rescaled to a fixed size, e.g. Fig. 5, for classifier training.

Foreground detection will return us some noisy and highly occluded foreground scenes. Occluded objects widely exist in the frames during the recording process of the camera. Since highly occluded objects can deteriorate the training at the current stage, we adopt an automatic selecting process to remove the images with occluded objects. This process is performed as follows. The sample S_i is selected if $S_l > w_l + \alpha_l$, $S_r < w_r - \alpha_r$, $S_t > h_t + \alpha_t$, $S_b < h_b - \alpha_b$ where w_r and h_b represent the camera resolution. In our implementation, 160×120 resolution is used, $w_r = 160$ and $h_b = 120$ in this case. w_l and h_b are set to be 0. $\alpha_l, \alpha_r, \alpha_t$ and α_b are arbitrary values designed manually to remove the occluded samples.

After cropping the foreground and removing the highly occluded foreground images, the remaining samples are rescaled to an average size. Finally, a positive dataset with a uniform image size will be obtained for training. Negative samples can be easily cropped from the frames when the foreground is not detected. Some samples from the auto-generated dataset are shown in Fig. 6 and Fig. 7. The negative samples consist of certain parts of the objects (in this case, humans). A positive sample is defined as the case where the main body of the object is included and detected in the image.

Fig. 3. Scene Captured with moving persons

Fig. 4. Corresponding foreground area

Fig. 5. Sample Foreground Extracted and rescaled

Fig. 6. Positive Samples extracted from the videos

Fig. 7. Negative Samples extracted from the videos

3.2 Convolutional Neural Network Structure

In this work, it is believed that a conventional convolutional neural network plus some overfitting-preventing and non-linear mapping techniques will be sufficient for learning simple and robust classifiers. Using other forms of CNNs, e.g. unsupervised feature learning to pre-train CNN [8] and deconvolutional neural network [22], may also provide good performance. However, The conventional CNN is relatively convenient in implementation. More importantly, only supervised training is required for the traditional CNN. As compared to the newly existing CNN, traditional CNN can save substantial training time if the pre-training process is not involved. The details of the CNN structure adopted to help robustly learn a human classifier will be discussed in the following of this section.

The CNN structure used in this paper is illustrated in Fig. 8. This is a simple and straight forward structure which can be easily implemented in our robotic system. It consists of 10 layers except the input and output layers. All neural weights and bias terms are randomly initialized. The input samples are in gray-scale with the size of 28×76.

Conv1 layer is a convolutional layer with 16 feature maps. A padding of 2 pixels with zero value is added to the edges of the input images to center the filters to the image edges. These augmented images are then convolved with 16 linear filters by Eq. 1, each filter with a size of 5×5. In this case, the output feature map has the same size as the input image, e.g., 28×76.

Pool2 layer is a sub-sampling layer where the max-pooling operation is performed on the 16 feature maps from the Conv1 layer. Overlapping pooling is used. This can be achieved by setting the kernelsize=3 and stride=2. The feature maps of Pool2 have half the row number and column number as those of Conv1, e.g. 14×38.

ReLU3 is a rectified linear unit layer, and there is no change in data dimension for this layer. Conv4 takes the output of ReLU3 as the input and filters the input with 36 kernels of size $5 \times 5 \times 16$. Similar to Conv1, a padding of 2 pixels is added. Feature maps of Conv4 are of size 14×38. On top of Conv4 is a rectified linear unit layer which is then followed by an average overlapping pooling layer, Pool6, with Kernelsize=3 and stride=2. Fully connected layer FC7 and FC9 have 10 and 2 neurons respectively. The dropout layer, Dropout8, has a dropout probability of 0.5. A softmax regressor layer is connected to FC9 to output the label of input samples.

Fig. 8. Full CNN structure for learning

Fig. 9. Carine Robot: Left) Overview; Right)Camera Location

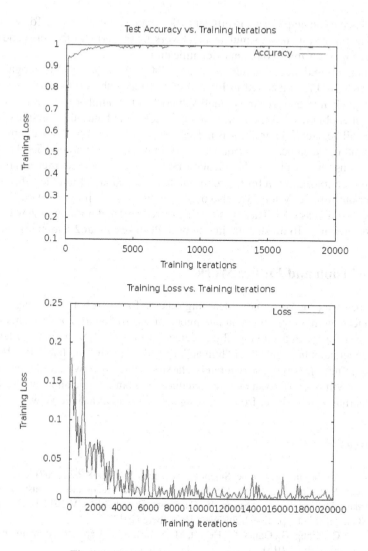

Fig. 10. Classification accuracy of the testing set

4 Experiments

The Carinet robot, as shown in Fig. 9, is a social bear developed at Social Robotics Laboratory, Interactive Digital Media Institute, National University of Singapore. The robot has one built-in camera, a 5-DOF exoskeleton inside, speech module and audio sensor running under a Linux Operating System. The objective of this robot is to closely interact with children via double channels, force interaction and face-to-face speech interaction. The vision sensor used in the robot is a simple webcam. The experiments were done on the data captured by the robot, in order to verify whether the proposed learning scheme can help Carine to efficiently learn an unknown object. To evaluate the proposed learning system, we simulate a real working scenario for the robot and test its learning/recognition ability on human identification.

With the proposed method, after training for 20000 iterations, 100% recognition accuracy is achieved as illustrated in Fig. 10. Fig. 10 also shows that the training loss converges gradually to an arbitrary small value when the number of training iterations increases. It is obvious that the more training is done, the better the classification performance will be. However, training is indeed a very time and power consuming task. A balance point between best performance and training time is at around 5000 iterations which gives us an acceptable classification rate of 99.4%, with shorter training time and less power consumption for the robot. In this case, about 54 minutes of training is needed for our robotic system. We also tried to train the same program with Caffe [23] on a GTX660 GPU as well. There is a significant reduction of training time without any lose of performance. To finish 5000 iterations, GPU takes about 2.5 minutes only.

5   Conclusion and Future Work

In this paper, we proposed a novel learning scheme for the robot to actively learn unknown objects without too much human intervention. To be more precise, this scheme consists of a foreground moving object detection plus a CNN learning to teach the robot to understand the concept of "human". Experiments showed that after about half an hour training, the robot can accurately classify scenes with/without human beings. Future work will be carried out on how to reduce the training time and how to allow the robot to learn a certain object from a moving view instead of a static view.

References

1. Gates, B.: A robot in every home. Scientific American 296(1), 58–65 (2007)
2. Fouhey, D.F., Collet, A., Hebert, M., Srinivasa, S.: Object recognition robust to imperfect depth data. In: Fusiello, A., Murino, V., Cucchiara, R. (eds.) ECCV 2012 Ws/Demos, Part II. LNCS, vol. 7584, pp. 83–92. Springer, Heidelberg (2012)
3. Romea, A.C., Xiong, B., Gurau, C., Hebert, M., Srinivasa, S.: Exploiting domain knowledge for object discovery (2013)
4. Hinton, G.E., Salakhutdinov, R.R.: Reducing the dimensionality of data with neural networks. Science 313(5786), 504–507 (2006)
5. LeCun, Y., Bottou, L., Bengio, Y., Haffner, P.: Gradient-based learning applied to document recognition. Proceedings of the IEEE 86(11), 2278–2324 (1998)
6. Zeiler, M.D., Taylor, G.W., Fergus, R.: Adaptive deconvolutional networks for mid and high level feature learning. In: 2011 IEEE International Conference on Computer Vision (ICCV), pp. 2018–2025. IEEE (2011)
7. Ouyang, W., Wang, X.: Joint deep learning for pedestrian detection. In: ICCV (2013)
8. Kavukcuoglu, K., Sermanet, P., Boureau, Y.-L., Gregor, K., Mathieu, M., Cun, Y.L.: Learning convolutional feature hierarchies for visual recognition. In: Advances in Neural Information Processing Systems, pp. 1090–1098 (2010)
9. Sermanet, P., Kavukcuoglu, K., Chintala, S., LeCun, Y.: Pedestrian detection with unsupervised multi-stage feature learning. arXiv preprint arXiv:1212.0142 (2012)

10. Sheikh, Y., Javed, O., Kanade, T.: Background subtraction for freely moving cameras. In: 2009 IEEE 12th International Conference on Computer Vision, pp. 1219–1225. IEEE (2009)
11. Lee, B., Hedley, M.: Background estimation for video surveillance
12. McFarlane, N., Schofield, C.: Segmentation and tracking of piglets in images. Machine Vision and Applications 8(3), 187–193 (1995), cited By (since 1996)163
13. Zheng, J., Wang, Y., Nihan, N., Hallenbeck, M.: Extracting roadway background image: Mode-based approach. Transportation Research Record (1944), 82–88 (2006), cited By (since 1996)14
14. Wren, C.R., Azarbayejani, A., Darrell, T., Pentland, A.P.: Pfinder: Real-time tracking of the human body. IEEE Transactions on Pattern Analysis and Machine Intelligence 19(7), 780–785 (1997)
15. Stauffer, C., Grimson, W.: Adaptive background mixture models for real-time tracking. 2, 246–252 (1999), cited By (since 1996)1864
16. KaewTraKulPong, P., Bowden, R.: An improved adaptive background mixture model for real-time tracking with shadow detection. In: Video-Based Surveillance Systems, pp. 135–144. Springer (2002)
17. Elgammal, A., Harwood, D., Davis, L.: Non-parametric model for background subtraction. In: Vernon, D. (ed.) ECCV 2000. LNCS, vol. 1843, pp. 751–767. Springer, Heidelberg (2000)
18. Heikkila, M., Pietikainen, M.: A texture-based method for modeling the background and detecting moving objects. IEEE Transactions on Pattern Analysis and Machine Intelligence 28(4), 657–662 (2006)
19. Nair, V., Hinton, G.E.: Rectified linear units improve restricted boltzmann machines. In: Proceedings of the 27th International Conference on Machine Learning (ICML 2010), pp. 807–814 (2010)
20. Krizhevsky, A., Sutskever, I., Hinton, G.E.: Imagenet classification with deep convolutional neural networks. In: NIPS, vol. 1, p. 4 (2012)
21. Hinton, G.E., Srivastava, N., Krizhevsky, A., Sutskever, I., Salakhutdinov, R.R.: Improving neural networks by preventing co-adaptation of feature detectors. arXiv preprint arXiv:1207.0580 (2012)
22. Zeiler, M.D., Krishnan, D., Taylor, G.W., Fergus, R.: Deconvolutional networks. In: 2010 IEEE Conference on Computer Vision and Pattern Recognition (CVPR), pp. 2528–2535. IEEE (2010)
23. Jia, Y.: Caffe: An open source convolutional architecture for fast feature embedding (2013), http://caffe.berkeleyvision.org/

Adaptive Control of Robotic Arm Carrying Uncertain Time-Varying Payload Based on Function Approximation Technique

Norsinnira Zainul Azlan[1] and Syarifah Nurul Syuhada Sayed Jaafar[2]

[1,2] Department of Mechatronics Engineering, Kulliyyah of Engineering, International Islamic University Malaysia, Jalan Gombak, 53100 Kuala Lumpur, Malaysia
sinnira@iium.edu.my, syarifahsyuhada90@gmail.com

Abstract. This paper presents a new adaptive controller based on the Function Approximation Technique (FAT) for a 2 degree of freedom (DOF) robot arm carrying uncertain time-varying payload. The mathematical model of the arm is expressed as the summation of the known and the uncertain terms in deriving the proposed control law. The time-varying uncertainty is expressed using an FAT expression, which avoids the need of linear parameterization of the mathematical model of the robotic arm. The expression also allows the update law to be easily derived by an appropriate choice of Lyapunov-like function. The stability proof of the controller is described in detail and computer simulation results are presented to demonstrate the effectiveness of the proposed technique.

Keywords: Adaptive control, Function Approximation Technique (FAT), unknown payload, time-varying uncertainties, robotic arm.

1 Introduction

Trajectory control of robotic arms is important in many applications including in service robots, humanoids, medical robots, teleportation and industry. In the service robots and humanoids applications, the robots are required to perform the same task as human and work in the same environment as human. The robot may need to carry various kinds of objects with time-varying weight. The complete information of the weight may not be known a priori in practice. For example, a service robot in a restaurant may need to carry a jug containing water and its weight changes as the water is being poured from it. A human arm can easily adapt with the unknown variation of the payload mass that it carries. However, for a robotic arm under the control of a model-based controller, an accurate dynamics model of the robotic arm is needed to achieve a high tracking performance. Nevertheless, the arm dynamics is influenced by the variation of the payload and this information may not be available in advance.

Several researches have been conducted in developing control laws to cater for the uncertain payload [1]-[7]. One of the effective techniques in dealing with uncertainties is the adaptive control. It has been one of the most important research areas since 1950's [1]. Cazalilla [2] implemented an adaptive control scheme for a 3-DOF

M. Beetz et al. (Eds.): ICSR 2014, LNAI 8755, pp. 390–399, 2014.

parallel manipulator. Lee [3] proposed an adaptive control scheme to provide the speed/position control of induction motors with lacking knowledge of some mechanical system parameters, such as the motor inertia, motor damping coefficient, and payload. Dai [4] et al. designed an adaptive control system for a quadrotor that delivers a cable-suspended payload without a prior knowledge of the payload mass. However, these controllers only focus on constant uncertain payloads.

Slotine and Li [5] presented the adaptive control of N degree of freedom (DOF) manipulators with time-dependent uncertainties by linearly parameterizing the manipulator dynamic model. Similarly, Pagilla and Zu [1] developed an adaptive control scheme for mechanical systems with time-varying parameters and disturbance by linearly parameterizing the manipulator model. However, in some cases, it is difficult to linearly parameterize the mathematical model of the robot [1]. The technique requires the knowledge of the complicated regression matrices [6].

Huang and Chien [7] introduced a regressor-free adaptive control for manipulators operating under time-varying uncertainties by the utilization of Function Approximation Technique (FAT). The uncertainties are expressed by FAT equations such as Fourier Series, Bessel and Taylor polynomials. Several researches have also been conducted in implementing neural network as an FAT [8]. With FAT, the uniform ultimate boundedness of the closed loop system can be achieved using Lyapunov stability theory. The work in [7] generalizes the unknown payload mass carried by a robot manipulator as the uncertainties in the inertia, coriolis and centrifugal and gravitational matrices of the robot. This method [8] results in a high number of adaptation gain matrices need to be adjusted. This makes the tuning effort of the control parameters to be time consuming.

This study proposes an FAT-based adaptive controller for controlling a robotic arm while operating under uncertain time-varying payload. The contribution of this paper is in the derivation of the FAT-based update law by specifically focusing on the uncertainties in the payload mass. The technique is capable of compensating time-varying payload and avoids the need to linearly parameterize the robotic arm's dynamic model. Only one adaptation gain is needed since the controller is designed based on the uncertain payload mass directly. The mathematical model of the robotic arm is divided into two parts, consists of the known parameters and uncertain payload term. The payload is represented by Function Approximation Technique (FAT) that is expressed as the multiplication of a constant weighting vector and a time-varying basis function

This paper is organized as follows, the dynamic model of a planar 2-DOF robotic arm are presented in Section 2. The FAT based adaptive control to compensate unknown time-varying payload and its stability proof are given in sections 3 and 4, respectively. Simulation results are discussed in Section 5, and finally conclusions are drawn in Section 6.

2   Dynamic Model of a 2-DOF Robotic Arm

The dynamic model of an N-DOF robotic arm can be represented using Lagrange equation as

$$M(q)\ddot{q}(t)+C(q,\dot{q})\dot{q}(t)+G(q)=\tau \tag{1}$$

where $q,\dot{q}$ and $\ddot{q}$ are the $N\times1$ vector of joint angular position, joint angular velocity and joint angular acceleration respectively, $M(q)$ is the $N\times N$ symmetric positive definite inertia matrix, $C(q,\dot{q})$ is the $N\times N$ matrix of Coriolis and Centrifugal torques, $G(q)$ is the $N\times1$ vector of gravitational torque and τ is the $N\times1$ vector of control input torque from the actuators.

For a 2-DOF planar robotic arm as illustrated in Fig.1, $M(q)$, $C(q,\dot{q})$ and $G(q)$ can be denoted as $M_2(q)$, $C_2(q,\dot{q})$ and $G_2(q)$ respectively. Since the unknown time-varying payload is carried by the end effector in the second link, its mass is regarded as a part of the mass of link 2. Therefore, the uncertain mass, $m_2(t)$ can be factorized from the inertial, coriolis and centrifugal matrices, and gravitational vector by expressing $M_2(q),C_2(q,\dot{q})$ and $G_2(q)$ as

$$
\begin{aligned}
M_2(q) &= M_{m1}(q)+m_2(t)P_m(q)\\
C_2(q,\dot{q}) &= m_2(t)P_C(q,\dot{q})\\
G_2(q) &= G_{m1}(q)+m_2(t)P_g(q)
\end{aligned}
\tag{2}
$$

and

$$q=\begin{pmatrix}\theta_1\\\theta_2\end{pmatrix},\dot{q}=\begin{pmatrix}\dot{\theta}_1\\\dot{\theta}_2\end{pmatrix},\ddot{q}=\begin{pmatrix}\ddot{\theta}_1\\\ddot{\theta}_2\end{pmatrix},M_{m1}(q)=\begin{pmatrix}\frac{1}{3}m_1l_1^2 & 0\\0 & 0\end{pmatrix},G_{m1}(q)=\begin{pmatrix}\frac{1}{2}m_1gl_1\cos\theta_1\\0\end{pmatrix},$$

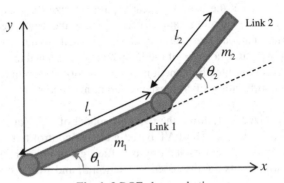

Fig. 1. 2 DOF planar robotic arm

$$P_m(q) = \begin{pmatrix} l_1^2 + \dfrac{1}{3}l_2^2 + l_1 l_2 \cos\theta_2 & \dfrac{1}{3}l_2^2 + \dfrac{1}{2}l_1 l_2 \cos\theta_2 \\[2mm] \dfrac{1}{3}l_2^2 + \dfrac{1}{2}l_1 l_2 \cos\theta_2 & \dfrac{1}{3}l_2^2 \end{pmatrix},$$

$$P_g(q) = \begin{pmatrix} \dfrac{1}{2}gl_1 \cos\theta_1 + \dfrac{1}{2}gl_2 \cos(\theta_1 + \theta_2) \\[2mm] \dfrac{1}{2}gl_2 \cos(\theta_1 + \theta_2) \end{pmatrix},$$

$$P_C(q,\dot{q}) = \begin{pmatrix} 0 & -\dfrac{1}{2}l_1 l_2 (\sin\theta_2)\dot{\theta}_2 - l_1 l_2 (\sin\theta_2)\dot{\theta}_1 \\[2mm] \dfrac{1}{2}l_1 l_2 (\sin\theta_2)\dot{\theta}_1 & 0 \end{pmatrix}, \qquad (3)$$

and l_i is the length of ith link, m_1 is the mass of the first link, g is the gravitational acceleration, $\theta_i, \dot{\theta}_i$ and $\ddot{\theta}_i$ are the angular displacement, velocity and acceleration of ith link respectively. Note that, in this model, $m_2(t)$ represents the total mass of link 2 and the uncertain payload since the unknown time-varying payload is regarded as a part of link 2. The following properties are utilized in the derivation of the adaptive control law:

- The inertia matrix, $M_2(q)$, is a symmetric positive definite matrix.
- The matrix $\dot{M}_2(q) - 2C_2(q,\dot{q})$ is skew symmetric.

3   Adaptive Control Based on Function Approximation Technique

In this section, an adaptive control strategy and its update law to compensate for unknown time-varying payload will be presented in detail.

Let the joint tracking error vector be $e_q(t) = q(t) - q_d(t)$, where $q_d(t)$ is the vector of the desired trajectory and is assumed to be twice differentiable. Setting the reference velocity error as $e = \dot{e}_q + \lambda e_q$ and the reference joint velocity as $\dot{q}_r = \dot{q}_d - \lambda e_q$, the adaptive control law can be described as [1]

$$\tau = \hat{M}_2(q)\ddot{q}_r + \hat{C}_2(q,\dot{q})\dot{q}_r + \hat{G}_2(q) - Ke \qquad (4)$$

where K is a 2×2 positive definite diagonal matrix, $\hat{M}_2(q),\hat{C}_2(q,\dot{q})$ and $\hat{G}_2(q)$ are the estimation of $M_2(q),C_2(q,\dot{q})$ and $G_2(q)$ respectively. From (2), these estimations can further be described as

$$\hat{M}_2(q) = M_{m1}(q) + \hat{m}_2(t)P_m(q),$$
$$\hat{C}_2(q,\dot{q}) = \hat{m}_2(t)P_C(q,\dot{q}),$$
$$\hat{G}_2(q) = G_{m1}(q) + \hat{m}_2(t)P_g(q).$$

(5)

It can be seen in (5) that only $m_2(t)$ is unknown, contributing to the uncertainties in $\hat{M}_2(q),\hat{C}_2(q,\dot{q})$ and $\hat{G}_2(q)$ matrices. Selecting the update laws as

$$\dot{\hat{W}}_{m_2} = -Q_{m_2}^{-1}\left[Z_{m_2}e^T P_m(q)\ddot{q}_r + Z_{m_2}e^T P_C(q,\dot{q})\dot{q}_r + Z_{m_2}e^T P_g(q)\right],$$

(6)

where $\dot{\hat{W}}_{m_2}$ is the vector of the constant weighting function of the FAT representation that is to be estimated, Z_{m_2} is the vector of the time-varying basis function of the FAT representation that can be chosen and Q_{m_2} is the adaptive gain matrix, will lead to the reference velocity error, e, the joint tracking error, e_q and the joint velocity tracking error, $\dot{e}_q$ asymptotically converge to zero as time, t approaches infinity. The stability proof for the proposed control strategy (4) and its update law (6) will be described in the following section.

4   Stability Proof

Substituting control law (4) into the mathematical model of the robot manipulator in (1) gives

$$M_2(q)\ddot{q} + C_2(q,\dot{q})\dot{q} + G_2(q) = \hat{M}_2(q)\ddot{q}_r + \hat{C}_2(q,\dot{q})\dot{q}_r + \hat{G}_2(q) - Ke \quad (7)$$

Subtracting ($M_2(q)\ddot{q}_r + C_2(q,\dot{q})\dot{q}_r$) from both sides of (7) yields

$$M_2(q)(\ddot{q} - \ddot{q}_r) + C_2(q,\dot{q})(\dot{q} - \dot{q}_r)$$
$$= [\hat{M}_2(q) - M_2(q)]\ddot{q}_r + [\hat{C}_2(q,\dot{q}) - C_2(q,\dot{q})]\dot{q}_r + \hat{G}_2(q) - G_2(q) - Ke$$

(8)

Utilizing (5) and replacing $\dot{q} - \dot{q}_r$ by e, (8) becomes

$$M(q)\dot{e} + C(q,\dot{q})e = \tilde{m}_2[P_m(q)\ddot{q}_r + P_C(q,\dot{q})\dot{q}_r + P_g(q)] - Ke.$$

(9)

$\tilde{m}_2$ is the estimation error representing the difference between the actual payload, $m_2(t)$ and its estimation $\hat{m}_2(t)$, where $\tilde{m}_2(t) = \hat{m}_2(t) - m_2(t)$. The actual and uncertain payload, $m_2(t)$ and $\hat{m}_2(t)$, can be expressed using FAT in terms of multiplication of constant unknown weighting function and the selected time-varying basis function, which can be described as

$$m_2 = W_{m_2} Z_{m_2} + \varepsilon_{m_2},\qquad(10)$$

$$\hat{m}_2 = \hat{W}_{m_2} Z_{m_2}.\qquad(11)$$

W_{m_2} is a $1 \times n_b$ vector of the true value of the weighting function, Z_{m_2} is an $n_b \times 1$ vector of the basis function, $\hat{W}_{m_2}$ is a $1 \times n_b$ vector of the estimated weighting function, n_b is the number of the basis function in the FAT representation and ε_{m_2} are the approximation error matrices which are assumed to be zero. The FAT representation, $W_{m_2} Z_{m_2}$ can be chosen as any orthonormal function such as Taylor Series, Fourier Series, Bessel functions and Legendre polynomials [9].

Let the estimation error of the weighting function denoted as $\tilde{W}_{m_2} = \hat{W}_{m_2} - W_{m_2}$ and substituting it into (9), the error dynamics becomes

$$M_2(q)\dot{e} + C_2(q,\dot{q})e = \tilde{W}_{m_2} Z_{m_2}[P_m(q)\ddot{q}_r + P_C(q,\dot{q})\dot{q}_r + P_g(q)] - Ke\qquad(12)$$

Define a Lyapunov-like function

$$V(e,\tilde{W}_{m_2}) = \frac{1}{2}e^T M_2(q)e + \frac{1}{2}\tilde{W}_{m_2} Q_{m_2} \tilde{W}_{m_2}^T\qquad(13)$$

Differentiating (13), the derivative can be obtained as

$$\dot{V}(e,\tilde{W}_{m_2}) = e^T M(q)\dot{e} + \frac{1}{2}e^T \dot{M}(q)e + \tilde{W}_{m_2} Q_{m_2} \dot{\tilde{W}}_{m_2}^T.\qquad(14)$$

Substituting the error dynamic (12) into (14) and utilizing the skew symmetric property of $\dot{M}_2(q) - 2C_2(q,\dot{q})$, the derivative becomes

$$\dot{V}(e,\tilde{W}_{m_2}) = -e^T Ke + \tilde{W}_{m_2} Q_{m_2} \dot{\hat{W}}_{m_2}^T + \tilde{W}_{m_2} Z_{m_2} e[P_m(q)\ddot{q}_r + P_C(q,\dot{q})\dot{q}_r + P_g(q)].\qquad(15)$$

Substituting the update laws (6) into (15), the derivative of the Lyapunov function can be finally reduced to

$$\dot{V}(e,\tilde{W}_{m_2}) = -e^T Ke\qquad(16)$$

Since (13) is positive definite and (16) is negative semi-definite, e and $\tilde{W}_{m_2}$ are bounded. Differentiating (16),

$$\dot{V}(e,\tilde{W}_{m_2}) = -2e^T K \dot{e} \tag{17}$$

From (17) and the error dynamics (12), it can be observed that $\ddot{V}$ is also bounded. Therefore, utilizing Barbalat theory, $\lim\limits_{t \to \infty} \dot{V} = 0$. From (16), this implies that $\lim\limits_{t \to \infty} e = 0$. Since $e = \dot{e}_q + \lambda e_q$, e_q and $\dot{e}_q$ will also asymptotically converge to 0. Therefore, under the application of control strategy (4) and update law (6), the actual trajectory of the robotic arm, $q \to q_d$ as time, $t \to \infty$ despite the time-varying uncertainty of the payload mass, $m_2(t)$. In this method, it can be seen that, only one adaptation gain matrix that is needed to be tuned in the update law, which leads to the simplification of the control law implementation.

5   Simulation Results

Simulation test has been conducted on the 2-DOF robotic arm using Matlab Simulink to investigate the effectiveness of the proposed method. The robotic arm is required to track the desired trajectories $\theta_{1d} = \theta_{2d} = \sin t$, under three different unknown time-varying payload conditions which are:

- Case 1: $m_2(t) = \sin t + 3$ (18)

- Case 2: $m_2(t) = \sin 30t + 3$ (19)

- Case 3: $m_2(t) = 3\sin 3t + 3$ (20)

- Case 4: $m_2(t) = \begin{cases} 1\,\text{kg} & 0s \leq t \leq 0.5s \\ 2\,\text{kg} & 0.5s \leq t \leq 3s \\ 4\,\text{kg} & 3s \leq t \leq 4s \\ 0.5\,\text{kg} & 4s \leq t \leq 5s \end{cases}$

$$\tag{21}$$

The sampling is chosen as 0.001 s and the known robotic arm parameters are $l_1 = 0.18\,\text{m}$, $l_2 = 0.26\,\text{m}$ and $m_1 = 0.8\,\text{kg}$. The controller parameters have been chosen as $K = \text{diag}[10 \quad 10]$, $\lambda = [100 \quad 100]^T$ and the adaptation gain has been tuned as $Q_{m2} = 10 I_{11}$. The uncertain payload in (11) have been approximated by the first 11 terms of Fourier series, where

$$\hat{W}_{m_2} = \begin{bmatrix} \hat{w}_{m_2 0} & \hat{w}_{m_2 1} & \hat{w}_{m_2 2} & \hat{w}_{m_2 3} & \hat{w}_{m_2 4} & \hat{w}_{m_2 5} & \hat{w}_{m_2 6} & \hat{w}_{m_2 7} & \hat{w}_{m_2 8} & \hat{w}_{m_2 9} & \hat{w}_{m_2 10} \end{bmatrix}$$

$$Z_{m_2} = \begin{bmatrix} \dfrac{1}{2} & \cos\left(\dfrac{\pi t}{5}\right) & \sin\left(\dfrac{\pi t}{5}\right) & \cos\left(\dfrac{2\pi t}{5}\right) & \sin\left(\dfrac{2\pi t}{5}\right) & \cos\left(\dfrac{3\pi t}{5}\right) & \cdots \end{bmatrix}$$

$$\cdots \quad \sin\left(\dfrac{3\pi t}{5}\right) \quad \cos\left(\dfrac{4\pi t}{5}\right) \quad \sin\left(\dfrac{4\pi t}{5}\right) \quad \cos\left(\dfrac{5\pi t}{5}\right) \quad \sin\left(\dfrac{5\pi t}{5}\right) \end{bmatrix}^T \tag{21}$$

$$\hat{m}_2(t) = \frac{\hat{w}_{m_2 0}}{2} + \hat{w}_{m_2 1}\cos\left(\frac{\pi t}{5}\right) + \hat{w}_{m_2 2}\sin\left(\frac{\pi t}{5}\right) + \hat{w}_{m_2 3}\cos\left(\frac{2\pi t}{5}\right) + \hat{w}_{m_2 4}\sin\left(\frac{2\pi t}{5}\right)$$

$$+ \hat{w}_{m_2 5}\cos\left(\frac{3\pi t}{5}\right) + \hat{w}_{m_2 6}\sin\left(\frac{3\pi t}{5}\right) + \hat{w}_{m_2 7}\cos\left(\frac{4\pi t}{5}\right) + \hat{w}_{m_2 8}\sin\left(\frac{4\pi t}{5}\right)$$

$$+ \hat{w}_{m_2 9}\cos\left(\frac{5\pi t}{5}\right) + \hat{w}_{m_2 10}\sin\left(\frac{5\pi t}{5}\right) \tag{22}$$

Fig. 2 illustrates the tracking response of joint 1 and joint 2 of the robotic arm under the influence of the proposed controller, tested under Case 1, where the unknown mass varies sinusoidally with an amplitude of 1 kg and period of 2π seconds. From the figure it can be observed that the robotic arm follows the desire trajectory accurately even though the exact value of the payload is not known exactly in advance. Fig. 3 shows the tracking error for both of the joints under Case 2, where the frequency of the change in the mass is 30 times higher than Case 1. From the results, it can be seen that the controller is effective in driving the robot joints to track the desired trajectory despite the rapid change in the uncertain parameter. High tracking error can be observed at approximately 1.6 seconds, 4.7 seconds, 7.8 seconds, 10.9 seconds and 14.2 seconds due to the high acceleration of the robot joints at these times. In Case 3, the amplitude and frequency of the variation of the mass of the payload is set to be higher and faster compared to Case 1. The result depicted in Fig. 4 (a) illustrates that the proposed control technique is still successful in driving the robotic arm to move according to the desired path although the change in the mass of the carried payload is higher and faster. It can also be seen that the tracking errors for both joints in this case are very low, which are between -0.012 rad until 0.006 rad. The simulation under Case 4 also shows small tracking error in the joints of the robotic arm when the mass of the payload is varied as a function of step input as illustrated in Fig. 4 (b). In this case the error varies between -0.006 rad to 0.003 rad. The results under these 4 cases prove that the proposed technique is capable of compensating unknown time-varying payload. The proposed technique is effective in controlling the robotic arm to track the desired trajectory in spite of the time-varying uncertainties in the payload mass.

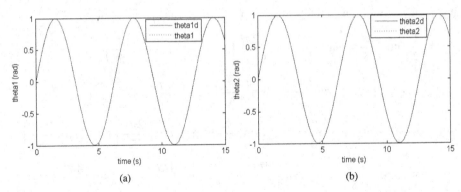

Fig. 2. Trajectory tracking response under Case 1 for (a) Joint 1 (b) Joint 2

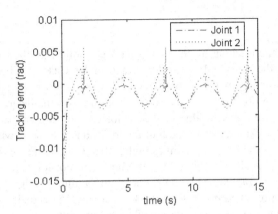

Fig. 3. Joint 1 and joint 2 tracking errors under Case 2

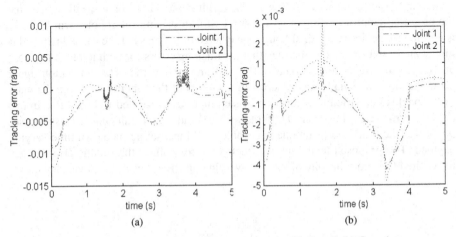

Fig. 4. Joint 1 and joint 2 tracking errors under (a) Case 3 (b) Case 4

6   Conclusion

This paper presents an FAT-based adaptive control strategy to compensate for uncertain time-varying payload in a 2-DOF planar robotic arm. The uncertain parameter in the mathematical model is factorized from the known terms in deriving the update law of the adaptive controller. The update law is designed based on FAT algorithm, which enables the estimation of unknown time-varying functions without the need to linearly parameterize the manipulator model. The controller design results in only one adaptation gain matrix to be tuned in the update law. The simulation outcomes under 4 mass variation cases verified that the proposed control technique has successfully control the robotic arm to track the desired trajectory in spite of the time-varying uncertainty of the payload that it carries. Future work involves the hardware experimental tests of the proposed control scheme, investigation on the extension of the proposed controller on robotic arms with a higher DOF and the formulation of the controller to compensate noise and unknown disturbances.

Acknowledgment. The authors would like to acknowledge Ministry of Higher Education Malaysia (MOHE) for supporting this research under the Research Acculturation Grant Scheme (RAGS).

References

1. Pagilla, P.R., Zhu, Y.: On the Adaptive Control of Mechanical Systems with Time-Varying Parameters and Disturbances. In: ASME International Mechanical Engineering Congress & Exposition, pp. 1–9 (2002)
2. Cazalilla, J., Vallés, M., Mata, V., Díaz-Rodríguez, M., Valera, A.: Adaptive Control of a 3-DOF Parallel Manipulator Considering Payload Handling and Relevant Parameter Models. Robotics and Computer-Integrated Manufacturing 30(5), 468–477 (2014)
3. Lee, H.T.: Adaptive Speed/Position Control of Induction Motor based on SPR Approach. International Journal of Control (2014)
4. Dai, S., Lee, T., Bernstein, D.S.: Adaptive Control of a Quadrotor UAV Transporting a Cable-Suspended Load with Unknown Mass, http://fdcl.seas.gwu.edu/paper/CDC14.3.pdf
5. Slotine, J.J.E., Li, W.: Applied Nonlinear Control. Prentice Hall, New Jersey (1991)
6. Kumar, N., Panwar, V., Borm, J.H., Chai, J.: Enhancing Precision Performance of Trajectory Tracking Controller for Robot Manipulators using RBFNN and Adaptive Bound. Applied Mathematics and Computation 231, 320–328 (2014)
7. Huang, A.C., Chien, M.C.: Adaptive Control of Robot Manipulators: A Unified Regressor-Free Approach. World Scientific Publishing Co. Pte. Ltd., Singapore (2010)
8. He, W., Ge, S.S., Li, Y., Chew, E., Ng, Y.S.: Impedance Control of a Rehabilitation Robot for Interactive Training. In: Ge, S.S., Khatib, O., Cabibihan, J.-J., Simmons, R., Williams, M.-A. (eds.) ICSR 2012. LNCS, vol. 7621, pp. 526–535. Springer, Heidelberg (2012)

Towards an Architecture for Knowledge Representation and Reasoning in Robotics

Shiqi Zhang[1], Mohan Sridharan[2], Michael Gelfond[3], and Jeremy Wyatt[4]

[1] Department of Computer Science, The University of Texas at Austin, USA
[2] Department of Electrical and Computer Engineering, The University of Auckland, NZ
[3] Department of Computer Science, Texas Tech University, USA
[4] School of Computer Science, University of Birmingham, UK
szhang@cs.utexas.edu, m.sridharan@auckland.ac.nz,
michael.gelfond@ttu.edu, jlw@cs.bham.ac.uk

Abstract. This paper describes an architecture that combines the complementary strengths of probabilistic graphical models and declarative programming to enable robots to represent and reason with qualitative and quantitative descriptions of uncertainty and domain knowledge. An action language is used for the architecture's low-level (LL) and high-level (HL) system descriptions, and the HL definition of recorded history is expanded to allow prioritized defaults. For any given objective, tentative plans created in the HL using commonsense reasoning are implemented in the LL using probabilistic algorithms, and the corresponding observations are added to the HL history. Tight coupling between the levels helps automate the selection of relevant variables and the generation of policies in the LL for each HL action, and supports reasoning with violation of defaults, noisy observations and unreliable actions in complex domains. The architecture is evaluated in simulation and on robots moving objects in indoor domains.

1 Introduction

Robots deployed to collaborate with humans in homes, offices, and other domains, have to represent knowledge and reason at both the sensorimotor level and the cognitive/social level. This objective maps to the fundamental challenge of representing, revising, and reasoning with qualitative and quantitative descriptions of uncertainty and incomplete domain knowledge obtained from different sources. As a significant step towards addressing this challenge, our architecture combines the knowledge representation and commonsense reasoning capabilities of declarative programming with the uncertainty modeling capabilities of probabilistic graphical models. The architecture has two tightly coupled levels with the following key features:

1. An action language is used for the system descriptions and the definition of recorded history is expanded in the high-level (HL) to allow prioritized defaults.
2. For any given objective, tentative plans are created in the HL using commonsense reasoning, and implemented in the low-level (LL) using probabilistic algorithms, with the corresponding observations adding statements to the HL history.
3. Tight coupling between the system descriptions enables automatic selection of relevant variables and the creation of action policies in the LL for any HL action.

M. Beetz et al. (Eds.): ICSR 2014, LNAI 8755, pp. 400–410, 2014.

In this paper, the HL and LL domain representations are translated into an Answer Set Prolog (ASP) program and a partially observable Markov decision process (POMDP) respectively. The novel contributions, e.g., histories with defaults and the tight coupling between the levels, support reasoning with violation of defaults, noisy observations and unreliable actions in large, complex domains. The architecture is evaluated in simulation and on robots moving objects to specific places in an indoor domain.

2   Related Work

Probabilistic graphical models such as POMDPs have been used to plan sensing, navigation and interaction for robots [13]. However, these formulations (by themselves) make it difficult to perform commonsense reasoning. Research in classical planning has provided many algorithms for knowledge representation and logical reasoning, but these algorithms require prior knowledge about the domain, tasks and the set of actions. Many such algorithms also do not support merging of new, unreliable information with the current beliefs in a knowledge base. ASP, a non-monotonic logic programming paradigm, is well-suited for representing and reasoning with commonsense knowledge [2]. It has been used to enable applications such as simulated robot housekeepers and natural language human-robot interaction [4,5]. However, ASP does not support probabilistic analysis, whereas a lot of information available to robots is represented probabilistically to quantitatively model the uncertainty in sensing and acting.

Researchers have designed architectures and developed algorithms that combine deterministic and probabilistic algorithms for task and motion planning on robots [8,9]. Examples of principled algorithms that combine logical and probabilistic reasoning include probabilistic first-order logic [7], Markov logic network [12], Bayesian logic [10], and a probabilistic extension to ASP [3]. However, algorithms based on first-order logic for probabilistically modeling uncertainty do not provide the desired expressiveness for commonsense reasoning, e.g., it is not always possible to express uncertainty and degrees of belief quantitatively. Other algorithms based on logic programming that support probabilistic reasoning do not support one or more of the desired capabilities such as: reasoning as in causal Bayesian networks; incremental addition of (probabilistic) information; and reasoning with large probabilistic components [3]. As a step towards these capabilities, our novel architecture exploits the complementary strengths of declarative programming and probabilistic graphical models, enabling robots to plan actions in larger domains than was possible before.

3   KRR Architecture

The syntax, semantics and representation of the transition diagrams of our architecture's HL and LL domain representations are described in an *action language* AL [6]. AL has a sorted signature containing three *sorts*: *statics*, *fluents* and *actions*. Statics are domain properties whose truth values cannot be changed by actions, fluents are properties whose values are changed by actions, and actions are elementary actions that can be executed in parallel. AL allows three types of statements:

$$a \text{ causes } l_{in} \text{ if } p_0, \dots, p_m \qquad\qquad \text{(Causal law)}$$
$$l \text{ if } p_0, \dots, p_m \qquad\qquad\qquad\qquad \text{(State constraint)}$$
$$\textbf{impossible } a_0, \dots, a_k \text{ if } p_0, \dots, p_m \qquad \text{(Executability condition)}$$

where a is an action, l is a literal, l_{in} is an inertial fluent literal, and $p_0, \dots, p_m$ are domain literals (any domain property or its negation). A collection of statements of AL forms a system description. As an illustrative example used throughout this paper, we consider a robot that moves objects of the sorts: *textbook*, *printer* and *kitchenware*, in a domain with four places: *office*, *main_library*, *aux_library*, and *kitchen*.

3.1 HL Domain Representation

The HL domain representation consists of a system description $\mathscr{D}_H$ and histories with defaults $\mathscr{H}$. $\mathscr{D}_H$ consists of a sorted signature (Σ_H) and axioms used to describe the HL transition diagram τ_H. Σ_H defines the names of objects, functions, and predicates available for use in the HL. The sorts in our example are: *place*, *thing*, *robot*, and *object*; *object* and *robot* are subsorts of *thing*. The sort *object* has subsorts: *textbook*, *printer* and *kitchenware*. The fluents of the domain are defined in terms of their arguments: $loc(thing, place)$ and $in\_hand(robot, object)$. The first predicate describes a thing's location, and the second states that a robot is holding an object. These predicates are *inertial fluents* subject to the laws of inertia. The domain has three actions: $move(robot, place)$, $grasp(robot, object)$, and $putdown(robot, object)$. The domain dynamics are defined using axioms that consist of causal laws such as:

$$move(Robot, Pl) \text{ causes } loc(Robot, Pl) \qquad\qquad (1)$$
$$grasp(Robot, Ob) \text{ causes } in\_hand(Robot, Ob)$$

state constraints:
$$loc(Ob, Pl) \text{ if } loc(Robot, Pl), \ in\_hand(Robot, Ob) \qquad\qquad (2)$$
$$\neg loc(Th, Pl_1) \text{ if } loc(Th, Pl_2), \ Pl_1 \neq Pl_2$$

and executability conditions such as:
$$\textbf{impossible } move(Robot, Pl) \text{ if } loc(Robot, Pl) \qquad\qquad (3)$$
$$\textbf{impossible } grasp(Robot, Ob) \text{ if } loc(Robot, Pl_1), \ loc(Ob, Pl_2), Pl_1 \neq Pl_2$$

Histories with Defaults. A dynamic domain's recorded history is usually a collection of records of the form $obs(fluent, boolean, step)$, i.e., a specific fluent observed to be true or false at a given step, and $hpd(action, step)$, i.e., a specific action happened at a given step; we abbreviate $obs(f, true, 0)$ and $obs(f, false, 0)$ as $init(f, true)$ and $init(f, false)$ respectively. *We expand on this view by allowing histories to contain (prioritized) defaults describing the values of fluents in their initial states.* We provide some illustrative examples below; see [6] for formal semantics of defaults.

Example 1 *[Example of defaults]*
Consider the following statements about the locations of textbooks in the initial state in our illustrative example. *Textbooks are typically in the main library. If a textbook is not there, it is in the auxiliary library. If a textbook is checked out, it can be found in the office.* These defaults can be represented as:

$$default(d_1(X)) \qquad\qquad default(d_2(X))$$
$$head(d_1(X), loc(X, main\_library)) \quad head(d_2(X), loc(X, aux\_library))$$
$$body(d_1(X), textbook(X)) \qquad\quad body(d_2(X), textbook(X)) \tag{4}$$
$$body(d_2(X), \neg loc(X, main\_library))$$

$$default(d_3(X))$$
$$head(d_3(X), loc(X, office))$$
$$body(d_3(X), textbook(X)) \tag{5}$$
$$body(d_3(X), \neg loc(X, main\_library)), \quad body(d_3(X), \neg loc(X, aux\_library))$$

where the literal in the "head" is true if all literals in the "body" are true. A history $\mathcal{H}_1$ with the above statements entails: $holds(loc(Tb_1, main\_library), 0)$ for textbook Tb_1. History $\mathcal{H}_2$ that adds observation: $init(loc(Tb_1, main\_library), false)$ to $\mathcal{H}_1$ renders default d_1 inapplicable; it entails: $holds(loc(Tb_1, aux\_library), 0)$ based on d_2. A history $\mathcal{H}_3$ that adds observation: $init(loc(Tb_1, aux\_library), false)$ to $\mathcal{H}_2$ should entail: $holds(loc(Tb_1, office), 0)$. History $\mathcal{H}_4$ that adds: $obs(loc(Tb_1, main\_library), false, 1)$ to $\mathcal{H}_1$ defeats default d_1 because if this default's conclusion is true in the initial state, it is also true at step 1 (by inertia), which contradicts our observation. Default d_2 will conclude that this book is initially in the *aux_library*; the inertia axiom will propagate this information to entail: $holds(loc(Tb_1, aux\_library), 1)$.

The following terminology is used to formally define the entailment relation with respect to a fixed $\mathcal{D}_H$. A set S of literals is *closed under a default d* if S contains the head of d whenever it contains all literals from the body of d and does not contain the literal contrary to d's head. S is *closed under a constraint* of $\mathcal{D}_H$ if S contains the constraint's head whenever it contains all literals from the constraint's body. A set U of literals is the *closure of S* if $S \subseteq U$, U is closed under constraints of $\mathcal{D}_H$ and defaults of $\mathcal{H}$, and no proper subset of U satisfies these properties.

Definition 1. *[Compatible initial states]*
A state σ of τ_H is *compatible* with description $\mathcal{I}$ of the initial state of history $\mathcal{H}$ if:

    1. σ satisfies all observations of $\mathcal{I}$,
    2. σ contains the closure of the union of statics of $\mathcal{D}_H$ and the set $\{f : init(f, true) \in \mathcal{I}\} \cup \{\neg f : init(f, false) \in \mathcal{I}\}$.

Let $\mathcal{I}_k$ describe the initial state of history $\mathcal{H}_k$. In Example 1 above, states compatible with $\mathcal{I}_1$, $\mathcal{I}_2$, $\mathcal{I}_3$ must contain $\{loc(Tb_1, main\_library)\}$, $\{loc(Tb_1, aux\_library)\}$, and $\{loc(Tb_1, office)\}$ respectively. Since $\mathcal{I}_1 = \mathcal{I}_4$, they have the same compatible states.

Definition 2. *[Models]*
A path P of τ_H is a *model* of history $\mathcal{H}$ with description $\mathcal{I}$ of its initial state if there is a collection E of *init* statements such that:
1. If $init(f, true) \in E$ then $\neg f$ is the head of a default of $\mathcal{I}$. Similarly, for $init(f, false)$.
2. The initial state of P is compatible with the description: $\mathcal{I}_E = \mathcal{I} \cup E$.
3. Path P satisfies all observations in $\mathcal{H}$.
4. There is no collection E_0 of *init* statements which has less elements than E and satisfies the conditions above.

We refer to E as an *explanation* of $\mathcal{H}$. Models of $\mathcal{H}_1$, $\mathcal{H}_2$, and $\mathcal{H}_3$ are paths consisting of initial states compatible with $\mathcal{I}_1$, $\mathcal{I}_2$, and $\mathcal{I}_3$; the corresponding explanations are empty. For $\mathcal{H}_4$, the predicted and observed locations of Tb_1 are different. Adding $E = \{init(loc(Tb_1, main\_library), false)\}$ to $\mathcal{I}_4$ resolves this problem.

Definition 3. *[Entailment and consistency]*
- Let $\mathcal{H}^n$ be a history of length n, f be a fluent, and $i \in (0,n)$ be a step of $\mathcal{H}^n$. $\mathcal{H}^n$ *entails* a statement $Q = holds(f,i)$ $(\neg holds(f,i))$ if for every model P of $\mathcal{H}^n$, fluent literal f $(\neg f)$ belongs to the ith state of P. The entailment is denoted by $\mathcal{H}^n \models Q$.
- A history which has a model is said to be *consistent*.

It can be shown that histories from Example 1 are consistent and that our definition of entailment captures the corresponding intuition.

Reasoning with HL Domain Representation. The HL domain representation is translated into a program $\Pi(\mathcal{D}_H, \mathcal{H})$ in CR-Prolog that incorporates consistency restoring rules in ASP [1,6]. ASP is based on stable model semantics and non-monotonic logics; it can represent recursive definitions, defaults, causal relations, and language constructs that are difficult to express in classical logic formalisms [2]. The ground literals in an *answer set* obtained by solving Π represent beliefs of an agent associated with Π; statements that hold in all such answer sets are program consequences. Algorithms for computing the entailment relation of AL, and for planning and diagnostics, reduce these tasks to computing answer sets of CR-Prolog programs. Π consists of causal laws of $\mathcal{D}_H$, inertia axioms, closed world assumption for defined fluents, reality checks, records of observations, actions and defaults from $\mathcal{H}$, and special axioms for *init*: $holds(F,0) \leftarrow init(F,true)$ and $\neg holds(F,0) \leftarrow init(F,false)$. Every default of $\mathcal{I}$ is turned into an ASP rule and a consistency-restoring (CR) rule:

$$holds(p(X),0) \leftarrow c(X), holds(b(X),0), not \ \neg holds(p(X),0) \quad \% \text{ ASP rule} \quad (6)$$

$$\neg holds(p(X),0) \xleftarrow{+} c(X), holds(b(X),0) \quad \% \text{ CR rule}$$

The CR rule states that to restore the program's consistency, one may assume that the default's conclusion is false. See [6] for more details about CR-rules and CR-Prolog.

Proposition 1. *[Models and Answer Sets]*
A path $P = \langle \sigma_0, a_0, \sigma_1, \ldots, \sigma_{n-1}, a_n \rangle$ of τ_H is a model of history $\mathcal{H}^n$ iff there is an answer set S of a program $\Pi(\mathcal{D}_H, \mathcal{H})$ such that:
  1. A fluent $f \in \sigma_i$ iff $holds(f,i) \in S$,
  2. A fluent literal $\neg f \in \sigma_i$ iff $\neg holds(f,i) \in S$,
  3. An action $e \in a_i$ iff $occurs(e,i) \in S$.

The proposition reduces: (a) computation of models of $\mathcal{H}$ to computing answer sets of a CR-Prolog program; and (b) a planning task to computing answer sets of a program obtained from $\Pi(\mathcal{D}_H, \mathcal{H})$ by adding the definition of a goal, a constraint stating that the goal must be achieved, and a rule generating possible future actions.

3.2   LL Domain Representation

The LL system description $\mathcal{D}_L$ has a sorted signature Σ_L and axioms that describe a transition diagram τ_L. Σ_L includes sorts from Σ_H and sorts *room* and *cell*, which are subsorts of *place* and whose elements satisfy static relation *part_of(cell,room)*. A static *neighbor(cell,cell)* describes the relation between cells. Fluents of Σ_L include those of

Σ_H, a new inertial fluent: $searched(cell, object)$—a cell was searched for an object—and two defined fluents: $found(object, place)$ and $continue\_search(room, object)$. Actions of Σ_L are viewed as HL actions represented at a higher resolution. The causal law:

$$move(Robot, Y) \textbf{ causes } \{loc(Robot, Z) : neighbor(Z, Y)\} \tag{7}$$

where Y, Z are cells, may (for instance) be used to state that moving to a cell in the LL can cause the robot to be in one of the neighboring cells. The LL includes a new action $search(cell, object)$ that enables robots to search for objects in cells; the corresponding causal laws and constraints are:

$$search(C, Ob) \textbf{ causes } searched(C, Ob) \tag{8}$$
$$found(Ob, C) \textbf{ if } searched(C, Ob), \ loc(Ob, C)$$
$$found(Ob, R) \textbf{ if } part\_of(C, R), \ found(Ob, C)$$
$$continue\_search(R, Ob) \textbf{ if } \neg found(Ob, R), \ part\_of(C, R), \ \neg searched(C, Ob)$$

The LL also has a defined fluent $failure(object, room)$ that holds iff the object under consideration is not found in the room that the robot is searching:

$$failure(Ob, R) \textbf{ if } loc(Robot, R), \neg continue\_search(R, Ob), \neg found(Ob, R) \tag{9}$$

In this action theory that describes τ_L, states are viewed as extensions of states of τ_H by physically possible fluents and statics defined in the language of the LL. Moreover, for every HL state transition $\langle \sigma, a, \sigma' \rangle$ and every LL state s compatible with σ, there is a path in the LL from s to some state compatible with σ'.

Unlike the HL, action effects and observations in the LL are only known with some degree of probability. The function $T : S \times A \times S' \rightarrow [0, 1]$ defines the state transition probabilities in the LL. Similarly, if Z is the subset of fluents that are observable in the LL, the observation function $O : S \times Z \rightarrow [0, 1]$ defines the probability of observing specific elements of Z in specific states. Functions T and O are computed using prior knowledge, or by analyzing the effects of specific actions in specific states (Section 4.1).

Since states are partially observable in the LL, reasoning uses *belief states*, probability distributions over the set of states. Functions T and O describe a probabilistic transition diagram over belief states. The initial belief state B_0 is revised iteratively using Bayesian updates: $B_{t+1}(s_{t+1}) \propto O(s_{t+1}, o_{t+1}) \sum_s T(s, a_{t+1}, s_{t+1}) \cdot B_t(s)$. The LL system description also includes a reward specification $R : S \times A \times S' \rightarrow \Re$ that encodes the relative *utility* of specific actions in specific states. Planning in the LL involves computing a *policy* that maximizes the cumulative reward over a planning horizon to map belief states to actions: $\pi : B_t \mapsto a_{t+1}$. We use a point-based approximate algorithm to compute this policy [11]. Plan execution uses the policy to repeatedly choose an action in the current belief state, and updates the belief state after executing that action and/or receiving an observation. We call this algorithm "POMDP-1".

Unlike the HL, the LL history only stores observations and actions over one time step. In this paper, the LL domain representation is translated automatically into POMDP models, i.e., data structures for $\mathscr{D}_L$'s components such that existing solvers can be used to obtain policies. One key consequence of the tight coupling between the LL and the HL is that the relevant LL variables for any HL action are identified automatically, significantly improving the efficiency of computing policies.

Algorithm 1. Control loop of the architecture

Input: The HL and LL domain representations, and the specific task for robot to perform.

  1 LL observations reported to HL history; HL initial state (s_{init}^H) communicated to LL.

  2 Assign goal state s_{goal}^H based on task.

  3 Generate HL plan(s).

  4 **if** *multiple HL plans exist* **then**

  5   Send plans to the LL, select plan with lowest (expected) action cost and communicate to the HL.

  6 **end**

  7 **if** *HL plan exists* **then**

  8   **for** $a_i^H \in$ *HL plan:* $i \in [1, n]$ **do**

  9   Pass a_i^H and relevant fluents to LL.

 10   Determine initial belief state over the relevant LL variables.

 11   Generate LL action policy.

 12   **while** a_i^H *not completed* **and** a_i^H *achievable* **do**

 13   Execute an action based on LL action policy.

 14   Make an observation and update belief state.

 15   **end**

 16   LL observations and action outcomes add statements to HL history.

 17   **if** *results unexpected* **then** Perform diagnostics in HL. ;

 18   **if** *HL plan invalid* **then** Replan in the HL (line 3). ;

 19   **end**

 20 **end**

3.3 Control Loop

Algorithm 1 describes the architecture's control loop. First, the LL observations obtained by the robot in the current location add statements to the HL history, and the HL initial state is communicated to the LL (line 1). The assigned task determines the HL goal state (line 2) and planning in the HL provides action sequence(s) with deterministic effects (line 3). If there are multiple HL plans, e.g., tentative plans generated for the different possible locations of a desired object, these plans are communicated to the LL; the plan with the least expected execution time is selected and communicated to the HL (lines 4-6). If an HL plan exists, actions are communicated one at a time to the LL along with the relevant fluents (line 9). For an HL action (a_i^H), the relevant LL variables are identified and the initial belief is set (line 10). An LL POMDP policy is computed (line 11) and used to execute actions and update the belief state until a_i^H is achieved or inferred to be unachievable (lines 12-15). The outcome of executing the LL policy, and the observations, add to the HL history (line 16). If the results are unexpected, diagnosis is performed in the HL (line 17); we assume that the robot can identify unexpected outcomes. If the HL plan is invalid, a new plan is generated (line 18); else, the next action in the HL plan is executed.

4 Experimental Setup and Results

This section describes the experimental setup and results of evaluating the architecture.

4.1 Experimental Setup

The architecture was evaluated in simulation and on physical robots. The simulator uses models that represent objects using probabilistic functions of features extracted from images, and models that reflect the robot's motion. The robot also acquired data (e.g., computational time of different algorithms) in an initial training phase to define the probabilistic components of the LL domain representation [14].

In each trial, the goal was to move specific objects to specific places; the robot's location, target object, and locations of objects were chosen randomly. An action sequence extracted from an answer set of the ASP program provides an HL plan, e.g., the plan to move textbook Tb_1 from the *main_library* to the *office* could be: *move(Robot, main_library), grasp(Robot, Tb_1), move(Robot, office), putdown(Robot, Tb_1)*. An object's location in the LL is known with certainty if the belief (in a cell) exceeds a threshold (0.85). Our architecture (with the control loop in Algorithm 1), henceforth referred to as "PA", was compared with: (1) POMDP-1; and (2) POMDP-2, which revises POMDP-1 by assigning high probability values to defaults to bias the initial belief. We evaluated two hypotheses: (H1) PA achieves goals more reliably and efficiently than POMDP-1; (H2) our representation of defaults improves reliability and efficiency in comparison with not using defaults or assigning high probability values to defaults.

4.2 Experimental Results

To evaluate H1, we first compared PA with POMDP-1 in trials in which the robot's initial position is known but the position of the object to be moved is unknown. The solver used in POMDP-1 is given a fixed amount of time to compute action policies. Figure 1(a) summarizes the ability to successfully achieve the assigned goal, as a function of the number of cells in the domain. Each data point in Figure 1(a) is the average of 1000 trials, and each room is set to have four cells (for ease of interpretation). PA significantly improves the robot's ability to achieve the assigned goal in comparison with POMDP-1. As the number of cells (i.e., domain size) increases, it becomes computationally difficult to generate good POMDP action policies which, in conjunction with incorrect observations (e.g., false positives) significantly impacts the ability to complete the trials. PA focuses the robot's attention on relevant regions (e.g., specific rooms and cells). As the domain size increases, the generation of a large number of plans of similar cost may (with incorrect observations) affect the ability to achieve desired goals—the impact is, however, much less pronounced.

Next, we computed the time taken by PA to generate a plan as the domain size (i.e., number of rooms and objects) increases. We conducted three sets of experiments in which the robot reasons with: (1) all available knowledge of objects and rooms; (2) only knowledge relevant to the assigned goal—e.g., if the robot knows an object's default location, it need not reason about other objects and rooms to locate the object; and (3) relevant knowledge and knowledge of an additional 20% of randomly selected objects and rooms. Figure 2 shows that PA generates appropriate plans for domains with a large number of rooms and objects. Using only the knowledge relevant to the goal significantly reduces the planning time; this knowledge can be automatically selected using the relations in the HL system description. Furthermore, it soon becomes

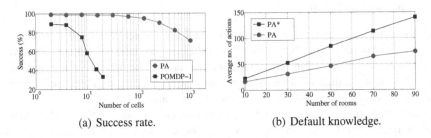

Fig. 1. (a) With a limit on the time to compute policies, PA significantly increases accuracy in comparison with POMDP-1 as the number of cells increases; (b) Principled representation of defaults significantly reduces the number of actions (and thus time) for achieving assigned goal

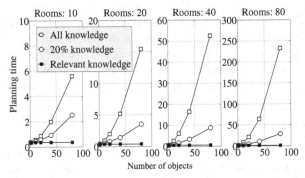

Fig. 2. Planning time as a function of the number of rooms and the number of objects in the domain—*PA* scales to larger number of rooms and objects

computationally intractable to generate a plan with POMDP-1 for domains with many objects and rooms; these results are not shown in Figure 2.

To evaluate H2, we first compared PA with *PA**, a version that does not include any default knowledge. Figure 1(b) summarizes the average number of actions executed per trial as a function of the number of rooms—each data point is the average of 10000 trials. We observe that the principled use of default knowledge significantly reduces the number of actions (and thus time) required to achieve the assigned goal. Next PA was compared with POMDP-2, which assigns high probability values to default information and revises the initial belief. The results with POMDP-2 can vary depending on: (a) the numerical value chosen; and (b) whether the ground truth matches the default information. For instance, *if a large probability is assigned to the default knowledge that books are typically in the library, but the book the robot has to move is an exception, POMDP-2 takes a large amount of time to recover from the initial belief.* PA, on the other hand, can revise initial defaults and encode exceptions to defaults.

Finally, PA was compared with POMDP-1 on a wheeled robot over 50 trials on two floors. Since manipulation is not a focus of this work, the robot asks for the desired object to be placed in its gripper once it is next to it. This domain includes additional places; the map is learned and revised by the robot over time. On the third floor, we considered 15 rooms, including offices, labs, common areas and a corridor.

To use POMDP-1 in such large domains, we used a hierarchical decomposition based on our prior work [14]. The experiments included paired trials, e.g., over 15 trials (each), POMDP-1 takes 1.64 as much time as PA to move specific objects to specific places; this 39% reduction in execution time is statistically significant; *p-value* = 0.0023 at 95% level of significance. A video of a robot trial can be viewed online: http://youtu.be/8zL4R8te6wg

5 Conclusions

This paper described a knowledge representation and reasoning architecture that combines the complementary strengths of declarative programming and probabilistic graphical models. The architecture's high-level (HL) and low-level (LL) system descriptions are provided using an action language, and the HL definition of recorded history is expanded to allow prioritized defaults. Tentative plans created in the HL using common-sense reasoning are implemented in the LL using probabilistic algorithms, generating observations that add to the HL history. Experimental results indicate that the architecture supports reasoning at the sensorimotor level and the cognitive level with violation of defaults, noisy observations and unreliable actions, and scales well to large, complex domains. The architecture thus provides fundamental capabilities for robots assisting and collaborating with humans in complex real world application domains.

Acknowledgments. The authors thank Evgenii Balai for his help with the ASP software used in the experimental trials. This work was supported in part by the U.S. ONR Science of Autonomy Award N00014-13-1-0766 and the EC-funded Strands project FP7-IST-600623. Opinions and conclusions in this paper are those of the authors.

References

1. Balduccini, M., Gelfond, M.: Logic Programs with Consistency-Restoring Rules. In: Logical Formalization of Commonsense Reasoning, AAAI SSS, pp. 9–18 (2003)
2. Baral, C.: Knowledge Representation, Reasoning and Declarative Problem Solving. Cambridge University Press (2003)
3. Baral, C., Gelfond, M., Rushton, N.: Probabilistic Reasoning with Answer Sets. Theory and Practice of Logic Programming 9(1), 57–144 (2009)
4. Chen, X., Xie, J., Ji, J., Sui, Z.: Toward Open Knowledge Enabling for Human-Robot Interaction. Human-Robot Interaction 1(2), 100–117 (2012)
5. Erdem, E., Aker, E., Patoglu, V.: Answer Set Programming for Collaborative Housekeeping Robotics: Representation, Reasoning, and Execution. Intelligent Service Robotics 5(4) (2012)
6. Gelfond, M., Kahl, Y.: Knowledge Representation, Reasoning and the Design of Intelligent Agents. Cambridge University Press (2014)
7. Halpern, J.: Reasoning about Uncertainty. MIT Press (2003)
8. Hanheide, M., Gretton, C., Dearden, R., Hawes, N., Wyatt, J., Pronobis, A., Aydemir, A., Gobelbecker, M., Zender, H.: Exploiting Probabilistic Knowledge under Uncertain Sensing for Efficient Robot Behaviour. In: International Joint Conference on Artificial Intelligence (2011)

9. Kaelbling, L., Lozano-Perez, T.: Integrated Task and Motion Planning in Belief Space. International Journal of Robotics Research 32(9-10) (2013)
10. Milch, B., Marthi, B., Russell, S., Sontag, D., Ong, D.L., Kolobov, A.: BLOG: Probabilistic Models with Unknown Objects. In: Statistical Relational Learning. MIT Press (2006)
11. Ong, S.C., Png, S.W., Hsu, D., Lee, W.S.: Planning under Uncertainty for Robotic Tasks with Mixed Observability. IJRR 29(8), 1053–1068 (2010)
12. Richardson, M., Domingos, P.: Markov Logic Networks. Machine Learning 62(1) (2006)
13. Rosenthal, S., Veloso, M.: Mobile Robot Planning to Seek Help with Spatially Situated Tasks. In: National Conference on Artificial Intelligence (July 2012)
14. Zhang, S., Sridharan, M., Washington, C.: Active Visual Planning for Mobile Robot Teams using Hierarchical POMDPs. IEEE Transactions on Robotics 29(4) (2013)

Author Index